D0085457

SCIENTIFIC CONTROVERSIES

SCIENTIFIC CONTROVERSIES

PHILOSOPHICAL AND HISTORICAL PERSPECTIVES

EDITED BY
Peter Machamer
Marcello Pera
Aristides Baltas

New York • Oxford
Oxford University Press
2000

Oxford University Press

Oxford New York
Athens Auckland Bangkok Bogotá Buenos Aires Calcutta
Cape Town Chennai Dar es Salaam Delhi Florence Hong Kong Istanbul
Karachi Kuala Lumpur Madrid Melbourne Mexico City Mumbai
Nairobi Paris São Paulo Singapore Taipei Tokyo Toronto Warsaw

and associate companies in
Berlin Ibadan

Copyright © 2000 by Oxford University Press, Inc.

Published by Oxford University Press, Inc.
198 Madison Avenue, New York, New York 10016

Oxford is a registered trademark of Oxford University Press.

All right reserved. No part of this publication may be reproduced,
stored in a retrieval system, or transmitted, in any form or by any means,
electronic, mechanical, photocopying, recording, or otherwise,
without the prior permission of Oxford University Press.

Library of Congress Cataloging-in-Publication Data
Scientific controversies: philosophical and historical perspectives / edited by Peter Machamer, Marcello Pera, and Aristedes Baltas.
p. cm.
Includes bibliographical references and index.
ISBN 0-19-511987-8
1. Science—Philosophy. 2. Science—Social Aspects.
3. Science—History. I. Machamer, Peter K II. Pera, Marcello, 1943– III. Baltas, Aristedes
Q175.S42323 1999
501—dc21 98-38234

9 8 7 6 5 4 3 2 1

Printed in the United States of America
on acid-free paper

Acknowledgments

We thank the many people who helped get this manuscript to completion. Peter Ohlin, the editor at Oxford, and before him Cynthia Read, have been most helpful and encouraging. Working with Catharine Carlin of Oxford has been a good thing. Georgianne Coleman worked on the manuscript and helped make it consistent and readable. Joe Piffaretti has virtually compiled the index.

Most of my gratitude for providing the courage to get through this editing task goes to Barbara Diven (and Tara and Michael Gainfort). She has provided the breath of life.

Pittsburgh
November, 1998

Peter Machamer

Contents

Contributors

Baltas, Aristides

Professor of Physics and Philosophy of Science at the National Technical University of Athens, with a Ph.D. in theoretical high energy physics, he has written (in English and Greek) on general philosophy of science and other sciences and purported sciences (for example, Marxism and psychoanalysis). His work is an effort to bring a mutually beneficial dialogue with Anglo-American and Continental philosophical traditions.

Barrotta, Pierluigi

Research Fellow in Philosophy of Science at the University of Pisa, he studied in Pisa, London, and Cambridge and has published a book on phlogiston controversy (*Dogmatismo ed eresia nella scienze: Joseph Priestley* [Angeli, Milano, 1994]. He is currently working on a book on scientific dialectics.

Freudenthal, Gideon

Editor of the journal *Science in Context* published by Cambridge University Press, he teaches at the Cohn Institute for the History and Philosophy of Science and Ideas, Tel-Aviv University. He is also the author of *Atom and Individual in the Age of Newton* (Reidel, 1986) and co-author of *Exploring the Limits of Preclassical Mechanics* (Springer-Verlag, 1992).

Gavroglu, Kostas

Professor in the Department of Methodology, History, and Theory of Science at the University of Athens. He wrote *Fritz London; A Scientific Biography* (Cambridge, 1995) and, with A. Simoes, *Concepts Out of Contexts: The Development of Low Temperature Physics 1881–1957* (Nihjoff, 1994).

Grandy, Richard

With a B.S. from the University of Pittsburgh (1963) and a Ph.D. from Princeton University (1968), he taught at Princeton University and the University of North Carolina at Chapel Hill prior to moving to Rice University, Houston, Texas, in 1980, where he is a member of the philosophy department, the Cognitive Sciences Program, and the Center for the Study of Science and Technology.

Hacking, Ian

University Professor at the University of Toronto, Canada, he includes among his publications *Mad Travelers: Reflections on the Reality of Mental Illnesses* (U of Virginia P, 1998) and *The Social Construction Of What?* (Harvard UP, 1999).

Kitcher, Philip

Formerly Presidential Professor of Philosophy at the University of California—San Diego, and now in the philosophy department of Columbia University, he has written several books and numerous articles on issues in the philosophy of mathematics, the philosophy of biology, and the philosophy of science.

Machamer, Peter

Professor of History and Philosophy of Science, University of Pittsburgh, he recently edited the *Cambridge Companion to Galileo* (Cambridge UP, 1998). He works on philosophy of psychology and the neurosciences as well as in sixteenth- and seventeenth-century history of philosophy and science.

Mamiani, Maurizio

Professor at the University of Ferrara, he previously taught at the Universities of Udine and Parma. He works extensively on 17th century topics, especially Newton and his contemporaries. Among his many books are *La mappa del sapere* (Angeli, 1983) and *Il prima di Newton* (Laterza, 1986).

Pera, Marcello

Formerly Professor at University of Pisa, he is currently (1998) a senator of the Italian Republic. He has worked on Aristotle, Galileo, and rhetoric and science.

Ruse, Michael

Professor of Philosophy and Zoology at the University of Guelph, Hamilton, Ontario, Canada, he is editor of the journal *Biology and Philosophy.* His most recent book is titled *Mystery of Mysteries: Is Evolution a Social Construction?* (Harvard University Press, 1999).

Salmon, Merrilee H.

Professor of History and Philosophy of Science at the University of Pittsburgh, she also holds an appointment in its department of anthropology. Her main research interest is in the philosophy of the social sciences, especially the fields of anthropology and archaeology.

Salmon, Wesley C.

With a Ph.D. from the University of California, Los Angeles in 1950, he is currently a University Professor of Philosophy at the University of Pittsburgh and author of two books, *Causality and Explanation* (Oxford UP, 1998) and *Explanation and the Causal Structure of the World* (Princeton UP, 1984).

Wallace, William A.

Professor Emeritus of Philosophy and History at the Catholic University, Washington, D.C., he teaches at the University of Maryland, College Park. He has published extensively on Galileo and philosophy of science, particularly on problems relating to causality and demonstrative proof.

SCIENTIFIC CONTROVERSIES

Scientific Controversies

An Introduction

A Paradox about Controversies

Aristotle disputed with his precursors and predecessors about atoms, void, space, movement, celestial spheres, and so on. Galileo argued against Ptolemy and contemporary seventeenth-century Aristotelians about the fundamental laws of motion, the structure of the universe, the causes of tides, floating bodies, and so on. Newton quarreled with Descartes, Hooke, Boyle, and many others about colors, light, and other topics. Einstein had controversies with Poincaré and Lorentz about absolute space and time, and with Bohr, Born, and many others about quantum mechanical laws.

Many major steps in science, probably all dramatic changes, and most of the fundamental achievements of what we now take as the advancement or progress of scientific knowledge have been controversial and have involved some dispute or another. Scientific controversies are found throughout the history of science. This is so well known that it is trivial.

What is not so obvious and deserves attention is a sort of paradoxical dissociation between science as actually practiced and science as perceived or depicted by both scientists and philosophers. While nobody would deny that science in the making has been replete with controversies, the same people often depict its essence or end product as free from disputes, as the uncontroversial rational human endeavor par excellence. Of course, neither scientists or philosophers have been unaware of controversies. Nevertheless, they have been reluctant to recognize when and how controversy plays a constitutive role in the development of scientific knowledge.

Most, if not all, philosophers have maintained either that, in science controversies arise from detectable, and so correctable, errors, or they should be relegated to nonepistemic, arational spheres, for they have no significant effect on the end product of knowledge. Sometimes this conviction has been so deep that the absence or resolution of disputes has been taken as the hallmark of scientific knowledge as compared to other disciplines or fields of experience. As Karl Popper has put it, "[S]cience is one of the very few human activities—perhaps the only one—in which errors are systematically criticized and fairly often, in time, corrected."[1] So, in this vein, it is typical of scientists and others to believe that any dispute or controversy is resolvable given further information or data, which, in their turn, necessarily will be forthcoming given enough time and money.

Controversies Should Not Exist

This view that science should be ultimately uncontroversial flourished at the time of the founders of modern science, although it dates back to Aristotle and to his idea of scientific knowledge as deduction from evident and necessary principles. On this view, controversies in science should not arise, and if they do it is because of inattention or lack of proper method (see chapters 3, 5, and 7). In his *Novum Organon,* Bacon writes that in science there is a method that "levels men's wit" like a rule or compass,[2] and therefore settles all disputes. In the *Assayer,* Galileo writes that there is nothing between necessary conclusions and paralogisms,[3] making explicit that scientific arguments are stringent and indisputable. Descartes, in his *Rules,* maintains that "whenever two persons make opposite judgments about the same thing, it is certain that at least one of them is mistaken, and neither, it seems, has knowledge."[4] Later on Leibniz advances the same view. Thanks to his own "philosophical calculus" or "mathesis universalis," he writes in a famous passage, "if controversies arise, there will be no more need for discussion between two philosophers than there is between two calculators. It will be enough for them to take a pen, sit at a desk, and tell each other (after calling a friend, if they like): let us calculate."[5] Kant follows in his predecessors' steps. In the preface to the second edition of his *Critique,* he turns the inability to resolve controversies into the criterion of demarcation between science and pseudoscience:

> [W]hether the treatment of such knowledge as lies within the province of reason does or does not follow the secure path of a science, is easily to be determined from the outcome. For if after elaborate preparations, frequently renewed, it is brought to a stop immediately it nears its goal; if often it is compelled to retrace its steps and strike into some new line of approach; or again, if the various participants are unable to agree in any common plan of procedure, then we may rest assured that it is very far from having entered upon the secure path of a science, and is indeed a merely random groping.[6]

The basic idea underlying this view is that in science one has to master nature, not opponents, as Bacon repeatedly puts it. This means that science is conceived as a sort of dialogue or interrogation in which one party (a scientist's inquiring mind) asks questions, while the other (nature) provides answers. Questions are theories or hypotheses or principles; answers are data, such as facts of experience, observational states of affair, and experimental outcomes. The procedure is impersonal, because only the objective content of theories and their match with the objective reality of data are at stake. And the results are compelling because the same means of interrogation and the same abilities to learn are available, in principle, to each and every person, globally conceived as a universal epistemic "I" (see chapter 5). Anyone can put questions to nature because nature is a book whose true meaning is manifest to those who read the book in the proper way. A particular reading is justified by appeal to a single, universal and independent scientific method (usually taken as a priori, or innate, corresponding to, or dictated by, the rules of thought or mind). Thanks to a good method, there is no need for discussion and no room for epistemic disputes: once questions have been raised and answers given, if controversies arise, method will settle them once and for all, and consensus will be reached (see chapter 5). In this respect science is like, or is part of, mathematics.

In this century, logical positivism and logical empiricism have developed and refined this idea. Here science continues to be seen as a sort of game between two players: a certain hypothesis h and certain empirical evidence *e,* where *e* is taken to support *h* or to be a test of *h.* This relation between *h* and *e,* too, like the early modern idea of the dialogue with nature, is taken to be objective and impersonal, because an organon exists that either allows *e* to be deduced from *h* (deductive logic, which provides the tools for the "logical construction of the world") or, if there is more than one *h,* establishes which one is made more probable by *e* (inductive logic, which offers the means for establishing the degree of confirmation of *h* or the basic information for choosing among competitive hypotheses or theories). Likewise, the outcome of this test is compelling and uncontroversial, as was the result of the old dialogue, because the organon relating *h* to *e* is a formal logic, free of any non-algorithmic factors. Thus, Rudolph Carnap joins Leibniz, because he too maintains that the relationship between *h* and *e* can be calculated. Popper joins both Bacon and Galileo because he too takes the dialogue with nature as regulated by methodological rules and as having objective criteria for success, although he stresses the fallibility of science and its tentative, conjectural character. In addition, he looks at scientific results as objective outcomes that we build up, step by step, in an independent "World 3," which is the rough equivalent of Galileo's *intensive* or God-like way of viewing scientific knowledge (see chapters 5 and 6). In general, the logical positivists agree with their predecessors that in science no epistemically relevant factors are involved other than facts and logic; in their hands Galileo's "sensible experiences and necessary demonstrations" or Descartes' "intuition and deduction" have been transformed into "protocol statements and logical inferences."

Yet Controversies Do Exist

In more recent times these views of epistemology and methodology have been called into question. Due to several important analytical results, which might be considered "theorems of limitation," the relationship between *h* and *e* has resisted logical analysis, while *e* has been recognized to be insufficient by itself for establishing even the most elementary semantical and methodological notions, such as reference, confirmation, and falsification. Regarding the *h–e* relationship, we now know that *h* is underdetermined by *e* (W. V. O. Quine); that a logic leading from *e* to *h* or reducing *h* to *e* does not exist (Karl Popper); that the degree of partial logical implication of *h* by *e* depends on a subjective free parameter (Carnap); that *h*'s final probability is a function of personal evaluations of prior probability (Bayes's theorem); that if *h* is falsified by *e* one can appeal the verdict by questioning auxiliary hypotheses (Duhem's thesis) or even by reconsidering the whole net of our background theories (O. Neurath's and W. V. O. Quine's holism); and so on. Concerning *e,* we know that, in order to describe even the most elementary bit of evidence (say, "This emerald is green"), one has to choose an appropriate language and, thus, decide which predicates are "projectible" and which are not (Nelson Goodman); that the reference of *e*'s descriptive terms is inscrutable and that two or more *h's,* observationally equivalent with respect to the same *e,* but incompatible among themselves, may exist (W. V. O. Quine); that *e* is theory laden (N. R. Hanson); and so on.

All these results can be considered criticisms internal to the logical positivist program. However, at the same time from the late 1950s onward, another way of looking at science came to the fore. Through the work of Stephen Toulmin, Hanson, Paul Feyerabend, Thomas Kuhn, and others (many of whom were influenced by the later Ludwig Wittgenstein), the history of science and, subsequently and concomitantly, the sociology and rhetoric of science have led to the recognition that historical, personal, cultural, and social factors enter into science. This approach focuses on new problems (theory change, incommensurability, scientific progress, the social construction of facts, the need to persuade, etc.). Recognition of such factors has made it evident that the very constitution of scientific knowledge is itself problematic, and therefore a likely subject for controversy. By their very nature these factors relate to one another and conflict with each other in ways that cannot be captured, let alone resolved or controlled, by logic alone. Accordingly, if the game of science is no longer an ideally logical one played between an objective *h* and an absolute *e* and regulated by a neutral arbiter, but a living one played by real scientists with concrete interests in actual research and cultural situations, then science can no longer aspire to a God-like view of the world. What science says the world is like at a certain time is affected by the human ideas, choices, expectations, prejudices, beliefs, and assumptions holding at that time. In this respect the image of science elaborated by historical philosophers and sociologists has done more justice to science as actually practiced than the logical positivists' rational reconstructions. The new program describes science better.

This is not to say, however, that the historical-social view of science has no shortcomings of its own. Indeed, it has; the pendulum has swung too far. By taking an anti-foundationalist and anti-methodological stand, that is, by replacing data with "social constructions" (Bruno Latour), method with "tricks" or "propaganda" (Feyerabend) or "routine conversation" (Richard Rorty), and theories with "social conventions" (David Bloor),[7] historicist philosophers and sociologists have transformed the well-ordered old dialogue with nature into a sort of battle whose outcome is determined by the forces present and the interests at stake at a given moment. This has left them facing a question symmetrical to that of the early moderns and logical positivists: while the latter had the problem of explaining why science is controversial, which should not be the case according to their own view, the former have the problem of explaining why everything is not enmeshed in never-ending controversy, how controversies determined by *social* factors lead to *natural* knowledge, and why there is so much agreement among practitioners. Saying that scientific objectivity is to be conceived as the resultant state of two vectorial forces, evidence and prior beliefs (Bloor) or that objectivity is to be considered as "solidarity" (Rorty) simply gives a name to the phenomenon, because the crucial challenge is to explain how this resultant state is obtained, and on what grounds such a solidarity is attained. Once granted that the voice of nature is only the voice as *we*—researchers *here and now*—understand it, what guarantees that nature's message is comprehended and that it is not purely an effect of the expectations and interests of the receivers?

In order to meet this challenge, we have to consider the idea of a controversy, its objects and origins, and the role of the many factors involved, as well as the ways controversies are conducted and the means by which they are settled.

Why Controversies Exist

To a large extent what can be disputed is isomorphic to the structure and factors that can be identified in the analysis of scientific theories and their functioning. Disagreements about truth, adequacy, relevance, and appropriateness can originate at or about any of them.

To start describing the objects and origins of scientific controversies, there is an obvious beginning. Whether one holds that the aim of science is to explain phenomena, to solve empirical or theoretical problems, or to describe regularities among data, there is no doubt that science, unlike fiction, art, or religion, is intended to comprehend the facts of the world. How well it comports with these facts is therefore the crucial test of the adequacy of scientific claims. Cosmologist Dennis Sciama has written that "in the middle of the debate we don't know the ultimate outcome and we must be guided by our own sense of the fitness of things"; however, "ultimately, the only test is the pragmatic one of whose ideas succeed the best."[8] Accordingly, it is the relationship between such ideas or cognitive claims and the relevant facts, or the question of whether the former match the latter, and to what extent, that are obvious sources of possible controversies.

Regarding the facts, the early moderns and the contemporary philosophers following them proved to be very optimistic. They held a sort of "facts-are-out-there" view. Sensible experiences, clear and distinct ideas, phenomena, observations, impressions, and sense data, or later, protocol sentences, basic statements, and *Konstatierungen,* were for them reliable building blocks provided by nature to any fair observer. They were not so naive as to not recognize that these blocks are affected by the ways the human mind perceives the world. Francis Bacon, for example, knew that observations could be "idol-laden," and Galileo did criticize the prejudices of the Aristotelians that, having been "imbibed with mother's milk," affected sense perception. However, they were confident that idols, prejudices, expectations, and the like could be expurgated and the mind left in a state to see clearly and think rationally. This is the main reason why they thought that controversies can be settled easily and completely.

Once we abandon this optimistic foundational epistemological basis, things become more complicated (Wilfrid Sellars). Facts, or the experiences that give rise to assertions of or about them, involve interpretation and, thereby, a certain degree of cognitive or social construction. Even the identification of what is to be explained is subject to various cognitive and social processes (see chapter 4).

This explains why in some cases the facts themselves can be controversial. A fact can be questioned as to its existence (see chapter 11), debated as to its inclusion in the domain of things to be explained (see chapter 12), or called into question as to its relevance in testing a theory (see chapters 8 and 9). And there is more. As some scientific fields develop, experiments become the way to ascertain facts, and these involve observation techniques, instrumentation, and experimental designs. Accordingly, all of the factors that are presupposed in assuming that an instrument is working properly, that an experimental device is appropriate, and that a design is adequate can become the subject of controversy. For example, there were those who argued that sunspots and the satellites of Jupiter were nothing but optical illusions created by Galileo's telescope.

In addition, the very nature of experiments can raise problems. Often experiments are conducted in group settings such as laboratories or depend on interactions among individuals and groups who exchange information about their results or techniques. These interactions and the appropriateness of such interchanges and borrowings can give rise to heated personal, nationalistic, or straightforwardly epistemological disputes (see chapter 10). One interesting type of dispute occurs when prior research results and the authority of the researchers who did the work are called into question or are themselves used as evidence in support of the claim (see chapter 13). The grounds for such questions are manifold and of distinctly different kinds, for example, fraud, inadequate experimental controls, contradictory new results, better experimental techniques, and so on.

In the philosophical and, sometimes, in the experimental literature as well, discussions about facts, phenomena, and observations and their relations to procedural techniques and instrumental devices are usually referred to the methodology of science. But methodology is not only concerned with the procedural rules and norms for observational and experimental practices. Methodology also consists of beliefs or metatheories about the evidential relations holding among empirical bases and scientific hypotheses or theories (see chapter 14). Thus, one could dispute whether a bit of data really is evidence for a given claim, what degree of support is provided by certain evidence, or what the overall strength of the evidence is for a given theory. Similarly, controversies can occur over whether the evidence warrants inferences to specific theoretical conclusions or predictions, for example, in the early 1960s in the debate over the quark model or "particle democracy," where the evidence for the quark model was scant indeed at the time. Besides empirical adequacy, methodology has also be taken to include other sorts of adequacy criteria for scientific theories, such as principles of symmetry, simplicity, comprehensivenss, explanatory power, fecundity, and so on. Such a debate occurred about the theory of special relativity and Einstein's focus on the "aesthetic" principle of symmetry rather than on the slim evidence that was available in 1905. Methodology has often been a source and object of scientific controversy.

For all these reasons, though a scientific controversy often appears to be about facts, phenomena, or method, it may reflect broader concerns about the theories or research programs and background assumptions that are implicated in the presentation of the relevant data (see chapter 2). Such concerns are quite complex, and often are not patent. They have been the source of many types of controversy and debate. The different traditions of research of the chemists and the physicists in the recent cold fusion controversy is an interesting example (see chapter 1).

If we now turn our attention to scientific theories, we can say that theories can be substantively described on the basis of the terms they include, their formal structures, and their explanatory force. First, the terms are supposed, minimally, to describe through their interrelations the observed regularities. Accordingly, controversies can involve the reference of terms, and the realistic commitments to such entities, while others focus on the exact nature of such terms or variables, for example, as idealizations, intervening variables, or hypothetical constructs. An example was the debate over the realistic status of the electron. Second, the structure of a theory can be the subject of debate: one may claim it to be a particular mathematical formalism, another a type or instance of a mechanical system, and still another a causal model. The controversies over the analytic formulations of classical mechanics turned on such considerations. More particularly, people can and have argued over the form of the equations or the type of mathematics that are proper for a given theory, and whether a given term ought be used (see chapter 14).

We should add that controversies concern not just the constitutive factors of a given scientific theory. Relations among theories, at a time or over time, also can bring problems to practitioners. The veracity of research traditions and the reliability of certain auxiliary theories have often been the cause of concern, and so of contention (see chapter 13). Further, individual theories are related to other, auxiliary theories whose results, explanatory framework, or structures they utilize in solving their own problems. Thus, some people have argued (wrongly) that the auxiliary theory of optics in Galileo's time could not be used to support his use of the telescope for making observations.

Even more complexity attends to theoretically based debates when one considers that theories, or even research traditions, do not stand isolated. The adherence to a theory or tradition always involves more than just that (see chapter 2). Choice of theories depends upon the education and training as well as on the interests and values, in varyingly broad senses, of the scientists—individually or in groups (see chapter 10). Even more abstractly, theories and traditions often reflect, sometimes unconsciously, higher level philosophical, ideological, or architectonic principles. In a way, such principles can be conceived as "external" factors, since they are borrowed from the surrounding culture. However, once they are made explicit and used in debate, they become "internal" and play a constitutive part in scientific practice or controversy.

Such background elements, as well as their implications, can be objected to in at least two ways. First, if they are conscious or are made conscious, they can be brought in as substantive parameters to a dispute. Second, at a higher level, the very relevance of such factors to the epistemic claims or procedural methods of a science can be debated (see chapter 7). For example, controversy can arise over whether feminist commitments are present in certain critiques of medical research, or over whether such "political" commitments are at all relevant. Or, again by way of example, people can argue about the nature of deterministic assumptions as formal constraints or even as regulative principles.

A few further dimensions of science that have been the object of controversy should finally be noted. First, scientists working with a theory and belonging to an experimental tradition do not cease to be part of broader social and cultural contexts. For example, it may be relevant that such a context involves small-group parameters such as the habits and traditions of a particular laboratory, or it may reflect national interests that have been established for overtly political reasons (see chapter 10). Second, even the context in which science is practiced can become the subject of controversy, as when, for example, a debate takes place about whether basic research can be carried on properly if it is supported by government or corporate sources (see chapter 13). Moreover, we should keep in mind that in every context the scientist fulfills many social roles. These roles may vary from that of an entrepreneur hoping for a money-making patent to that of a timorous lab assistant afraid to offend the boss. Factors of such kinds do affect the manner and substance of scientists' work in various ways. Perhaps most abstract are those debates, such as the Velikovsky incident or the Lyschenko research program, that ultimately touch on whether a theory or program is really science or just pseudoscience (see chapter 12).

On the more concrete end, the technological or social effects of developing a theory or of working on a project may become controversial. Questions of utility, value, interest, and righteousness attend upon the actual or perceived effects of a given project or tradition (see chapter 2). Such effects can be the subject of controversy, both scientific and political. The controversies surrounding the Manhattan Project and the atomic bomb are but one well-known example. Finally, the social or political and the epistemological come together in the deceptively simple question of whether or not something is worth knowing or just a waste of research time.

How Controversies Are Settled

Suppose we now take for granted that, in actual practice, science can be controversial, and scientific controversies may involve, or arise from, each of the factors that constitute it: theories, facts, experiments, epistemic values, philosophical or ontological assumptions, ways of thought, ideological commitments, social environment, and the like. Then three questions remain to be addressed. First, are controversies essential to science? Obviously, it is one thing to say that scientists do argue, and it is another thing to say that they must argue, that is to say, that controversies are necessary to the growth of science. Second, regarding this growth, how is a claim that emerges from a controversy transformed into scientific knowledge? Here, too, it is one thing to say that a party has overcome another in a debate, and it is a different matter to maintain that the outcome of a debate is a step forward in the knowledge of nature. Finally, how are scientific controversies settled? Once again, to say that one party has scored a victory over another in a debate is different from saying that the controversy has been settled and that the claim of the winning party is to be considered better than the one of the rival party.

As the last question seems to provide the key for the others, let us start with it. If science were a zero-sum game with only two players—nature and the inquiring mind—regulated by a finite set of procedural rules, there would be one way alone for settling scientific controversies: to establish whether the former has checkmated the latter. This was precisely the office of method according to the early modern theorists (see chapters 3, 5, and 6), and the function of logic and evidence according to the logical positivists. However, once we realize the facts themselves can be disputed in different ways, that the relationship between theories and facts is not straightforward, that "theory" is a label covering a cluster of different entities or a set of factors of different kinds, that no purely methodological rules can represent what actually happens in discussion, and, as well, that no universal epistemic "I" can replace the personal actors, the two-player game needs to be modified.

The typical situation can be better represented in terms of a dialectical game where one party is committed to refute the claim of another in front of judges. Under this construal, a controversy is like a trial, where the accuser and the defendant are engaged in a dispute in front of a judge or jury. As in a trial, where the judge is asked to establish which view is preferable given the admitted relevant evidence, so in a scientific controversy the judges, that is to say, the scientific community, are called to establish which claim is better given the relevant factors adduced by both parties. Aristotle had already resorted to this judiciary model when he had remarked that in scientific inquiry, one has to "start with a review of the theories of the others," because "to give a satisfactory decision as to the truth it is necessary to be rather an arbitrator than a party to the dispute,"[9] as well as when he wrote that "he who has heard all the contending arguments, as if they were parties to a case, must be in a better position for judging."[10]

For a lot of good reasons, in the last instance boiling down to the degenerate, scholastic use of dialectic, which had become an art of verbal disputations about Aristotle's writings, the early modern theorists abandoned the model of the trial and embraced the methodological one. As Descartes wrote, representing the view of all of them, "[O]rdinary dialectic is of no use whatever to those who wish to investigate the truth of things. Its sole advantage is that it sometimes enables us to explain to others arguments which are already known."[11] This move was successful, and present-day science is to a large extent the outcome of it. Yet, by so doing, and thanks to their epistemological optimism, the view dismissed the role played by the community and, along with it, by the interplay of all the factors, including evidence, of which the community is interpreter and preserver. In this century, Popper has resumed the model of the trial, but he has been careful to distinguish the "verdict of the jury," whose "convictions may be called the 'motives' of the decision" and which "are not justifications," from the "judgment of the judge," which "needs, and contains, a justification."[12] Yet like the positivists, Popper was careful to exclude any role of the community from the validation of scientific claims: from his point of view, as from theirs, science remains a game with two players, one who advances hypotheses and another who only says "Yes" or "No," because, as he wrote repeatedly, "what ultimately decides the fate of a theory is the result of a test, i.e. an agreement about basic statements,"[13] or because "in the natural sciences . . . experience [is] the impartial arbiter of their controversies."[14]

If we take the model of the trial, or an arbitrated dispute, seriously, then dialectical and rhetorical argumentation—taken in the technical sense of the art of attacking a view and defending it, as well as of convincing or persuading an audience of the merits of a view—helps provide an answer to our question.

In the first place, we recognize that scientific controversies are conducted in terms of dialectic and rhetoric. This means that the parties not only advance arguments (appeal to *logos*), but also present their arguments in a certain way in order both to create a favorable state of mind in the community (appeal to *pathos*) and to give their own views a certain authority (appeal to *ethos*). Arrangement and style, which are the ways of such presentations, are thus neither illegitimate means of persuasion nor mere verbal ornaments or embellishments that might be expurgated by the practice of science (see chapters 3 and 9). More specifically, a style of reasoning, especially if taken in the sense that has been recently investigated by A. C. Crombie, is not an external, dispensable, linguistic dress of a proof that is already given and the same for everybody, everywhere, and at any time, but it is precisely what constitutes a proof for a certain community in a certain period. In actual practice, there is not a process of proving a claim and a logically or chronologically different process of presenting it or making it acceptable or palatable. Dialectical and rhetorical processes merge into a single activity, which ends up with the argumentative victory of a group over another and the persuasion of the community, and on which the acceptance of the claim, that is, its very constitution as a scientific claim, depends. Take this dialectical and rhetorical process away and the the public character of a scientific claim vanishes into thin air.

Here is where dialectical and rhetorical analysis proves to be more flexible and adequate than the methodological one. Unlike method, at least the method as conceived by the early moderns, styles of reasoning have to be taken as changing over the time; even logic itself has changed over time. Efficient in certain circumstances, certain methods may be inefficient or even inapplicable in others. For example, "arguing from evident principles" by using a deductive model of reasoning is an effective means of conducting a controversy in a community that takes geometry as the paradigm of science, but it is quite ineffective for another community that works according to Baconian standards. And the same holds with such styles as "deducing from phenomena" or "deriving consequences from hypotheses" or their modern cognitive science counterparts (see chapter 4).

In the second place, if controversies are conducted in terms of dialectics and rhetoric, then the "logic" of refutation and persuasion establishes the way they are settled (see chapters 1 and 3). As Aristotle put it, to refute an adversary, one has to produce "a deduction to the contradictory of the given discussion;"[15] but not just any deduction performs the job. What is necessary in addition is that the deduction provided by one party be a sound or convincing *argument* for the other and for the community. By itself, a deduction is not a convincing argument. It becomes so if the party that provides it argues correctly (that is, according to a valid logical form) either from certain concessions made by the latter party or from certain bits of knowledge, such as facts, theories, assumptions, and values, that are accepted by the community (Aristotle's *endoxa*). We may then say that, when a party has scored a dialectical victory over another by producing a convincing argument, a controversy is settled in favor of that party (see chapter 3). Here "convincing" is not to be taken in a psychological or sociological sense, but in a logical or, more precisely, argumentative sense. The notion is normative: if an argument proceeds validly from concessions or from endoxa accepted by the community, then it is such that it *ought* to convince, no matter whether the party to which it has been addressed actually changes its mind or not (see chapters 3 and 14).

This logic of refutation and persuasion helps explain why the settlement of a controversy can hardly be definitive. In the first place, during a controversy endoxa may be questioned as to their relevance to the case in question (see chapters 2 and 7). This is typically the case with controversies about theories during radical changes. Second, even when the relevant endoxa are recognized and shared, they may be challenged and changed, or they may be given different weight or different interpretations and configurations (see chapter 11). Conflicts of values, for example, conflicts about how simplicity is to be understood, are a case in point. Third, when a party has provided a refutation, the other may try to repair the alleged inconsistency by using one or another of the several moves that are typical of the situation or by inventing new ones. One may admit an inconsistency but deny it is serious, or withdraw a concession, or resort to higher levels of discussion, or retort that the rival view is also affected by difficulties and inconsistencies (see chapter 8). Thus, algorithmic decision procedures for closing a controversy do not exist.

This does not mean that controversies have no end. In actual practice they do, although the closure point cannot be fixed by general logical or methodological rules. What we can say from a logico-philosophical point of view is that the participants in a controversy have at their disposal an escape tree that they can use to avoid the inconsistencies and difficulties they may happen to be embedded in (see chapter 1). Only when one has explored all the branches of such a tree unsuccessfully—for example, because one is not able to continue with the debate, because there happens to be no answer to the objections, or because any answer advanced meets greater difficulties—can one be considered rationally defeated and the controversy rationally closed. Of course, one may still resist, but in such cases resistance is psychological and the community has the right to dismiss it. If rationality is linked with argumentation, then the notion of rational attitude is as normative as that of a convincing argument. Rational is what *ought* be done (for example, accepting or rejecting a theory) if the arguments for it are convincing (see chapter 1).

When Controversies Are Transformed into Knowledge

This leads us to the second question. When a controversy is closed, the winning party is the one that has resisted most if not all attacks and, although often not immediately, persuaded the community. In Aristotle's terms, it becomes an *endoxon* claim, that is, one that is accepted or held to be true by most if not all the members of the community or, among them, by the most influential or reputable (the leaders or the experts). But how can such a claim be considered a bit of scientific knowledge? How can it to be said to refer to nature? After all, it emerges as persuasive from a dispute, and as Plato had already remarked in his polemics against rhetoric, given an appropriate influence on the "psychology of the masses," people can be convinced of anything, for example, even that a monkey is a horse.[16] We have already remarked that this is a crucial point, especially for social constructivists, and it is certainly a challenge for any philosophy of science that acknowledges the role of the community in the formation of scientific knowledge (see chapters 4 and 13).

Once the Paradise of the early moderns and the logical positivists is admitted to be lost because science is recognized as having more dimensions than logic and evidence, and the symmetrical Hell of the social constructivists is also recognized to be inadequate because scientific knowledge needs to be something firmer and more durable than a cultural fashion, the middle way between these two extremes may be alluded to, but has not been found yet. It is not easy to see where to go. Some hints at a promising direction, however, can be detected.

Plato's objection against rhetoric is attractive but it can be rejected. The introduction of dialectic and rhetoric into scientific discourse does not require that it is not objective, in the sense of being both intersubjectively testable or agreed upon and in touch with nature. Indeed, a scientific claim acquired through a dialectical and rhetorical process is objective in the former sense, because it is compelling even for those who resist it but have no convincing counterarguments to oppose it. If a claim, after an open exchange of arguments and counterarguments, becomes endoxon, the resistance of the losing party can still retain any degree of psychological or emotional strength, but the community has the right to ignore it. As we have shown, rationally speaking, regarding the arguments advanced in the dispute, if the claim is justified, the losing party cannot but retreat, unless the fairness of the dispute itself is questioned, in which case the burden of proof shifts to the questioner. Einstein's strong conviction against the logical and philosophical feasibility of the "orthodox" interpretation of quantum mechanics is a good case in point. When he wrote to Born that he could not "base this conviction on logical reasons," but could only produce his "little finger as witness,"[17] he implicitly recognized that, in terms of arguments, he was defeated.

A scientific claim acquired through a dialectical and rhetorical exchange is also objective in the latter sense. Plato's objection, which has been repeated many times by a host of philosophers, including Hobbes, Locke, Kant, Schopenhauer, and in this century Chain Perelman, stems from two presuppositions: a dichotomy exists between the verbal world of human communication, or beliefs or convictions, and the real world of nature "out there"; and a privileged organon (method) also exists that allows us to discover what that world of nature is like in itself. The latter presupposition is the underlying view that is now so difficult to maintain that it has been rejected almost unanimously. The former stems from an extreme form of metaphysical and epistemic realism that is also difficult to defend.

Our discourses in science are not about a "world of paper," as Galileo was afraid they would be once we deprive them of the guidance of a fixed method; they are about the world of nature. And such they remain even if we recognize that scientific discourse, like any other discourse, has dialectical and rhetorical dimensions. This is so because scientific dialectic and rhetoric do not dismiss empirical evidence, which is our privileged access to nature; they simply combine it with other factors. Admitting that scientific knowledge is not based on empirical evidence alone, but is the product of a set of historical, cultural elements, including evidence—because evidence is not out there present to our senses as an apple on a tree ready to be picked—is tantamount to denying that it is a sort of mirror, but does not require maintaining that it is a mere construction. Dialectic and rhetoric, once they are recognized as the means of formation of scientific claims, compel us to understand realism in a different way, not to dismiss it, although it is an open philosophical question to outline that way properly and explicate it analytically.

Whether Controversies Are Essential

The last question, about the necessity of controversies, has now to be addressed. Obviously, the necessity of controversies cannot be inferred from any historical record. Moreover, the history of science witnesses both that there have been important theory changes that have been controversial (Copernicus, Darwin, etc.) and that there have been other major achievements (for example, the unification of electricity with magnetism) that have not been. There are also cases in which controversies are explicitly avoided (see chapter 14). What, then, can be concluded?

Sociologically speaking, science is sometimes an antagonistic game, and sometimes a cooperative enterprise. In actual practice, scientists aim both to collaborate, through the interchange of information, and to compete, with an exchange of criticism. Collaboration and criticism, however, are not mutually exclusive. Collaboration does not exclude criticism, because criticism can be constructive and offered in the spirit of mutual pursuits. This is the case, for example, with the spirit behind the peer review system of scientific publication or with the idea behind the request of the reproducibility of results. In another sense, too, criticism and competition do not exclude collaboration, since they need not be acrimonious or hostile. Although personal idiosyncrasies or group interests may play a role in triggering a competition (see chapters 10 and 13), scientists, in fact, believe that the final result to be reached is public, accessible to anyone, and such that anyone can agree to it, independently of any personal, social, cultural factors. Despite the many possible individual or social differences, the ethics of the profession is communitarian, and this is the reverse of the common view that the epistemology of science is universal. It is an empirical question, requiring further empirical research, to establish, case by case, how and when the differences become reconciled, and how and when they prove to be irreducible.

From a philosophical point of view, one general lesson can perhaps be drawn about controversies. Again, it is linked with the notion of the epistemic *endoxa,* that is, the set of all the factors in terms of which scientific discourses are conducted and settled. As the accepted endoxa at a given time constitute the body of scientific knowledge or what is called science at that time, they are, or act as, the touchstones of any new claim. There is then no room for controversies in at least three situations. First, when a new claim fits with the accepted endoxa, it is immediately absorbed into the body of knowledge (see, for example, the refinements of parameters, the detection of new phenomena, etc.). Second, when the new claim does not fit with the endoxa but scientists do not want to change or cannot change them, for example, because of their merits in many fields, the claim is rejected as absurd, or neglected as unimportant, or maintained as a strange anomaly (for example, the anomalous perihelion of Mercury in astronomy, the paradoxes in quantum mechanics, the lack of fossil records or of intermediate stages of species in evolutionary biology, etc; see chapters 11 and 13). Third, there is no room for controversy when the new claim alters the accepted endoxa, but the alteration is recognized to be useful, for example, because it rearranges them in a more aesthetic way, because it unifies them, or because it allows scientists to discover that one or more of them are redundant (see the synthetic theory of evolution, Maxwell's electromagnetism, etc.).

Controversies are fated to arise when the new claim does not fit with the accepted endoxa and the community cannot neglect it. This creates a problem of consistency that cannot be solved unless the set of accepted endoxa is altered in some way or another. Rival parties then emerge with different solutions as to how a remedy can be found for the anomoly, and which endoxa need to be altered or modified. The deeper and wider the scope and effects of the needed alterations, the greater and more important the controversy. Scientific revolutions are controversial not because of the dramatic changes in the content of the body of scientific knowledge they introduce, but because of the dramatic alterations they bring about in the traditional, accepted endoxa that are necessary to allow for such changes.

Although this may be considered a general pattern, nothing specific can be drawn from it. Neither the phenomenon of controversies nor its actual ways of practice can be inferred from a general philosophy of the history of science. Philosophy can analyze the phenomenon; help understand its several dimensions; make clear its sources, origins, and forms; and reflect upon the tools used, but it is up to the history of science to examine, case by case, how controversies in actual practice go, and why they go the way they do. These examinations need to use tools and concepts from sociology and other empirical disciplines, for example, discourse analysis, economics, and so on. Philosophy, history, and the sociology of science work better if the one does not neglect the results of the others.

About This Volume

The contributors to this volume have different views about different subjects, but all of them agree on the last point of the introduction. This, besides their kindness, patience, and spirit of collaboration for which the editors are greatly indebted to them, has prevented them from engaging in a meta-controversy about scientific controversies. They have addressed the issues from the points of view of philosophy, logic, argumentation theory, history, cognitive science, and sociology. This interaction among them is an assurance that mutual cooperation is possible.

Some volumes that issue from conferences and congresses deal with many and varied topics, while others deal only with the single issue for which they have been devised. This volume, despite its diversity, belongs to the latter category. This does not diminish but rather exalts the role played by the Istituto Italiano per gli Studi Filosofici, Naples, which provided to the contributors to this volume the unforgettable atmosphere they had in Vico Equense, 23–26 June 1993. All of the contributors express their gratitude to the Istituto.

Notes

Larry Laudan in his *Science and Values* (Berkeley: University of California Press, 1984) has told a story similar to ours about controversies and consensus in science. Of course, Laudan has a different idea about how to make the story end happily.

1. K.Popper, Conjectures *and Refutations,* Routledge and Kegan Paul, London 1963, p. 216.
2. F. Bacon, *Novum Organon,* in J. Spedding, R. Ellis, and D. D. Heath (eds.), *The Works of Francis Bacon* vol. 4, Longman, London 1860, p. 63.
3. See G. Galilei, *Il Saggiatore,* in *Le Opere di Galileo Galilei* (20 vols.), national edition, ed. A. Favaro, vol. 6, p. 269: "Therefore, just as there is no objective half-way between truth and falsehood, likewise in necessary demonstrations, either one draws a compelling conclusion or one ends up in inexcusable paralogisms, leaving no room for using limitations, distinctions, distortions of the meaning of the words, or other tricks in order to stand on one's own feet."
4. R. Descartes, *Rules for the Direction of the Mind,* in J. Cottingham, R. Stoothoff, and D. Murdoch (eds.), *The Philosophical Writings of Descartes* (2 vols.), vol. 1, Cambridge University Press, Cambridge 1985. p. 11.
5. G. W. Leibniz, *Philosophische Schriften* (7 vols.), ed. C. I. Gerhardt and G. Olms, Hildesheim, Germany 1961, vol. 7, p. 237.
6. I. Kant, *Critique of Pure Reason,* ed. Norman Kemp Smith, MacMillan, London 1929, p. B vii.
7. The *loci classici* of these views are, in particular, B. Latour and S. Woolgar, *Laboratory Life: The Construction of Scientific Facts,* Princeton University Press, Princeton, N.J., 1986 (2nd ed.); P. K. Feyerabend, *Against Method,* New Left Books, London 1975; R. Rorty, *Philosophy and the Mirror of Nature,* Princeton University Press, Princeton, N.J., 1980; and D. Bloor, Knowledge *and Social Imagery,* Routledge and Kegan Paul, London 1976.
8. D. Sciama, "Cosmological Models," in *Cosmology Now,* British Broadcasting Corporation, London 1973, p. 56.
9. Aristotle, *On the Heavens,* 279b 5–12, in *The Complete Works of Aristotle* (2 vols.), ed. J. Barnes, Princeton University Press, Princeton, N.J., 1984.
10. Aristotle, *Metaphysics,* 995b 2–4, in *Complete Works.*
11. R. Descartes, *Rules,* p. 37.
12. K. Popper, *The Logic of Scientific Discovery,* Hutchinson, London 1959, pp. 109–10.
13. Ibid., p. 109.
14. K. Popper, *The Open Society and Its Enemies* (2 vols.), Routledge and Kegan Paul, London 1945, vol. 2, p. 218.
15. Aristotle, *Sophistical Refutations,* 165a 3, in *Complete Works.*
16. Plato, *Phaedrus,* 260c, in *The Collected Dialogues,* ed. E. Hamilton and H. Cairns, Princeton University Press, Princeton, N.J., 1987.
17. A. Einstein and M. Born, *The Born-Einstein Letters,* trans. I. Born, Walker, New York 1971, p. 158.

Part I

THE STRUCTURE OF SCIENTIFIC CONTROVERSIES

1

Patterns of Scientific Controversies

PHILIP KITCHER

Current thinking about scientific controversies is dominated by two models. One, popular among logical empiricists and their intellectual descendants, proposes that scientific controversies are ultimately settled by experimentation, evidence, and the exercise of reason. The principal rival to this rationalist model, which I shall christen the anti-rationalist model, supposes that context-transcendent canons of reason and evidence have no power to resolve major scientific controversies. Both models come in simplified and more sophisticated versions, and in recent years much ink has been spilled by sophisticates on one side deriding the oversimplifications made by the other.

The most primitive type of rationalism proposes that scientific controversies are resolved by designing and carrying out crucial experiments.[1] Rival protagonists agree that each theory has determinate implications about what will be observed in a particular experimental setup; the experiment is run, nature speaks, and at least one of the disputants retires defeated. Unfortunately, as has become commonplace since Pierre Duhem, nature typically does not offer clear indictments. More sophisticated forms of rationalism attempt to show how assignments of credit and blame can be localized, perhaps by considering sequences of theories within a research program (Lakatos 1970), batteries of bootstrap tests (Glymour 1980), or posterior probabilities (see the recent accounts of Bayesianism offered in Howson and Urbach, 1989, or Earman, 1992).[2]

Anti-rationalists often suppose that rationalists are committed to the simplest position, and that mere invocation of Duhem or W. V. O. Quine is checkmate.[3] The simplest types of anti-rationalist model suggest that the gap between the evidence and the resolution of a controversy is bridged by the interests of (some of) the disputants. More subtle versions emphasize the role that decisions about the social order play in the specification of canons of reason and evidence. Instead of thinking of scientists as responding in some automatic fashion to their interests, anti-rationalists allow that, in a certain sense, controversies are resolved through the appeal to reason and evidence, denying only that there are universal standards of evidence. Claiming that problems about the natural order are intertwined with problems of social order, they suggest that the ultimate explanation for the resolution of a dispute in a particular way will invoke factors that go beyond any canons of universal rationality.[4]

I believe that our understanding of the issues between rationalists and anti-rationalists will be advanced if we begin by understanding the variety of courses that scientific controversies take *before* they are resolved. In what follows, I shall try to defend a Tolstoyan thesis: each resolution of a scientific controversy proceeds in much the same way; all unresolved scientific controversies are unresolved in their own way. I hope that my development of this thesis will capture the genuine insights both of rationalists and anti-rationalists.

1

Many discussions of scientific controversies (and, I suspect, almost all rationalist discussions) present a stereotype. Two participants are involved, at least one of whom is a single figure whose views will ultimately triumph. Thus, the chemical revolution is about the struggle between Lavoisier and Priestley; the Copernican controversy centers on the dispute between Galileo and Pope Urban VIII (or, perhaps, some anonymous Church Aristotelians, real-life counterparts of Simplicio). But many scientific disputes *evolve.* The position that is finally adopted may not be among those that sparked the initial controversy (so, for example, the Great Devonian Controversy was concluded with consensus on a position that had initially seemed quite impossible).[5] Moreover, many controversies involve advocates of many different positions that are interrelated in complex ways (there is important variation among sixteenth-century astronomers and physicists about the proper relations of the disciplines, the character of astronomy, and, of course, the status of Copernicanism).[6] Finally, the focus of a controversy is not always some statement, hypothesis, or theory. In some instances, what is in dispute is the significance of particular questions or the reliability of an instrument (a current controversy in evolutionary theory concerns the significance of questions about adaptation, and the debate between Boyle and Hobbes centered on the legitimacy of the air pump).[7]

In understanding the variety of scientific controversies, I propose to start with the idea of a scientific community in which there is both a consensus practice and a diversity of individual practices.[8] Practices are multidimensional, involving other things besides statements. Specifically, I take an individual practice, viewed as a representation of a scientist's cognitive commitments, to contain a language, a set of questions taken to be significant, a set of explanatory schemata, a set of accepted statements about the natural phenomena that are the community's distinctive concern, experimental techniques, instruments, assessments of authority, methodological canons, and statements about how the community's project relates to human well-being.[9] Now, many scientific controversies surely do focus on some disputed statement about the natural world, but others are concerned with different components of practice. Besides the examples already noted, there are controversies about legitimate scientific language (current disputes in systematics), explanatory schemata (worries about final causes in the seventeenth century), and the implications of some part of science for human well-being (the sociobiology controversy, the debate about the human genome project).[10]

How does the consensus practice of the community relate to the individual practices of the community members? The obvious first answer is to declare that the consensus practice of a community is a multidimensional entity each component of which contains exactly those elements universally shared within the community. I suggest that this answer identifies the *core* of the community's commitments, but that it is extended by adding elements based on the assessments of authority within the community. Thus, if all members of the community recognize a certain class of community members as authoritative on a particular range of questions, then any answers to such questions that are shared within the pertinent class belong to the consensus practice.

Individual practices plainly outrun consensus practice. Hence, the basis for controversy is always present.[11] A controversy erupts when some individual (or group of individuals) within the community proposes that a particular component of consensus practice should be modified to become more closely aligned with the individual practice(s) of the proposer. Although an individual's practice may extend consensus practice in many ways, it is typically impossible for any scientist to engage in many controversies at once. Moreover, because the attention of the community is limited, only a few proposals for modifying consensus practice can be seriously considered at any one time.

The first question that we should ask is, How do controversies begin? I understand this question as inquiring into the causal factors that underlie the emergence of some differences in individual practice as matters of public debate. Here, there is surely no doubt that prior assessments of the significance of questions play some role.[12] Given the established importance to astronomy of questions about the positions of planets, it was hardly surprising that, when Copernicus proposed what he claimed to be a superior scheme for computing those positions, he received the attention of the astronomical community. However, other factors may contribute to the outbreak of a controversy. So, for example, it is likely that communitywide assessments of the talents of those who propose the modification of consensus practice influence whether or not the proposal is seriously discussed: Pons and Fleischmann began a controversy about cold fusion, at least in part, because both were highly regarded within the electrochemical community.[13] Similarly, in the controversies between Newton and Leibniz and in the debate about N-rays, considerations of national allegiance played important parts.[14]

If we were looking down on the history of science from a more detached, ethereal vantage point, we would like to see controversies as fruitful. We would hope that controversies would regularly lead communities to states in which they were closer to achieving their scientific aims. For present purposes I shall be blunt and dogmatic about the nature of these aims.[15] Unlike many realists, I do not think of the aim of inquiry as the construction of the complete true story of the world.[16] Unlike antirealists, I do not conceive of that aim as simply that of "saving the phenomena" or "solving problems" or simply as "continuing the conversation."[17] Part of the aim of science is to achieve accurate representations of parts of nature, but the importance of representing those parts accurately derives from our desire for a picture of how the world works. Lots of true statements about nature are intrinsically insignificant because they contribute nothing to our total explanatory scheme. Science, I suggest, aims at significant truth, and sometimes is prepared to sacrifice truth in the interests of significance. The pertinent notion of significance is grounded in the goal of systematic explanation of as many facets of nature as possible. So, to put my cards completely on the table, I take the aim of inquiry to be the provision of a maximally unified set of explanatory schemata that will generate the largest possible set of true instantiations.[18]

Scientists' assessments of significant questions are based on their views about how this unified explanatory scheme will be achieved, and where the current gaps are. The recognition of gaps points the way to the significant open questions, often organizing those questions into hierarchical structures.[19] *If* their evaluations are correct, and *if* the community's attention is typically directed to disagreements about significant questions, then we might expect that scientific controversy will generally prove fruitful (although direct attempts to tackle the most significant questions might be premature, and a well-functioning scientific community should surely have institutions for terminating unprofitable debate on such issues). By contrast, if the community's attention is often diverted to side issues of minimal importance—perhaps because there are deep divisions along national lines, perhaps because of an overdeferential attitude to certain people with high prestige but narrow concerns—then the investigative effort of the community may be expended in epistemically worthless ways. In general, we may inquire after the extent to which the factors that cause scientific controversy to erupt in a community are likely to lead the community to improve its epistemic state.

The issues that arise in understanding the eruption of scientific controversies are plainly questions in *social* epistemology. To resolve them requires us to identify better or worse distributions of investigative effort, and to examine the effects of various types of forces and conditions in moving the community closer to or farther from a good distribution. Picking out examples of poor distributions is not hard: if all members of the community campaign for their own pet modifications of consensus practice, none listening to the proposals of others, there is effectively no controversy and the community dissolves; but if all members of the community are obsessed with some proposal for relatively trivial modification, matters are not much better. Much harder is the task of finding general principles that characterize good or bad distributions, and analyzing the general effects of various considerations. Are national allegiances always detrimental? At this stage of philosophical inquiry, for all the attention lavished on scientific controversies, we still lack satisfying answers to such basic questions.[20]

2

Many of the most famous scientific controversies persist for a substantial period of time. More exactly, the time interval between the first public discussion of a proposal for modifying consensus and the community's coalescence on a new consensus with respect to the issues raised in that proposal is often measured in decades. The resolution of the questions raised by Copernicus took a century; Wegener's proposal about the motions of the continents was debated, off and on, for fifty years; Lavoisier argued with champions of the phlogiston theory for almost twenty years; and even the relatively small-scale Great Devonian Controversy involved intensive discussion among all the British geological elite for ten years. Rationalist models that emphasize the epistemic achievements of the heroes frequently try to argue that, from the beginning, there were compelling reasons for preferring the ultimately triumphant position.[21] But as Thomas Kuhn and numerous writers after him have made abundantly clear, we cannot suppose that controversy persisted for many years because those who were ultimately defeated were just too blind (or prejudiced) to recognize the decisive considerations that favored the opposing views.[22]

Rationalist accounts face the problem of accounting for the *persistence* of controversy. If there are rational grounds for favoring one point of view over another relatively early in the controversy, then some explanation must be given of why apparently intelligent people continue to resist. If there are no such grounds, then it will be important to explain why the appropriate attitude is not one of agnosticism, and why those involved do not take this attitude. From the community's perspective, of course, it is a good thing for scientists to explore ideas that have very little support—and, if belief is a necessary concomitant of serious exploration, it is desirable that there should be people in the community who firmly believe novel claims that leap beyond the available evidence.[23] The epistemic well-being of the community may be promoted because the clash of proposals eventually results in a superior line of reasoning or even in a position superior to any that had been available at earlier stages.

Let me make this discussion more concrete by considering one of the most well-studied examples in the history of science. After working on the problem of the motion of the planets for nearly four decades, Copernicus received, on his deathbed, the first copy of his *De Revolutionibus Orbium Coelestium* (1543). Much philosophical discussion has focused on features that might have made the system of *De Revolutionibus* superior to Renaissance versions of Ptolemaic astronomy. The ground rules are that there is a geocentric system that is observationally equivalent to Copernicus's heliocentric system, and that the Copernican system reduces the number of epicycles and eccentrics but by no means eliminates them. Give that the dispute could not have been decided by appeals to accuracy or to the most obvious types of simplicity, the philosophical question has thus been, What was the rational basis for preferring Copernican astronomy in 1543?

Now, this question strikes me as profoundly misguided, because it presupposes that we should introduce the rational/irrational dichotomy and apply it in favor of Copernicanism at the very beginning of a protracted controversy. We can find our way to a better idiom by considering the strains of one of the most sensitive answers to the traditional philosophical question. Clark Glymour (1980, ch. 6) has deployed his account of bootstrap testing to argue that there are important differences between the Copernican system and any observationally equivalent Ptolemaic version. Although Glymour is careful to refrain from attributing his evaluation of the situation to Copernicus, it is noteworthy that his reconstruction attends to just those "mathematical harmonies" that Copernicus made an important feature of his case.

Glymour argues that there are parts of the Copernican system that are testable, whose counterparts in the Ptolemaic system are not testable. Among his examples are Copernican theses about relative distances and claims about periodicities in the planetary motions. However, we ought to ask two questions. (1) What is the connection between the testability of parts of the Copernican system and the rational preferability of Copernicanism as a whole? (2) Are there testable constituent claims of the Ptolemaic system whose counterparts are not testable within the Copernican system? Now, it does follow from Glymour's account of testing that Copernican astronomy offers a type of information that its Ptolemaic rival does not, and that, assuming the correctness of other parts of the system, this information is tested and confirmed. Whether this should be construed as a situation in which Copernicanism has stuck its neck out to make more definite commitments that have survived potential defeat at the hands of experience, or whether it simply testifies to the fact that the ancillary claims of Copernicanism are well designed to support its faulty views about such matters as the relative distances of the planets, needs to be argued.

More important, however, when we treat Copernicanism realistically, as some but by no means all of Copernicus's followers did, it appears that the answer to the second question is that there are indeed parts of an observationally equivalent Ptolemaic system whose counterparts within the Copernican framework are not testable. Consider the Ptolemaic claim that the annual velocity of the earth is zero. Given the Ptolemaic hypothesis that the distance to the fixed stars is not large in comparison with the earth-sun distance, we can use the observed stellar parallax to compute the velocity of the earth, thus testing and confirming the Ptolemaic claim that the velocity is zero. Of course, the Copernican method of responding to the absence of observed stellar parallax is to retract the assumption about the distance to the fixed stars, effectively leaving it impossible to compute the annual velocity of the earth. Should we count this as a triumph for Ptolemy? Perhaps our post-Copernican inclination is to say that this alleged "calculation from the phenomena" only reveals the false coherence of parts of the Ptolemaic system. But if we do say this, question (1) arises again to haunt us, for, as I have noted, it is possible to take an exactly parallel attitude toward the alleged advantages of the Copernican system.

The moral of this story is that the epistemic situation, as of 1543, is far more confused than Glymour makes it out to be—and this, I believe, is a problem that Glymour shares with other rationalists. There are important considerations that favor retention of the geocentric approach and other considerations that support heliocentrism. In such situations, it is important for the community that the decision not be made too early, that both perspectives be thoroughly explored. Now, it is possible that this could be accomplished through a process of convening the potential participants, allocating them to one line of investigation or another, and having them pursue either heliocentrism or geocentrism with an attitude of studied agnosticism and detachment. Whether this is *psychologically* or *socially* realistic is quite another matter. Thus, there arises another important question, How are scientific controversies sustained?

Miriam Solomon (1992) has argued that welcome cognitive diversity can be sustained within scientific communities by the propensity individuals have to overrate the importance of particular salient effects. Thus, in the genuinely mixed predicament of the astronomical community in the second half of the sixteenth century, some scholars might have been moved to pursue Copernicanism in a thoroughly undetached, committed, fashion because they were struck by the "mathematical harmonies" of the system and were quite oblivious to the real difficulties that it faced. Others might have been convinced by the Aristotelian arguments against the possibility of the earth's motion and have been impervious to the charms of heliocentrism. Elsewhere, I have suggested that distributions of cognitive effort might be achieved in quite different ways, if scientists are driven by desires to achieve fame and fortune. For, when the epistemic situation is unclear, an underrepresented point of view will appear to offer avenues for the advancement of a career. Thus, advocates or opponents of heliocentrism might have been motivated by the desire for advancement—and, I believe, such motives can be identified in the cases of Giordano Bruno, in some of the Dominicans who denounced Galileo, and, perhaps, even in Tycho Brahe.[24]

Valuable division of cognitive labor could be achieved either in Solomon's preferred way, or in mine. But, of course, both suggested explanations abstract from the complex decision-making processes of real people. Even if some "salience heuristic" can lead scientists to overvalue certain kinds of considerations, we might expect that the rough-and-tumble of scientific debate would force them to come to terms with the factors they have slighted. However, once having staked one's reputation to the defense of one point of view, motivations of career advancement might reinforce the initial decision. In general, we might anticipate that the distribution of scientific effort will reflect complicated interactions among a variety of factors: curiosity about the phenomena, desire to preserve or enhance one's reputation, judgments about the importance of various types of evidence, the need to respond to those who disagree. I doubt very much that scientific communities ever reach optimal distributions, but, on the basis of analyses I have offered elsewhere, I would claim that complete homogeneity is frequently a very poor distribution in terms of advancing the community's epistemic state, and that the complex forces of human motivation and evaluation typically lead it to avoid this unsatisfactory outcome in favor of something substantially better.[25]

Philosophical attempts to make the ultimately triumphant position rationally preferable even at early stages in the controversy seem to me to be doubly unfortunate. In the first place they disguise interesting issues about the ways in which cognitive variation is kept alive. Second, they distort the character of the decision problem facing the individual scientist. Given two or more imperfect rivals, scientists have to commit themselves on the basis of inconclusive reasons.[26] The situation is analogous to one in which one wants to find one's way out of a wood and recognizes various possible paths through the trees. Standing pat achieves nothing. One must commit oneself and pursue a particular course. Even though there are no guarantees—maybe even no high probabilities—of success, commitment is hardly irrational.

3

Rationalists can, I believe, concede all the points that I have been making. What surely concerns them most is that controversies are *closed* by reason and argument. Reinvoking Hans Reichenbach, they might concede that the emergence and maintenance of scientific controversies take place in all types of ways and involve all types of causal factors. But if errors in reasoning and the drive to advance one's career play a valuable role in the initiation and sustaining of scientific controversies, can we really suppose that these causal factors unaccountably disappear when the controversy comes to an end?[27]

Anti-rationalists deny that controversies are closed by appeal to evidence and argument because they think that there are no ways to formulate universal standards of good reasoning that will license the acceptance of proposed modifications of consensus practice. The issue between the sophisticated rationalist and the sophisticated anti-rationalist is fought over the following thesis:

> [R]: In major scientific controversies, change in consensus practice is accomplished because those deemed by members of the community to be authoritative with respect to the range of topics in questions concur in recommending the change.[28]

The explanation for this agreement consists in the authorities' undergoing processes that are superior (in the sense of promoting acceptance of significant truth) to any that generated cognitive commitments at earlier stages of the controversy.[29]

The anti-rationalist counterproposal is that, at the times of closure, there are always alternatives to the accepted modification that could be adopted with equal reason.

In many circles, it is currently popular to defend this claim by invoking the underdetermination thesis. Once we have achieved the insight that "any statement can be held true come what may," we should see that maintaining any (consistent?) alternative proposed at an earlier stage of the controversy is an open possibility. If communities do reach consensus, it is because they find that one particular "way of going on" is sanctioned in the "form of life" that they have antecedently chosen—or, perhaps, because they use the occasion of the controversy to fashion a new social order (as has been claimed for the resolution of the Boyle-Hobbes debate).

Now, there are many different underdetermination theses on the contemporary philosophical landscape, and it is important to be explicit about which of them is to be invoked in arguing against [R]. One thesis stems from Duhem (1906). When we make theoretical predictions about what will be observed in a given situation, there are typically many hypotheses in the derivation of the report of expected observation. Thus, if the expected result is not found, deductive logic by itself only tells us that at least one of the bundle of hypotheses must be mistaken. But Duhem himself did not draw the conclusion that the innocence of any hypothesis we choose can be preserved, contending instead that conflicts with the deliverances of experience are resolved by the *bon sens* of scientists.

A major source of more ambitious underdetermination theses is the doctrine of space-time conventionalism, founded in the recognition that there are alternative observationally equivalent ways of constructing a physics-plus-geometry. Historically, the doctrine has been tied to a distinction between observation language and theoretical language (as well as verificationist approaches to the meaning of statements in the theoretical language). Quine's influential versions of the underdetermination thesis result from denying the distinction between theoretical language and observation language (Quine 1952 §6). Space-time conventionalism proposed that there is a class of observation statements with which space-time theories must be consistent. Generalizing, and abandoning the idea that observation statements are privileged, we obtain one form of the underdetermination thesis: there are any number of consistent total theories that will be compatible with "the totality of stimuli." What is it to declare that a theory is "compatible with the totality of stimuli"? On one reading of Quine, the reading encouraged by his remarks about maintaining statements about brick houses on Elm Street, the "stimuli" exert no constraints whatsoever. So the underdetermination thesis reduces to the truism that, given any inconsistent set of statements, we can always find a consistent subset containing any chosen element of the original set that is itself consistent. (Moreover, in light of Quine's claims about the revisability of logic, it is not even clear whether the consistency requirement does any work.)

Quine's own writings oscillate between this extremely weak formulation, and much more exciting theses. He sometimes supposes that underdetermination will persist, given an "ideal organ of scientific method," where I take this organ to include rules for responding to the stimuli (thus enabling the stimuli to exert some kind of pressure on the total system) as well as the methodological canons of simplicity and conservatism that Quine frequently mentions.[30] This far stronger version of the underdetermination thesis is the source of the anti-rationalist claim that only in the light of antecedent decisions about the social order are there rules that dictate a particular way of responding to experience.

I have characterized scientific controversies as taking place on a *field of disagreement* in which alternative individual practices compete as candidates for the modification of consensus practice. All participants, I have claimed, subscribe to a common goal, the production of a maximally unified set of explanations for the broadest possible class of phenomena. Rather than thinking of their proposals as faced with only one type of difficulty, the avoidance of inconsistency, we should recognize the competing claims of the search for unifying explanation. Precisely because these claims are so powerful, the recurrent response of practicing scientists to underdetermination arguments is that the alleged proliferation of alternative is specious: the problem is typically that there is *no* available practice that does everything that the community desires. Controversies arise and linger, not because there are large numbers of perfect solutions, but because there are alternative *imperfect* solutions, possible modifications of practice that point toward some future systematization that will navigate between the Scylla of lost, or disunified, explanations, and the Charybdis of inconsistency.[31] Different scientists weigh the perils and promises differently, and, I have suggested, it is a good thing for the community that they do so.

The real challenge posed by appeals to underdetermination stems from what we might appropriately call Kuhnian underdetermination, that is, from the claim that this need for trade-off between the achievement of explanatory systematization and the avoidance of inconsistency is a permanent feature of the scientific condition. I shall try to resist this conclusion in three different ways. First, I sketch reasons for thinking that, given what we understand about the factors that sustain scientific controversies, the proposed anti-rationalist explanation for the closure of controversies is itself threatened by the alleged phenomenon of Kuhnian underdetermination. Second, I indicate a general account of the closure of scientific controversies. I conclude by describing very briefly how this account might apply to two major examples in the history of science.

During a scientific controversy, the different prospects and promises of rival modifications of consensus practice offer "career niches" for the ambitious young scientist. If we now assume Kuhnian underdetermination, then alternative niches will still be available at the time of closure. Anti-rationalists thus owe an explanation of why such niches are no longer occupied. Unless we suppose that the competitive pressures on individual scientists are somehow overridden by some type of authoritarian constraint, there will no more be an anti-rationalist sociological explanation for the achievement of consensus than the type of account that rationalists prefer. It is important to note the symmetry between this challenge and that leveled by anti-rationalists. The latter typically demand that their rationalist opponents should specify the constraints on scientific decision that enable scientists to close controversies. If my general account of the eruption and continuation of scientific controversies is correct, then we can ask, with equal justice, what overrides the social pressure to occupy vacant career niches.

Now, in some instances there may be satisfying social explanations. Consider Wesley Salmon's example of the noncontroversy about quasars (presented in chapter 14). Apparently, there is a vacant niche that we might expect ambitious members of the astrophysical community to occupy. In discussion of the example, Ian Hacking suggested a possible explanation for the vacancy: perhaps there is a "trading zone" between theoreticians and observers, and participation in this zone involves meeting conditions that preclude questioning the orthodoxy about quasars. Hacking's proposal points to a line of social constructivist response to my challenge.[32] The idea that community members will distribute themselves so as to occupy all available career niches presupposes that there are not *social* barriers to the occupancy of niches that are *epistemically* available. In some types of scientific communities, the interdependence of some classes of workers or the needs of workers for scarce resources may create such barriers. If a theorist cannot develop new models without collaborating with an observer/experimentalist, or if research cannot be undertaken without involving a subcommunity that controls operation of a certain type of equipment, then niches that might seem epistemically available—and attractive—to the aspiring theorist will be closed off because of the orthodoxies that hold sway among observers/experimentalists or in the pertinent subcommunity. (We might imagine that these orthodoxies are in place because of the salience to the relevant people of certain types of evidence, as envisaged in Solomon, 1992.)

The idea that some subcommunities might exert pressure on others, thus blocking off career niches, is an important one, but it does not completely answer my challenge. For a full account would need to show why the epistemic availability of a niche does not lead members of *both* (or all) subcommunities to occupy it. Why are there not theoreticians and observers prepared to work together to defy the rules of the "trading zone"? Perhaps the response is that the *coordinated* break with orthodoxy is too hard to accomplish. Or perhaps, as some of Salmon's documents seem to indicate, members of the community believe that the niche in question has been adequately explored.[33]

Even if my challenge to the social constructivist explanation should be successful, *tu quoque* arguments are not ultimately satisfying, and I believe that an adequate defense of [R] should identify the constraints that govern the closure of scientific controversies. I begin with the thesis that scientists in the thick of a controversy face two types of predicaments: those of inconsistency and those of explanation. Inconsistency predicaments are generated by the fact that some of the claims they make are incompatible with others; most obviously, commitment to a general explanatory schema may lead one to claims about what will be observed under specified conditions. When those conditions are realized, the impinging stimuli incline one to incompatible claims—and there is no resisting this response to the stimuli except by calling into question some generally accepted thesis about the proper making of observations. Explanatory predicaments stem from the difficulty of extending accepted schemata to cover phenomena that cannot currently be brought within their scope. There is an obvious reciprocal relationship between these two kinds of predicament. Very frequently, one can clear up an inconsistency by abandoning or restricting some explanatory schema, which greatly diminishes the unifying power of the practice. By the same token, an explanatory schema can sometimes be extended to cover new cases, or can be generalized, at the cost of introducing inconsistencies.

These two notions provide an abstract account of the field of disagreement, which reveals the character of much scientific debate. To defend a particular proposal for modifying consensus practice, one must show, constructively, that it has the potential to find solutions to the predicaments that it faces and, adversarially, that rival proposals lack resources. Given a predicament (of either type), we can envisage a multiply branching structure, the *escape tree,* that encompasses the various possibilities for responding to that predicament. For an inconsistency predicament, for example, there will be various ways of deleting elements of the inconsistent set. Each of these possibilities will threaten an explanatory loss, thus generating a new predicament. Response to that explanatory predicament may now involve the provision of new instantiations of old schemata, with potential inconsistencies, or the introduction of new schemata, with loss of unification. Hence, in exploring one predicament, one may swiftly be led to others, and difficulties may ramify. Alternatively, it may be possible to find a path through the escape tree along which no further predicaments arise. Any path that cannot be pursued without generating further predicaments is blocked; paths that offer release from the initial predicament without generating new predicaments are open.[34]

I claim that much of the debate in major scientific controversies consists in showing that one can find open paths in the escape trees generated by one's own predicaments, and that one's opponents' escape trees are blocked. One of the most dramatic instances occurs in Galileo's *Dialogue Concerning the Two Great World Systems* (1632), where Salviati (aided by Sagredo) responds to Simplicio's complaints by presenting lines of escape from the predicaments delineated, and shows how Simplicio is constantly faced by predicaments whose escape trees are blocked. By the end of the dialogue, the set of unsolved (and apparently unsolvable) predicaments for geocentrism is so vast in comparison with those for heliocentrism that, unless some alternative geocentric modification of astronomical-physical practice can be found, there is no sense in which geocentrism remains a live option. This is, I believe, typical of the resolution of scientific controversies. Although it may be possible to attain consistency by truncating one's commitments in the ways in which Simplicio is forced to do, the explanatory losses involved in doing so are overwhelming.[35]

The systematic closing of options may affect different subfields differently. So, for example, a particular subcommunity may find that the predicaments it faces are resolvable by one of the contending modifications of consensus practice, while for other subcommunities the issues remain wide open. If this occurs, then there may be a piecemeal closing of controversy, as first one subgroup and then others find that a particular way of escaping their principal predicaments is available. Arguably, this occurred in the transition in geological practice that culminated with the acceptance of plate tectonics, and Miriam Solomon has used this example to suggest that piecemeal closure is typical (Solomon, 1994).

Let me develop Solomon's approach by combining it, informally, with both the pressures for cognitive diversity proposed in Kitcher (1990) and those she favors in Solomon (1992). Imagine a community broken into subcommunities, for each of which a particular type of evidence is salient. We may assume that pooling these types of evidence covers all the problems that a controversial proposal must address. As the controversial claim successfully tackles some of these problems, it will become the orthodoxy within particular subcommunities, those for whom the problems solved constitute the salient evidence. But within other communities, for which salient problems remain unaddressed, there will be continued pressure to explore alternatives: as argued in Kitcher (1990; 1993, ch. 8), the opportunity to make a reputation by subverting the nascent orthodoxy within other communities will beckon the aspiring scientist. Hence, the controversy will be closed *across the entire community* only when the whole range of problems that the proposal should address has been adequately tackled. At this stage, there will be an ideal argument for closing the controversy, one that deals even-handedly with all the evidence, even if it is never formulated and is entirely irrelevant to the piecemeal processes of decision making that have occurred.[36]

My own view is that there is likely to be a range of cases. Sometimes a community is sufficiently small, or its members are all sensitive to a sufficiently broad range of predicaments, that closure only occurs when an overarching line of solution has been developed—often through exchanges that make publicly available a line of reasoning far superior to anything that the individual contributors have antecedently achieved. Major examples in the history of science seem to me to reveal this social forging of communitywide lines of reasoning: the acceptance of Darwin's thesis of descent with modification, the triumph of Lavoisier's "new chemistry," and the closure of the Great Devonian Controversy are marked by public presentation of arguments that emerge from interactions between proponents and critics.[37] But the translation to plate tectonics does show the possibility Solomon proposes, as biogeographers, paleometeorologists, students of paleomagnetism, researchers on the sea floor, and other geologists sequentially fall into line.

4

I close with two examples of the closure of scientific controversies. Consider, first, the chemical revolution.[38] In the first phase, which occurred in the 1770s, Lavoisier was able to show that there is a weight gain in the calcination of metals, and he demonstrated that the weight gain equals the weight loss of the air. This did not settle the issue between him and the phlogistonians, because it was still possible to maintain that the absorption of a substance from the air was accompanied by the emission of phlogiston from the metal. Both sides agreed that the task of explaining chemical reactions consisted in identifying the reactants and their proportions by weight in a way that would satisfy the "principle of the balance." The challenge for participants in the controversy (various stripes of phlogistonians and anti-phlogistonians) was to provide a set of schemata that could be instantiated to cope with combustion, acid-metal reactions, the decomposition of water, and so forth. Lavoisier's attack on the phlogiston theory consisted in showing that no consistent, unified set of schemata could be given, subject to the constraint that phlogiston is emitted in calcination and combustion.

To appreciate fully the force of this attack would require understanding how all kinds of possibilities were systematically blocked off by the experiments performed by Lavoisier and his coworkers. I shall use a small piece of his campaign as an illustrative example.

In the 1780s, Henry Cavendish offered an explanation of the received observations about the calcination of metals and the decomposition of water by proposing the following principles:

metal calx = (metal – phlogiston) + water

inflammable air = phlogiston

inflammable air + vital air = water

Using these principles, Cavendish was able to provide explanations of the general form favored by all parties to the controversy for a wide class of reactions. However, his approach ran into difficulties with Lavoisier's "gun-barrel" experiment, in which iron filings placed in a gun barrel are heated to red hot, and water is passed through the barrel. The products of the experiment are the black calx of iron and inflammable air. Cavendish and his supporters supposed that the inflammable air (phlogiston) is released from the iron, leaving the "basis of the iron" (iron – phlogiston), and that this basis combines with water to yield the black calx. Lavoisier interpreted the experiment as involving the decomposition of the water to form vital air and inflammable air. The former combines with the iron to form the black calx; the latter is released. At a purely qualitative level, both succeed.

Lavoisier was able to show that an amount of vital air (oxygen) whose weight equaled the gain in weight of the calx would combine with the inflammable air collected in the experiment, yielding an amount of water whose weight was exactly that of the water lost in the experiment. His favored interpretation can explain this *quantitative* result. By contrast, on the phlogistonian account, there is no reason why the precise amount of phlogiston released by the metal should combine with vital air to yield the amount of water lost. How could phlogistonians extend their preferred schemata to deal with this explanatory problem?

The problem arises from the fact that the phlogistonian account of calcination depends on keeping two processes *independent*. Phlogiston is supposed to leave the metal, and water is supposed to be absorbed. The gun-barrel experiment reveals that these processes are connected. It cannot be proposed that the phlogiston in the water displaces the phlogiston in the metal, for that would violate a principle about chemical affinities needed to explain a vast array of reactions. Nor can it be suggested that metals can only absorb as much water as they release, for an essential part of the phlogistonian explanation is that the basis of the metal is common to all calxes, and it is recognized that some metals (such as iron) are able to form more than one calx, and that the differences depend on the amount of substance absorbed. Cavendish is thus bereft of resources for explaining the coincidence in values.

For my second example I turn to a controversy whose focus was an instrument. Between 1610 and 1613, Galileo was able to convince almost all his contemporaries that his telescope gave accurate information about both terrestrial and celestial phenomena. The terrestrial part was easy, and we have contemporary records of how the elder statesmen of Venice puffed up the stairs to turn the telescope on incoming vessels, confirming their observations when the ships reached port. But Galileo's claims about additional stars in well-known constellations, the surface of the moon, and the satellites of Jupiter could not be checked in this way. Here, then, we seem to have a clear case of underdetermination: one could either accept the veridiality of Galileo's observations in the heavens or denounce the telescope as inadequate to fathom the characteristics of the celestial realm.

Let us leave Galileo for a moment, and consider Quine's provocative remark about maintaining the statement "that there are brick houses on Elm Street" by pleading hallucination (Quine 1952). The obvious reply is that we accept some general ways of accounting for people's perceptual responses, and the plea of hallucination is at odds with these. To be sure, we could modify our explanatory schemata by introducing some special condition about the happenings on Elm Street, but, since we do not make exceptions for Elm Street events across the board, this would produce an unmotivated disunification of our explanatory schemata.[39]

Now, the important point about the case of Galileo and the telescope is that orthodox Aristotelians already had in place a set of explanatory distinctions for terrestrial and celestial events. To the blunt suggestion that the workings of the telescope on the earth would suffice to show its accuracy in the heavens, the Aristotelians could *justifiably* plead that Galileo's alleged new phenomena were deceptions of the instrument. Galileo's success depended on his ability to show both that the invocation of the distinction was problematic and that the distinction itself could not be sustained.

The first part of his strategy consisted in stressing the continuity between naked-eye observations and the deliverances of the telescope. Just as the use of the telescope on the earth typically revealed more detail in observational situations than had previously been seen with the naked eye, so too Galileo's comparison of pre- and post-telescopic observations of the moon and of the Pleiades presented to Aristotelians the predicament of explaining why it seems that the telescope adds more detail to the ordinary observational picture. (I conjecture that Galileo could have, and perhaps did, reinforce this argument by showing how increasing detail is obtained by using telescopes of increasing power.)[40] Second, in a large body of writings from the 1600s through the 1630s, Galileo argued that the celestial/terrestrial distinction was untenable both because of such phenomena as the novae and because of internal incoherence. The character of this argument is that the Aristotelian response to the telescope presupposes claims about the heavens that face inconsistency predicaments; attempts to escape these inconsistency predicaments generate explanatory losses.

I have at best gestured at the pattern exemplified in the closure of these two important scientific controversies. Nonetheless, I hope to have provided some motivation for pursuing further my central claims. Scientific controversies have beginnings, middles, and endings. In their beginnings and middles, they are far more various than traditional forms of rationalism have suspected. Yet in their ending there is some unity, as the community collectively realizes that all but one of its options faces insuperable obstacles. Anti-rationalists paint for us pictures of scientists free and unconstrained, on whose musings the world has little impact. By contrast, given the perspective I have adopted, the task of science—the task of producing a unified account of a complex world—is so difficult that it is hardly surprising that the continuing press of experience can batter even the most ingenious maverick into submission.

Notes

I thank the participants in the conference on scientific controversies, not only for their comments on an earlier version of the manuscript, but also for the stimulating discussions that took place in Vico Equense. In light of these exchanges, I have articulated or modified several points.

1. Given these experiments, it is supposed that "simple rules" will decide the debate. For lucid formulation of the ideas of the "Founding Fathers" of the rationalist model, see Marcello Pera's chapter 3 of this volume.

2. Latter-day developments of rationalism forsake the Fathers' optimistic view that the rules of method are simple, in favor of much more complex considerations. How these more complex means of evaluation relate to the intuitive judgments of scientists is an important question.

3. See, for example, Collins (1985).

4. The most developed anti-rationalist approach along these lines is that of Shapin and Schaffer (1985).

5. In my judgment, Martin Rudwick's narrative of the Great Devonian Controversy provides the richest account we have for any scientific controversy. See Rudwick (1985) and, for summary discussion, Kitcher (1993, pp. 211–218).

6. A brief taxonomy is provided in Kitcher (1993, pp. 207–209).

7. In surveying a range of possibilities, I do not want to give the impression that scientific controversies can arise with respect to *any* number of the categories I mention. I concur with Gideon Freudenthal's insightful identification of some necessary conditions on the objects of controversy in chapter 7 of this volume.

8. For more extensive discussion of these notions, see Kitcher (1993, ch. 3).

9. Although this last constituent was not identified in the account of practices given in Kitcher (1993), I believe it to be crucial for understanding some controversies. Precisely because of judgments about the relation between a proposed project and human well-being, controversies have erupted over human sociobiology (see the articles in Caplan 1978) and parts of behavior genetics (see the articles in Block and Dworkin 1976).

10. The last controversy involves a wide range of issues and types of considerations. Some opponents of the project believe that mapping and sequencing the human genome is unjustified because the information that will be provided will be mostly useless. Others think that the project will distort the social structure of biology, hampering valuable types of biological research. Still others claim that the project will distort biological education, producing people who will lack the skills to solve the truly significant biological problems. These perspectives tend to be offered in informal conversations. *Public* opposition to the project tends to center on

its social consequences—concerns about genetic discrimination, misidentifications in forensics, and so on. See Hubbard and Wald (1993) for trenchant presentation of these arguments.

11. But it is important that the individual practices are rival developments of a shared consensus practice. Here my kinship with Freudenthal's claims about relation to a common conceptual scheme should be evident.

12. So, for example, in discussing the Human Genome Project, an eminent medical geneticist was quite forthright about the significance of the project, characterizing it as the culmination of a half-century's work on the molecular basis of heredity. An equally eminent molecular geneticist and developmental biologist was equally critical, claiming that the project was irrelevant to the really important problems in contemporary biology.

13. See Kitcher (1992; 1993, ch. 8) for discussion.

14. Kostas Gavroglu's insightful discussion of episodes in the history of physical chemistry reveals clearly how complex are the notions of national and professional affiliation. See his chapter 10 in this volume. These affiliations are often reinforced by national (or regional) self-images, and broader philosophical conceptions. Thus, I would argue that large themes of pragmatism, as well as the nineteenth-century notion of "manifest destiny" and the image of the pioneer, all play a role in the distinctive character of the physics and chemistry pursued at the nascent Californian institutions (Caltech and UC Berkeley) in the early decades of this century.

15. A somewhat more detailed account of these aims is given in Kitcher (1993, ch. 4).

16. Thus I diverge from Hilary Putnam's characterization of the (metaphysical) realist's account of the aims of inquiry. See, for example, Putnam (1990).

17. For these rival approaches, see van Fraasen (1980), Laudan (1977), and Rorty (1979).

18. Here I link the account of the aims of science offered in Kitcher (1993) to the view of explanation proposed in Kitcher (1981, 1989), thus going beyond the more ecumenical view of Kitcher (1993). I do so here because, as will become apparent in section 3, I regard the principle of unification as playing an important part in the closure of controversies. One obvious way of accounting for that role is to make the notion of unification somehow *constitutive* of the aims of inquiry. Perhaps those like Wesley Salmon who maintain a stronger type of realism, will be able to resist this assimilation either by showing how to give an alternative account of the closure of controversies, which does not involve the appeal to unification, or by forging some other link between the maxim of unification and the aims of inquiry. But I do not see how to do this.

19. Here I attempt to extend Thomas Kuhn's classic discussion of the ways in which puzzles arise in normal science. For further analysis, see Kitcher (1993, pp. 114–115) and, for an example, Culp and Kitcher (1989).

20. In Kitcher (1993, ch. 8) I attempt to show how complex the problems are and to provide some precise solutions in some special instances. For a more general defense of this conception of social epistemology, see Kitcher (1994). Richard Grandy's chapter 4 of this volume obviously shares a great deal with the perspective I favor, and his analysis serves as a paradigm for the approach to the problems that I would commend.

21. For Pera's "Fathers," these reasons are "simple." Hence, opposition to the triumphant position must be seen as based on dogmatism, prejudice, or just sheer stupidity.

22. The locus classicus is Kuhn (1962, ch. 12). For an extremely sensitive account of one of Kuhn's central characters, Joseph Priestley, see Pierluigi Barrotta's chapter 9 in this volume. I have more to say about the chemical revolution and the treatment of it by both Kuhn and Barrotta below.

23. For defense of these claims, see Kitcher (1990; 1993, ch. 8). I should note explicitly that the gambit of supposing that welcome diversity can be sustained by supposing that we can pursue options in a detached way, remaining properly agnostic, seems to me to be decisively refuted by facts about our psychology.

24. Some of Bruno's contemporaries portray him as a vulgar careerist whose wild pronouncements are made to obtain attention. The documents printed in Finocchiaro (1990) give a very clear picture of the motivations of Galileo's accusers. Tycho's desire for personal glory seems to me a likely cause of the suppression of other sources of the "compromise" between

geocentrism and heliocentrism. For a revealing account of a murky story, see Gingerich and Westman (1991).

25. But it should be noted that one simple way of combining Solomon's account of cognitive pressures for diversity with the social drive for career distinction shows interference, with *lowering* of diversity. See Kitcher (1993, pp. 374–378). I suspect that this may be the result of using too crude a model.

26. This seems to me to diverge in important respects from the Fathers' conception of the conduct of the epistemically responsible scientist, who is bound by rules not to "go beyond the evidence." Thus, the rationalism I articulate in this chapter is very much a human rationalism that takes seriously both the cognitive limitations of the subject and the subject's immersion in the community. Here, I believe, I am in deep agreement with Grandy (see chapter 4).

27. For forceful presentation of this challenge as a problem for the account given in Kitcher (1990), see Solomon (1994).

28. There are serious questions about the size and character of this decision-making set. How many people ought to concur for the community to reach consensus? I discuss a simplified version of this problem in Kitcher (1993, pp. 382–387). Grandy's chapter 4 in this volume explores the same questions in an elegant way.

29. Here I take for granted a standard for evaluating reasoning in terms of the expected attainment of epistemic goals. For much more discussion, see Kitcher (1993, ch. 6).

30. These become far more prominent in Quine (1960) and his later writings.

31. Here, I believe, my proposals are close to those of Aristides Baltas and Gideon Freudenthal in their chapters 2 and 7, respectively, in this volume. Faced with the task of explaining the length of scientific controversies, each of us resists the Fathers' solution (the ignorance, blindness, or prejudice of the losers) and the relativist-constructivist solution (underdetermination rules, and the controversy can only be resolved through mobilizing powerful social resources), proposing that the resolution of controversies involves complicated trade-offs that tax the judgment of the most sophisticated reasoners. Freudenthal's way of working out this approach seems especially close to my own.

32. It has obvious affinity with the account of the social structure of the community of high-energy physicists offered in Pickering (1984). By adapting that account to the project of explaining the *non*eruption of controversies, Hacking (chapter 12) points the way to an appropriate engagement between the approach to sociology of science offered in Kitcher (1993, ch. 8) and that favored by many professional sociologists of science. The abstract models I develop should be corrected and refined by recognizing the ways in which the particular structures of communities affect their epistemic projects. The opposition in *style* between the impressionistic reports of contemporary sociology of science and the formal decision-theoretic models I recommend should not blind us to the possibility of mutual enrichment on matters of *substance*. The example of "trading zones" and the possibility of resultant barriers to occupancy of "career niches" makes this point extremely clear.

33. Thus, the explanation of the noncontroversy may be epistemic: the alternative has been pursued and has led nowhere. I do not claim that this is the right explanation but simply offer it as an alternative to Hacking's intriguing hypothesis.

34. Ian Hacking's study of putative instances of multiple personality (see chapter 12) provides an example of cases in which further interactions with nature modify the character of the escape tree. New examples of people displaying a variety of symptoms call into question the explanatory successes that were supposedly achieved at earlier phases of the controversy. Hence, the "multiple personality debate" has an overall cyclic character and can be seen as a drama with a number of "acts," each of which has the structure I take to be typical of a scientific controversy.

35. Although geocentrism becomes untenable, it might appear that Tycho's compromise might survive and that Galileo was somehow cheating by confining his attention to *two* "Great World Systems." But I think that Galileo's title is profoundly significant. There were only two great world *systems*. In the wake of Kepler's work and Galileo's incipient thoughts about mechanics, the Tychonic model had no power to explain the motions of heavenly bodies. By the early seventeenth century, there is growing support for the idea that control of satellites depends

on proximity and the size of the central body. In these terms, there is no coherent way of developing Tychonic explanations for the motions of the bodies in the solar system. For one cannot square the ability of the earth to serve as the center of rotation for the sun with its *inability* to control the motions of bodies (such as Mars) that are sometimes closer to it. For Tycho's astronomy there is thus a direct choice between bald inconsistency and extreme explanatory loss.

36. Obviously this scenario needs to be analyzed more precisely than I am able to do here. Perhaps the informal expectations will not be borne out by a more formal treatment. It is also possible, of course, that for reasons discussed in connection with Salmon's example and Hacking's suggestion about it, the "ideal" operation of piecemeal closure should be modified. Barriers may close off certain paths of continued resistance in subcommunities so that there is communitywide agreement before all the problems have been solved. Obviously the relations of interdependence of various subcommunities are crucial here.

37. I discuss these examples in Kitcher (1993). Barrotta's treatment of the role played by Joseph Priestley in the transition to the new chemistry is a lucid demonstration of how even the most apparently stubborn critics can play an important constructive role in the delineation of argument. See his chapter 9 of this volume.

38. I draw on a far more detailed account offered in Kitcher (1993, pp. 272–290). That account seems to me to be compatible with the conclusions drawn by Barrotta in his own exemplary study of the episode (chapter 9).

39. Although I shall not attempt a precise analysis here, it seems to me that the notion of adhocness is best understood in terms of the introduction of exceptional cases that do not correspond to causal factors picked out in our background explanatory schemata. The explanatory schemata identify the kinds of factors relevant to the presence or absence of certain types of states, events, or properties, and a distinction we draw among cases is ad hoc just when we sanction no explanatorily relevant factors for making distinctions among the pertinent events, states, or properties.

40. As far as I know, nobody has investigated Galileo's use and distribution of increasingly powerful telescopes. But the sequence of observations made with more powerful instruments must surely have reinforced observers' convictions that they were seeing the same things in more and more detail.

References

Block, Ned, and Dworkin, Gerald (1976), The IQ Controversy, New York: Pantheon.

Caplan, Arthur (ed.) (1978), The Sociobiology Debate, New York: Harper and Row.

Collins, H.M. (1985), Changing Order, London: Sage.

Culp, Sylvia, and Kitcher, Philip (1989), "Theory Structure and Theory Change in Contemporary Molecular Biology," British Journal for the Philosophy of Science, 40, 459–483.

Duhem, Pierre (1906), The Aim and Structure of Physical Theory, Princeton, N.J.: Princeton University Press.

Earman, John (1992), Bayes or Bust? Cambridge, Mass: MIT Press.

Finocchiaro, Maurice (1990), The Galileo Affair, Berkeley: University of California Press.

Gingerich, Owen, and Westman, Robert (1991), "The Wittich Connection," Transactions of the American Philosophical Society, 78, part 7.

Glymour, Clark (1980), Theory and Evidence, Princeton, N.J.: Princeton University Press.

Howson, Colin, and Urbach, Peter (1989), Scientific Reasoning, LaSalle, Ill.: Open Court.

Hubbard, Ruth, and Wald, Elijah (1993), Exploding the Gene Myth, Boston: Beacon.

Kitcher, Philip (1981), "Explanatory Unification," Philosophy of Science, 48, 507–531.

——— (1989), "Explanatory Unification and the Causal Structure of the World," in P. Kitcher and W. Salmon (eds.), Scientific Explanation, Minneapolis: University of Minnesota Press.

——— (1990), "The Division of Cognitive Labor," Journal of Philosophy, 87, 5–22.

——— (1992), "Authority, Deference, and the Role of Individual Reason in Science," in E. McMullin (ed.), The Social Dimensions of Scientific Knowledge, Notre Dame: University of Notre Dame Press.

——— (1993), The Advancement of Science, New York: Oxford University Press.
——— (1994), "Contrasting Conceptions of Social Epistemology," in F. Schmitt (ed.), Social Epistemology, New York: Rowman and Allanheld.
Kuhn, Thomas S. (1962), The Structure of Scientific Revolutions, Chicago: University of Chicago Press.
Lakatos, Imre (1970), "Falsification and the Methodology of Scientific Research Programmes," in I. Lakatos and A. Musgrave (eds.), Criticism and the Growth of Knowledge, Cambridge: Cambridge University Press, 91–196.
Laudan, Larry (1977), Progress and Its Problems, Berkeley: University of California Press.
Pickering, Andrew (1984), Constructing Quarks, Chicago: University of Chicago Press.
Putnam, Hilary (1990), "A Defense of Internal Realism," in J. Conant (ed.), Realism with a Human Face, Cambridge, Mass.: Harvard University Press.
Quine, W.V. (1952), "Two Dogmas of Empiricism," in From a Logical Point of View, New York: Harper, 20–46.
——— (1960), Word and Object, Cambridge, Mass.: MIT Press.
Rorty, Richard (1979), Philosophy and the Mirror of Nature, Princeton, N.J.: Princeton University Press.
Rudwick, Martin J. S. (1985), The Great Devonian Controversy, Chicago: University of Chicago Press.
Shapin, Steven, and Schaffer, Simon (1985), Leviathan and the Air-Pump, Princeton, N.J.: Princeton University Press.
Solomon, Miriam (1992), "Scientific Rationality and Human Reasoning," Philosophy of Science, 59, 439–455.
——— (1994), "Social Empiricism," Nous.
Van Fraassen, Bas (1980), The Scientific Image, Oxford: Oxford University Press.

2

Classifying Scientific Controversies

ARISTIDES BALTAS

Despite its importance, the study of scientific controversies has not developed as much as the study of other aspects of science. This is, I believe, is one more of the aftereffects of Thomas Kuhn's work. In one sense, Kuhn's *The Structure of Scientific Revolutions* (1962) did, of course, place some notion of scientific controversy at the center of attention. But, in another sense, most of the relevant issues were rarely pinpointed and discussed in their own right. Instead, they tended to remain unclear and indistinct, covered under the all-embracing—or so it appeared at the time—paradigm change. Moreover, much of post-Kuhnian philosophy of science forced itself to respond to what it considered Kuhn's overall challenge to rationality, rather than occupy itself with the finer details of what it is for scientists to disagree with one another and to fight out such disagreements. This, then, implies that to study scientific controversies we have to start from the beginning. Traditionally, to attempt classification forms precisely such a beginning.

Of course, in philosophy as in everything else, no beginning is ever virgin; all starting points cannot but be sullied by the set of presuppositions peculiar to them—and this in more senses than one. For one thing, one cannot but be indebted in the case at hand to what the debate over the rationality of science (at least as related to the issue of paradigm change) came up with. For example, Kitcher (1993) speaks of "dissolving" rationality and tries to lay to rest the whole issue without succumbing to the presently dominant social constructivism. In chapter 1 of this volume, he applies his views to the analysis of the patterns that scientific controversies tend to follow, trying to capture the genuine insights of both parties of the old debate, rationalists and anti-rationalists alike.

The other sense in which the beginning in question cannot be innocent is more logical than historical: it is evident that one cannot talk of *scientific* controversies without relating what one has to say on a conception of science; and one cannot talk of *classifying* scientific controversies without a classification principle (or a coherent set thereof) predicated on this conception. I believe that this is not as trivial as it appears. Although most of us tend to take for granted that the demarcation problem is not an issue any longer (see, for example, Laudan 1992), I am convinced that a sustained discussion of what the different sciences are, and of why they are sciences, would shed some new light on numerous philosophical puzzles anxiously waiting to be more adequately handled. I am, of course, aware that this is not the place to argue for this conviction. Accordingly, I will let the issue rest and turn immediately to what I have to say about the classification principle in question.

The idea I have in mind permitting, I believe, the formulation of an interesting principle for classifying scientific controversies develops from an insight shared by many philosophers who, however, have not exploited its capacities in full. The idea is that the interpretation tying a scientific conceptual system to our experience at large harbors inevitably what we can call *background assumptions*.

On Background "Assumptions"

It practically goes without saying that the effective functioning of any science involves a set of particular processes of inquiry regarding the science's object. Such processes cannot proceed unless they keep fixed two kinds of things. The first is a set of overtly postulated premises, a list of general and specific axioms, on the basis of which testable claims can be formulated. Without entering too much into details, we can say that, as these premises are overtly postulated, they present themselves as in principle open to doubt and, at least indirectly, as subject to confirmation, adjustment, or refutation. The second item that the functioning of a science keeps fixed has received much less attention in philosophy of science proper. I am referring to the amorphous plethora, the indefinitely deep and broad ocean, of *background "assumptions"* that any process of inquiry cannot but carry along with it, at least for the most part silently, inexpressibly, blindly.

As Wittgenstein (1972) shows, these are not proper assumptions—hence the quotes. They are neither a priori and indubitable nor a posteriori and open to doubt. In normal conditions, they cannot be the object of any kind of justificatory move. Rather, they constitute quasi-logical, or rather *grammatical*, conditions allowing the concepts involved in the inquiry to make sense. As Wittgenstein (1972) himself phrases it (*On Certainty*, 341–343; see also Morawetz 1980), they are the "hinges" that have to stay put for the door of inquiry (its questions and its doubts) to open. For example, under normal conditions, the zoology of cats is not obliged first, to inquire whether our purring companions grow in trees like bananas, whether they are extraterrestrials in disguise, and so forth; second, to run the appropriate tests; and finally, to present explicit arguments to the effect that they are not. That our pets are no such things, the zoology of cats takes silently for granted as a matter of course. In other words, the "assumption" that they are not belongs to the amorphous set of grammatical conditions allowing the concept "cat" to make the kind of sense necessary for our employing it with confidence in the relevant inquiries. However, that such "assumptions" are not a priori signifies that, under very particular circumstances, any one of them can become "illuminated" and emerge from the background. It will appear, then, as a proper assumption (without quotes) that had been "unjustifiably" taken for granted without our being aware of it, necessitating thereby an investigation as to its warrants. The completely baffling and totally unaccountable behavior of some particular cats may thus indeed force us in the long run to consider an extravagant hypothesis like the above. When this happens, such an "assumption" will appear post hoc as an illegitimate prejudgment, as a bias, as a prejudice, as an unwarranted presupposition.

Since, now, the background of any of our practices is variously determined by our experience at large, the fact that such "assumptions" belong to the background of a process of inquiry signifies that they tie, so to speak, this process to that experience. That their function is grammatical signifies that such work consists in partly fashioning the semantical dimension of the corresponding concepts, as this dimension is again determined by our experience at large. Francis Bacon's "idols," Paul Feyerabend's "natural interpretations," Imre Lakatos's "hidden lemmas," Larry Laudan's "ontological commitments," Thomas Kuhn's "changeable Kantian categories" constitute many attempts to name the "assumptions" in question and to cope theoretically with their inescapable existence.

In forming grammatical conditions allowing the concepts involved in a scientific inquiry to make sense, these "assumptions" reside *in the interpretation* of the corresponding conceptual system (and by means of it, to everything this interpretation determines as, for instance, the sense in which the corresponding experimental transactions and their results are to be taken). More specifically, they are involved in the pictures, the analogies, the metaphors, and so forth, that scientists draw on in order to understand the concepts their own work produces, and to communicate their results. As such, they are not mere Wittgensteinian "ladders," to be thrown away after the system is established, nor Fregean elucidations that do not belong to the conceptual system proper (Joan Weiner 1990). At least some of the connotations of those pictures and analogies tarry, as ineradicable vestiges, on the concepts involved, latently forming part of their meaning. In this way, they surreptitiously close the horizon of that meaning, channeling correspondingly the relevant investigations. Only a subsequent upheaval of the science in question can bring to the fore some of the "assumptions" involved. For example, the "assumptions" involved in the concept of wave, understood as designating the propagation of a medium's disturbances, required an ether as the carrier of electromagnetic waves; or the "assumptions" involved in the Newtonian concept of motion required that all moving bodies, however small, could not but follow trajectories. It was only the coming to being of relativity theory and of quantum mechanics which showed that the meaning of the concepts in question could be "extended"—or radically changed—beyond what these "assumptions" silently dictated.

Although the space they dwell in is amorphous and indeterminate, we can distinguish four kinds, or rather four levels, of background "assumptions." Thus, going from the "deep" to the "superficial," and from the blindly "accepted" (in quotes because no real acceptance can be involved here) to the almost consciously adopted, we have the following.

Constitutive Background "Assumptions." These are the "assumptions" that are instrumental in setting up a scientific conceptual system; they assure its coherence and they determine its identity in the sense that their effective disclosure transforms it radically. To stay with physics, the background "assumptions" involved in the concepts of space and of time, dictating that simultaneity does not depend on relative motion or that position and momentum are independent attributes of all moving bodies, can be considered as constitutive of the conceptual system of Newtonian mechanics.

Interpretative Background "Assumptions." These are the "assumptions" determining the way a scientific conceptual system is "naturally" interpreted when it is first established. Their disclosure does not transform radically the conceptual system; it only shows that the capacity to account cognitively for phenomena that hitherto appeared as totally unaccountable in its terms inheres in the system itself. The inclusion of Leibniz's "conservation" results within the structure of Newtonian physics is the result of disclosing the particular interpretative "assumptions" that prevented this from happening before. Freudenthal (1986) discusses this episode extensively and in a way that shows admirably the complex interplay between the cognitive and the social dimensions without, for that matter, collapsing the one onto the other. Or, to give another example, physics coming to grips with low-temperature phenomena amounts to the disclosure of some such interpretative "assumptions": in describing this process, Gavroglu and Goudaroulis (1989) isolate a mechanism that leads the relevant concepts out of the context of their then reigning interpretation.

Participation Background "Assumptions." These are the "assumptions" determining the particular scientific traditions, national or otherwise, and/or the particular "styles of reasoning" scientists tend to follow. For example, scientists implicated in an "empiricist" tradition or style of reasoning may be hampered by participation "assumptions" from recognizing the significance of some theoretical result for the pursuit of their own work. Conversely, a "rationalist" tradition or style of reasoning, focusing on high theory and axiomatics, harbors participation "assumptions" that may blind the scientists implicated as to the relevance of some particular experimental result for their own theoretical work. Hacking (1982, 1983), Gavroglu (1990), and Gavroglu and Simoes (1994) discuss some historical episodes that exemplify the role of such "assumptions" in a very illuminating manner, albeit from slightly different angles.

Preference Background "Assumptions." These are the background "assumptions" involved in the criteria scientists employ in choosing the research program they are going to work on out of the available alternatives. For example Geoffrey Chew's proposal of particle "democracy" attracted many high-energy physicists in the late 1960's, those despairing of making sense of the proliferation of "elementary" particles on the basis of the traditional "building block" schemata.

It should be clear that carrying out any scientific game implicates all these categories of background "assumptions": scientists work with a constituted conceptual system that cannot but be already interpreted; they belong to a tradition and they follow a style of reasoning; they choose—or are forced into—the research program they will serve. Moreover, these categories cannot be distinguished sharply from one another. In residing, as I said, within the amorphous and indeterminate background of the corresponding investigation, they form together a continuum, extending from the more deeply entrenched, that is, the "assumptions" that only a radical theoretical break can reveal, to those lying nearer the "surface," that is, those that can be disclosed and discarded relatively easily, if strong enough reasons are adduced. Finally, I should add that background "assumptions" may be performing their work at the substantive, at the methodological, or at both these dimensions of the relevant process of inquiry.

Given the foregoing small analysis, I can now state what scientific controversies amount to and why they occur as well as formulate, on this basis, the principle required for their classification.

What Are Scientific Controversies?

At first glance, scientific controversies are simply disagreements among the practitioners of a given science regarding some aspect of their practice. They are disagreements over the way to tackle a given problem or puzzle, over what an adequate solution to such a problem or puzzle amounts to, or even over the criteria to be employed for such an assessment. A scientific controversy, however, is not a mere disagreement: it is one that cannot be readily settled by resorting to the commonly accepted disciplinary canons for conducting the relevant inquiry, as these have been developed up to that time. The disagreement may even be profound enough to implicate the assessment of those very canons. As the reader should have gathered by now, my main thesis is that scientific controversies occur when disagreeing scientists *do not share background "assumptions."*

Let me elaborate somewhat on this. First of all, background "assumptions" exist inescapably, as noted above, in any process of inquiry while such a process is addressed to the world that exists independently of it and of the background "assumptions" it involves. This is to say that the world need not conform to what these "assumptions" silently dictate, while the resistance it may correspondingly manifest appears in the form of a problem, or puzzle, which the inquiry in question cannot solve or dissolve, at least for the moment (Baltas 1997). Beyond the manifestation of such a resistance, the world remains, however, mute: it cannot point out by itself which of the "assumptions" involved are at fault. Thus, in front of a problem or puzzle that dumbly resists its resolution with the means at hand, those conducting the inquiry have to resort to some strategy of their own design for coping with the situation. In some occasions, different such strategies may seem to different scientists as worth pursuing.

Now, each such strategy is an inquiry in its own right, involving its own set of background "assumptions." Background "assumptions" derive, as I implied above, from the positions in the widest possible sense (social, professional, ideological, etc.) of the scientists implicated. That different such sets of background "assumptions," and hence that different such strategies, are in principle available derives from the fact that these positions constitute an indefinitely rich variety. That these strategies are different implies that the "assumptions" they involve are not shared. As background "assumptions" are not proper assumptions, explicitly stated and overtly accepted, their precise role and function regarding the strategy they determine remain hidden from view. This is to say that, in pursuing their different strategies, scientists are constrained by something they do not share and are in no position to lay bare on the table of discussion. Their disagreement amounts to a controversy because they debate an issue without rendering explicit the very factors whose silent existence precludes their all resorting at the same moment and in the same manner to the same set of criteria, norms, or canons.

If this is indeed what a scientific controversy amounts to, the general pattern of its resolution should run as follows. First of all, one of the parties implicated, while pursuing its particular strategy, must come up with a *new* scientific result that solves—or at least appears to solve—the problem and/or dissolves the puzzle lying at the root of the controversy. On the basis of what I have said, it follows that the new scientific result achieves this to the extent, and only to the extent, that its coming to being *effects the disclosure* of one (or more) of the background "assumptions" that had been silently at work during the controversy. However, this may not be immediately sufficient in itself to end the controversy. The other parties implicated need not acknowledge that the new result indeed achieves what the first party claims, and the controversy may linger. Our analysis of background "assumptions" permits us to isolate the grammatical condition—as opposed to the merely sociological and/or psychological—that will allow the new result eventually to win the day, such resistance notwithstanding.

The disclosure of an background "assumption" accomplishes simultaneously two things. On the one hand, it adds new grammatical possibilities to the conceptual system of the corresponding science: for the party having effected the disclosure, the horizon of inquiry is no longer closed by the mute existence of this "assumption," and new avenues of research are thus opened. These will eventually lead to the establishment of additional scientific results. On the other hand, the disclosure creates *a new vantage point* from where the party in question can look back at the particulars of the controversy. The misconstruals due to the work that the disclosed "assumption" had been silently performing, and had prevented the new scientific result from being attained before, can thus be located. The existence of additional scientific results, in conjunction with the localization of the previous misconstruals, places the party whose strategy was successful in this manner at an *objectively* superior position (Baltas 1992) regarding its opponents, who continue to be blinded by the work mutely performed by the "assumption" in question, and arms him or her with important rhetorical ammunition. By employing such ammunition the party who has effected the disclosure will make his or her point of view prevail in the longer or shorter run.

Given that different "levels" of background "assumptions" can be distinguished, it follows that scientific controversies can be classified according to those "levels." However, I should stress that the existence of a background "assumption," its particular "level," and the role it effectively plays at any juncture of a science's development can be determined only ex post facto, *after* the "assumption" has been disclosed *and* from the *new* vantage point created by this disclosure. In other words, it is impossible to arbitrate a scientific controversy before such a disclosure has been effected and/or from a vantage point that pretends to remain neutral regarding the two parties implicated, namely, the one who has effected the disclosure and the one who continues to be blinded by its silent existence.

Kinds of Scientific Controversies

As the space allowed does not encourage indulging in too fine an analysis, let me distinguish *three major kinds* of scientific controversies. Reversing the order above, these can be described as follows.

First, there are controversies in which the parties implicated share both the constitutive and the interpretative background "assumptions" involved in the inquiry but *not* the participation and/or the preference "assumptions." In other words, this kind of controversy involves only disagreements stemming from participating at different traditions, from following differing "styles of reasoning," and from entertaining different subjective preferences as to what is worth pursuing, either at the substantive or at the methodological level. No incommensurability of any sort is involved here, and disagreement does not hamper communication. Each party understands almost perfectly what the other is doing and why it is being done; the "assumptions" they entertain lead them only to consider their own line of investigation as more promising.

The almost forgotten controversy between the physicists supporting, with Chew, "particle" democracy" and those advocating "particle hierarchy," or the still ongoing controversy between the proponents of a steady-state cosmological model and those championing the idea of an expanding universe (Pera 1987) belong to this category. Moreover, priority disputes and fights over the distribution of credit belong here, too. For obvious reasons, we can call controversies of this first kind *surface controversies*.

To help come to grips with the deeper aspects of the cognitive dimension of science, surface controversies are not as probative, or as consequential, as those belonging to the following two categories, for, concerning precisely the cognitive level, controversies of this kind are quasi-immediately resolved after the new scientific result has been established. However, for understanding the social dimensions of the scientific endeavor, their significance is decisive: resolution at the cognitive level does not necessarily entail immediate resolution at the social level.

Differences in participation and preference "assumptions" cannot be telling, in the sense of giving rise to scientific controversies in their own right, unless the "assumptions" lying "deeper" in the background are shared. Accordingly, they will not concern us further in developing this classification.

The second kind of scientific controversy comprises those in which the parties implicated share the constitutive "assumptions" involved in the inquiry but *not* all the relevant interpretative "assumptions." Some, the more benign, forms of incommensurability may be encountered here, as well as some communication problems. As I implied above, the Newton-Leibniz dispute, analyzed by Freudenthal (1986), as well as the debates over the adequate explanation of the baffling (in terms of the then reigning interpretation of the conceptual system of quantum mechanics) low-temperature phenomena, analyzed by Gavroglu and Goudaroulis (1989), are cases in point. Another example is the Bohr-Einstein exchanges over the interpretation of the quantum mechanical formalism.

That all parties participating at controversies of this second kind share the constitutive "assumptions" of the conceptual system involved constrains the deployment of such controversies decisively. Incommensurability phenomena and communication issues become really serious only to the extent that *this conceptual system in itself* is not yet developed far enough, which is to say that the constraints it imposes on the controversy are not yet sufficiently clear to the disputants. Independent of sociological and subjective considerations, which undoubtedly played their own role, this can explain the *cognitive* dimension of the readily apparent differences in exacerbation between the Newton-Leibniz conflict, on the one hand, and the more recent disputes, on the other: in the first case, the relevant conceptual system was less developed, and hence less constraining, than it was in the others.

The resolution of a scientific controversy belonging to this second kind, as it is effected through the disclosure of the interpretative "assumption(s)" implicated, either leaves intact the conceptual system involved (quantum mechanics showed that it had the capacity, as it stood, to account for low-temperature phenomena) or develops that system in a way that renders it capable to accommodate post hoc both parties of the controversy, with relatively minor adjustments of their initial positions. The Newton-Leibniz case and, for that matter, the controversy over the wave or the matrix formulations of quantum mechanics are good examples here. As said above, we are entitled to say that the controversy was indeed about unshared interpretative "assumptions" only *after* it has been resolved and the interpretation of the relevant conceptual system has been correspondingly clarified.

Finally, the third kind of scientific controversy comprises what we can call *deep controversies*. In scientific controversies of this kind, the parties implicated do not share some of the constitutive background "assumptions" involved in the inquiry. Without, again, paying too much attention to details, we can distinguish here two major subcategories.

First, there are the controversies implicating some of the background "assumptions" assuring the coherence and determining the identity of a conceptual system that is *already well constituted* within an *already existing science*. This is to say that the cognitive perspective on the world defining that science—together with all the substantive and methodological ingredients that go with it (Baltas 1997)—*is not at issue*. Important incommensurability questions and grave communication problems arise in such cases, but the fact that all parties in the dispute work within the same perspective not only constrains in important ways the deployment of the controversy, but also restricts substantially the areas where incommensurabilities appear and communication breaks down. The Lorentz-Einstein dispute or that between the proponents of the new quantum physics and those of classical mechanics are cases in point.

The second subcategory of deep controversies comprises those occurring when the very perspective defining a science is in the process of being carved out. The background "assumptions" at issue in such cases are *not* those constitutive of the new conceptual system that is in the process of being established. Rather, they are those determining the identity and assuring the coherence of the particular *old* theories that the new conceptual system constitutively challenges. Since the controversy is not deployed within the disciplinary confines of an established science, it is, as a rule, less focused than it is in the cases we encountered above, and in that sense it tends to spill all over the place. Almost any part of common wisdom, with the deeply entrenched background "assumptions" it harbors, may be called to the rescue against the threat, or the scandal, represented by the new, highly counterintuitive concepts. The dispute of Galileo with the Aristotelians exemplifies best this subcategory (Damerow et al.

1992), while the controversy opposing Volta to Galvani (Pera 1991) or that between Lavoisier and Priestley, which Pierluigi Barrota analyzes in this volume (chapter 9), can with qualifications be classified here, too.[1]

Obviously, deep controversies are the most significant from the cognitive point of view. They are revolutionary processes whose stake is the establishment of a radically novel conceptual system. To repeat, this either inaugurates a new science, and hence opens a wholly new perspective on the world, or it challenges some highly confirmed theory within the confines of an already existing science. In both cases, the new view appears almost incomprehensible to those still holding to the old theories, while the reverse is not true. Those supporting the novel conceptual system are objectively in the position to know which are the weaker links in their adversaries' strategy and thence to fight them better. The final resolution of a deep controversy is tantamount to the total victory of the new conceptual system. This is a victory showing no mercy: from the vantage point of the victors, and if sufficient time has passed, the defeated views appear no better than plainly irrational. Hard work is required in order to make the Aristotelian viewpoint sound less than ridiculous even to our undergraduate physics students, for example.

This last point implies that, on the present analysis, the existence, the deployment, and the outcome of scientific controversies of any kind do not involve in principle, and need not question the rationality of any of the participants. Once we accept that any inquiry and any debate cannot but rest on a background of "assumptions," all the parties implicated, prospective winners and losers alike, can be as rational (or as irrational) as our pet theories of rationality might suggest without this impinging on any aspect of a scientific controversy. In the present volume, Philip Kitcher (chapter 1) and Marcello Pera (chapter 3) defend this view and elaborate on it from different angles, while Barrota (chapter 9) makes the corresponding case very well on behalf of Priestley.

And this allows me to conclude. An attempt at classification as the present one, that is, an exercise in pure description with no normative import on issues of rationality and the like, would be senseless unless it aspires to be adequate to the subject matter it sets out to organize. It follows that the attempt hangs literally in the air unless concrete controversies, as many and as varied as possible, are studied in the detail required. The many excellent chapters in this volume, which do precisely this, are invaluable to me not only for their intrinsic worth but also because they literally place in our hands the very means for assessing what precedes, for filling in its all too obvious lacks, for remedying its no less apparent deficiencies and shortcomings, as well as for criticizing its governing ideas.

Notes

1. *Post festum*, we can say that what was at issue in the Priestley-Lavoisier controversy was the conceptual system of the new science of chemistry, and at least one of issues in the Galvani-Volta case was the separation of physiology from physics. As both these controversies involved the opening of the corresponding new perspective (chemistry and physiology), I rank them as deep controversies of the second subcategory. However, these two new perspectives were not as radically novel and as all-encompassing as the one at issue in the dispute of Galileo with the Aristotelians: Galileo's work opened the way for all the natural sciences, as we conceive them today. Moreover, as a matter of fact, both Priestley and Lavoisier, and Galvani and

Volta fought each other *within* that wider perspective. One could then object to my grouping and maintain that I should have classified either or both of these controversies under our first subcategory of deep controversies, or even under the second category. This objection is well taken. However, to defend the classification proposed or to refine it by adding further subcategories would oblige me to enter deeply into the details of each controversy, something I cannot do here.

References

Baltas, A. (1992), "Shifts in Scientific Rationality and the Role of Ideology," in M. Assimakopoulos, K. Gavroglu, and P. Nicolacopoulos (eds.), *Historical Types of Rationality, Proceedings of the First Greek-Soviet Symposium on Science and Society*. Athens, Greece: National Technical University of Athens.

——— (1997), "Constraints and Resistance: Stating a Case for Negative Realism," in E. Agazzi (ed.), *Realism and Quantum Physics*, Rodopi, Amsterdam-Atlanta, Ga.

Damerow, P., Freudenthal G., McLaughlin P., and Renn J. (1992), *Exploring the Limits of Preclassical Mechanics*. New York: Springer.

Freudenthal, G. (1986), *Atom and Individual in the Age of Newton*. Dordrecht: Reidel.

Gavroglu, K. (1990), "Differences in Style as a Way of Probing the Context of Discovery," *Philosophica* 45, 53–75.

Gavroglu, K., and Goudaroulis, Y. (1989), *Concepts Out of Context(s): The Development of Low Temperature Physics 1881–1957*. Dordrecht: Nijhoff.

Gavroglu, K., and Simoes, A. (1994), "The Americans, The Germans, and the Beginnings of Quantum Chemistry: The Confluence of Divergent Traditions" *Historical Studies in the Physical and Biological Sciences*. 25, 47–110.

Hacking, I. (1982), "Language, Truth and Reason," in M. Hollis and S. Lukes (eds.), *Rationality and Relativism*. Cambridge, Mass.: MIT Press.

——— (1983), *Representing and Intervening*, Cambridge: Cambridge University Press.

Kitcher, P. (1993), *The Advancement of Science*. Oxford: Oxford University Press.

Laudan, L. (1992), "The Demise of the Demarcation Problem," in R. S. Cohen and L. Laudan (eds.), *Physics, Philosophy and Psychoanalysis. Essays in Honor of Adolph Grunbaum*. Dordrecht: Kluwer.

Morawetz, T. (1980), *Wittgenstein and Knowledge*. Sussex: Harvester.

Pera, M. (1987), "From Methodology to Dialectics: A Post-Cartesian Approach to Scientific Rationality," in A. Fine and P. Machamer (eds.), *PSA 1986*, vol. 2. East Lansing, Mich.: Philosophy of Science Association.

——— (1991), *The Ambiguous Frog*. Princeton, N.J: Princeton University Press.

Weiner, J. (1990), *Frege in Perspective*. Ithaca, N.Y.: Cornell University Press.

Wittgenstein, L. (1972), *On Certainty*. New York: Harper and Row.

3

Rhetoric and Scientific Controversies

MARCELLO PERA

The Sins of the Fathers and the Regrets of the Sons

The philosopher-scientists who gave rise to modern science, let us call them the Founding Fathers or, for short, the Fathers, were deeply attached to the idea that science is uncontroversial. This is not because in science there are no controversies, which, especially for those who were struggling against Scholastic doctrines, it would have been ironic to maintain, but because in science there are ways of solving conflicts of opinion definitively.

The origin of this idea is to be found in three other views or acts of faith to which the Fathers were also attached, namely, (1) that nature has its own order that is independent of us, that is, that nature is a book or a text; (2) that there are means for discovering such an order, that is, that there is a method for deciphering the sense of the book of nature; and (3) that these means are universal, that is, that they are shared by everybody as parts of the very structure of the human mind. Descartes wrote that "whenever two persons make opposite judgements about the same thing, it is certain that at least one of them is mistaken, and neither, it seems, has knowledge."[1] Galileo took the verdicts and laws of nature as stringent and inexorable; therefore, he wrote that

> just as there is no objective half-way between truth and falsehood, likewise in necessary demonstrations, either one draws a compelling conclusion or one ends up in inexcusable paralogisms, leaving no room for using limitations, distinctions, distortions of the meanings of words, or other tricks in order to stand on one's own feet; but it is necessary with few words and at the first attack to be either Caesar or nothing.[2]

In the same vein, Bacon wrote that his own new organon "leaves but little to the acuteness and strength of wits, but places all wits and understandings nearly on a level."[3] Later on, Leibniz improved on this organon with his "universal analysis" and "scales of proofs," thanks to which "all truths can be discovered by anybody and with a secure method"[4] or the degrees of probability of propositions can be estimated. As he wrote, once we have this analysis at our disposal, "if controversies arise, there will be no more need for discussion between two philosophers than there is between two calculators. It will be enough for them to take a pen, sit at a desk, and say (after calling, if they like it, a friend): let us calculate."[5]

We, the Sons, have lost this faith and believe that the Fathers, though great, were philosophically optimistic and sometimes naive. Although a few of us still stick to some sort of realism, all of us have gotten rid of the foundationalistic idea that there are pure, neutral ways to access nature, be they empirical or rational, and we have completely abandoned the view that there is a universal method that, as an organon, allows us to use those ways of access or, as an arbiter, allows us to establish whether they have been used rightly. The Cartesian "reliable rules which are easy to apply," the Baconian rules that act like a "rule or compass," the Leibnitzian analysis that transforms reasoning into calculus, as well as their up-to-date versions, are for us shattered dreams.

This is why we are faced with difficult problems. We have lost the earthly Paradise but we still hanker for it. If there is no independent, impartial arbiter, how can we say a controversy is settled? And when it is settled, how can we say the winning party has reached the truth or progressed in relation to its rival?

I will say something about the second problem, but it is the first that will concern me most here. Obviously, a controversy is settled by a victory of one party over another. But mere victory is not enough, because one may happen to win with wrong ways or bad reasons or irrelevant means. What we need is at least an *honest* victory, a victory with the *right* means. But what does "right" mean? One of the misbelievers has come to maintain that the very idea of being right or rational makes no sense for lack of a supreme tribunal (be it God's eye, the Human Mind, Method, or the Universal Community). In this I detect the voice of despair.

Although we must admit that no such tribunal exists or, as Richard Rorty, the misbeliever in question, has put it, that we cannot explain the success of science because there is no "epistemologically pregnant answer"[6] to the question as to why one party in a dispute is right and the other wrong, we must also recognize that people make *decisions* about one theory being better than another. Thus the idea of being right or rational in a controversy cannot be "out of place" merely because no one, from the outside, can compare our claims and nature itself, or because no one can be absolutely right or rational, or right or rational with respect to an ahistorical, neutral code. The very fact of decision involves reasons, and reasons involve the distinction between good and bad, valid and invalid, sound and unsound, right or wrong, which, in its turn, involves the question of the standards with respect to which reasons can be judged to be so. When Rorty writes that "Europe did not *decide* to accept the idiom of Romantic poetry, or of socialist politics, or of Galilean mechanics,"[7] he simply does injustice to many of the struggles and discussions that took place at the time, to many people who used up their energies and lives in these discussions and struggles, and to the communities that examined the different options, pondered them, and finally made up their minds.

However, the fact remains that, for us, in scientific controversies there is no impartial arbiter. If we want to combine this fact with the fact of decisions, I see no other more promising way than elaborating upon the idea of an "honest victory without an impartial arbiter". I shall call it "dialectical victory." I come later to the question of dialectics as well as rhetoric. For the moment, let us take "dialectical victory" in the intuitive sense of a victory carried off through argumentative means. My first concern is with these means, that is, the terms in which controversies are conducted and victories won.

Controversies and Dialectical Victory

Consider two interlocutors, or parties in a dispute, A and B, with different views, respectively T and T', as to how to solve a scientific problem, say, a fact recalcitrant to explanation, an anomaly, a conceptual tension or inconsistency. A controversy then arises between A and B.

Since, for the Fathers, according to their three acts of faith, (1) A and B address nature directly, (2) there are means for revealing its order, and (3) the use of such means is regulated by universal rules, they had an easy way (easy to define, if not always easy to apply) of settling this controversy.

A wins a victory over B = A proves T or falsifies T'

In this context, "victory" is a *metaphoric* concept. Actually, for the Fathers, A does not defeat B—A defeats nature itself. This is so because, as Bacon put it, the aim of science is "to overcome, not an adversary in argument, but nature in action."[8] Hence, for the Fathers, all dialectical means of arguing are to be banned from science and to be relegated to the realm of "altercations," to use one of Galileo's favorite expressions against the dialecticians. If nature speaks for itself or if, by mastering it, we may induce it to reveal its true structure, no real discussion with interlocutors is needed, unless it is to overcome their resistance, to clear their vision by ridding it of all sorts of idols or preconceived ideas that prevent them from reading the book of nature accurately. In the same way, for the Fathers, no rhetorical persuasion is needed, because no persuasion can be more efficient than that imposed by the language of nature itself.

For us, the Sons, the situation is different. As our loss of faith has induced us to maintain that there is neither a nature "out there" speaking for itself nor neutral means or universal rules for discovering its structure, we have no other means of settling the controversy between A and B than the resources offered by the ongoing discussion between them. Our way, then, is like this:

A wins a victory over B = A refutes B's arguments

Here "victory" is *literal*—it is a dialectical victory. It refers to the fact that A engages in a real discussion, a real exchange of arguments and counterarguments, with B, until A refutes B. It is this inner dialectical victory, not the outer verdict of nature, that decides what nature's voice is. Thus, in this view, refutation and persuasion are needed in an essential, constitutive way, that is, not in order to overcome resistance to an answer, but in order to construe the answer itself.

All this has several consequences. The first from which all the others stem, is that we, the Sons, have to question the anti-dialectical reaction of the Fathers and perhaps go back to Aristotle's view of science and scientific controversies. Granted, the Fathers had good reasons for rejecting dialectics (namely, its Scholastic degeneration), and we have to be grateful to them because our science is mainly the ripened fruit of such a rejection. But while the Fathers did better, I think Aristotle was farther sighted.

According to Aristotle, at least to my interpretation of Aristotle, science makes use of demonstration (demonstrative syllogism or didactic argument) insofar as it is a body of already acquired knowledge to be organized in order to be transmitted. For him, demonstration is the logical structure of the *static* of science. But insofar as science is a work in progress, an ongoing inquiry, a research, it does not use demonstration, or it does not use it in the same way; it uses dialectical syllogism. The logical structure of the *dynamics* of science, that is, of science in the making and not in the teaching, is dialectics. Since dialectics is not a science itself, but serves all sciences and arts, this means that, for Aristotle, scientific research possesses no special organon, but rather uses the organon common to all sorts of discussions in which distinctions are made regarding good or bad, right or wrong, rational or irrational arguments.

It is not my intention here to examine Aristotle's position or to take him as a protective shield for my own view. I merely start from the assumptions that the Fathers have committed philosophical sins, as far as their metaphysical realism, foundationalism, and methodologism are concerned; that these sins cannot be expiated by simply finding a remedy for this or that defect of their view; and that a new philosophical image of scientific research is needed. In other words, I believe that we have to change the boat handed down to us by the Fathers rather than simply repair its sails. What I suggest is a dialectical view. If it proves promising and is accepted at least as a research program, it will confirm once again that the irony of history exists, because when the Fathers found themselves confronted with the dilemma of Scholastic dogmatism or scepticism, they decided to get rid of precisely that dialectics I am suggesting to use to get out of a similar dilemma, the one between positivistic dogmatism and irrationalism, now affecting the philosophy of science.

The idea of a dialectical victory raises two preliminary questions. The first concerns the factors of scientific dialectics, as I call them, that is, those factors in terms of which the dialectical exchanges of arguments and counterarguments are conducted. The second refers to the rules regimenting such exchanges. Elsewhere, I have examined the former question extensively and touched upon the latter.[9] Here I shall elaborate this further. However, a brief look at the factors is needed.

The list of these factors of scientific dialectics includes established facts, accepted theories, admitted values, shared assumptions, and presumptions. These are the substantive elements that given interlocutors or parties in a controversy appeal to in given situations in order to reach an agreement or weaken or reinforce consensus over a given claim. Notice that these elements are *typical* of, or *proper* to, science because they establish its way of proceeding, constitute its domain, define its tradition. As far as logic or the ways of arguing are concerned, Rorty is right when he says that science is a sort of "routine conversation." Aristotle's view was no different. But regarding its substantive factors, science is different from any other intellectual or practical activity, because the factors of science are not the same as the factors of other activities.

Due to this, the factors of scientific dialectics are both *internal* and *epistemic*. These distinctions are notoriously rather troublesome. My view is that "internal" denotes all those factors that, at a given time, are admitted by the community as pertinent to the dialogue or dispute concerning a given cognitive claim. "Epistemic" denotes all those factors (not only facts, therefore, but also values, assumptions, etc.) that, at a given time, are considered by the community as relevant to the cognitive value of a given claim. Epistemic factors are then internal, because no factor is considered by a community to be relevant to the cognitive value of a claim if it is not admitted by that community to be pertinent to the dialogue in which such a claim is discussed. And the other way round: internal factors are epistemic, because no factor is admitted by a community to be pertinent to the discussion of a cognitive claim if it is not considered by the same community to be relevant to its cognitive value. In other words, "internal" means "internal to the dialogue or dispute"; "epistemic" means "relevant to the merit of the claim." Since whatever is internal to the dialogue influences its course and conclusion and since the merit of the conclusion, for lack of other independent ways of establishing it, depends on the dialogue, internal and epistemic coincide.

A first consequence of this view is that external (and therefore nonepistemic) factors do not count in science, not because they may not happen to influence it, but because when they do, they become internal and epistemic. Thus, the cult of the Sun and of its eminence may have been an external cultural factor in the sixteenth and seventeenth centuries, but when it influenced the birth of modern astronomy, it became a value or assumption, that is, an internal epistemic factor. Another consequence is that the closure of scientific controversies always depends on internal factors. Again, this is not because external factors may not become decisive, but because when they do, the controversy is not really scientific, but rather a controversy between science and something else, for example, the Church in the Galileo affair or a political party in the Lyssenko affair.[10]

With the list of substantive factors in mind, let us now examine the argumentative techniques in terms of which controversies are conducted and victories can be won.

Refutation Schemes

The participants in a controversy attempt a dialectical victory in the sense roughly explicated above of refuting each other. Consider a controversy between A, the proponent, and B, the opponent. To refute B, A has to

1. acquire a set of departure premises,
2. find certain proper bridge or covering premises, and
3. use appropriate inference rules in order to derive the desired conclusion.

As for item 1, departure premises are either those propositions that are conceded by B or those propositions that both A and B must admit, for example, because they refer to states of affairs or already established conclusions everybody in the community agrees upon. Aristotle made a distinction between dialectical arguments proper (from reputable opinions, *endoxa*) and peirastic arguments (from concessions).[11] However, on the one hand, he maintained that a concession cannot be any proposition whatsoever, because "no one in his senses would make a proposition of what no one holds" for "no one would assent" to it.[12] On the other hand, he thought that a mere concession that would not be generally accepted would make the argument contentious.[13] Therefore, as concession does not hold if it is not reputable and reputable opinion does not work dialectically if it is not accepted by the opponent,[14] we can group concessions and reputable opinions in the same class. I shall call such a class "accepted premises."

As for item 2, the bridge premises are those propositions that, together with the accepted premises or by themselves, lead to the conclusion. This is mainly the function of Aristotle's *topoi*;[15] in my view, this is the function of the substantive factors of scientific dialectics. The rules of logic constitute item 3. But this is a delicate question examined carefully later.

Considering these three elements, the ideal, highly schematized, structure of refutation may be depicted like this:

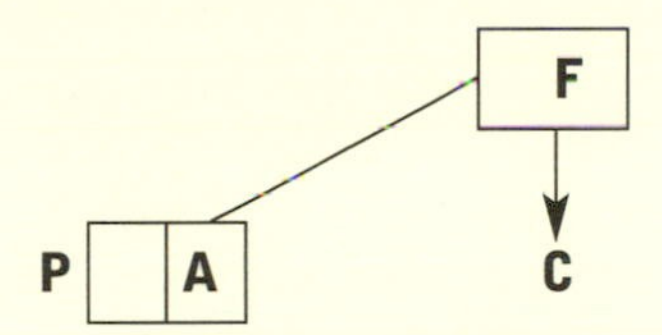

where P is the class of accepted premises, A that subclass of those accepted premises which is actually used by the proponent and usually explicitly conceded by the opponent, F the set of substantive factors holding in the specific dialectical situation in which the controversy takes place, and C the conclusion.

Notice that all the elements of this structure can be questioned, and a controversy may involve different levels. A good illustration is provided by the controversy between Galileo and Francesco Ingoli about the Copernican system. When Ingoli puts forward the objection that "the earth is in the center of the eighth orb because the stars, whatever position they have on the horizon, appear to have the same magnitude,"[16] Galileo replies by questioning the premises. He says, first, that to assume that the stars of the firmament are all placed on the same sphere "is so doubtful that neither you nor anyone else will ever prove it." Second, the claim that a given star always appears to be of the same magnitude "is full of difficulties, which make it very uncertain," because (a) "very few stars can be seen when they are near the horizon," (b) "their apparent magnitude is always altered in various ways by vapors and other impediments," (c) no "natural eye can ever detect a very small change that might take place in two or three or four hours," and finally (d) "the same authors who place the earth at the center claim that, because its radius is insignificantly small compared to the great distance of the stellar sphere, therefore stars do not appear larger near the middle of the heavens than near the horizon."[17]

When Ingoli argues that the earth must be at the center because it "is denser and heavier than the solar body" and this "can be proved by the authorithy of the Philosopher and all Peripatetics who say that all celestial bodies have no gravity,"[18] Galileo questions the factors in terms of which his opponent conducts the discussion by replying that "in natural phenomena human authority is worthless," because nature "makes fun of constitutions and decrees of princes, emperors, and monarchs, and at their request it would not change one iota of its laws and statutes."[19]

Finally, Galileo sometimes questions Ingoli's way of reasoning. For example, Ingoli asserts that "if the Sun were at the center of the universe, its parallax would be greater than the Moon's," because "the farther bodies are from the first mobile, where astronomers put their places, the greater their parallax is" and because "the Sun, according to Copernicus himself, is farther than the Moon from the first mobile since the latter is out of the center while the former is in the center, and the center is the farthest place from the periphery."[20] To this Galileo replies that the argument contains an error because "it is not the distance of a star from the firmament (or from anything else which you may place as a boundary for the parallax) which makes it greater, but the vicinity of the star to the eye of the observer, namely the earth."[21] And this error reveals a paralogism (*petitio principii*), because the proposition that the greater the distance of a body from the firmament, the greater its parallax, presupposes that the earth is at the center of the universe, which is precisely what Ingoli wants to show.

Later on, Galileo charges Ingoli with more flawed reasoning. Ingoli's syllogism—denser and heavier simple bodies are closer to the center; the earth is denser than the Sun; therefore, the earth is closer to the center[22]—contains a paralogism (*quaternio terminorum*) because in the first premise the center is taken in the sense of the center of the earth, while in the conclusion it is taken in the sense of the center of the universe.

Let us leave these complications aside for the moment and examine the logic of refutation. According to Aristotle, refutation is "a deduction to the contradictory of the given conclusion."[23] This is what may be called "strict refutation" or "refutation in a proper sense." But Aristotle himself admitted there are other ways of carrying off a dialectical victory,[24] which we may call "refutation in a broad sense." Here are a few typical refutation schemes.

1. *Refutation due to falsification of the consequences of the opponent's hypothesis,* that is,

$$((H \therefore O) \,\&\, \text{not-}O) \therefore \text{not-}H.$$

This is the best way of winning a scientific controversy. If the proponent shows that the opponent's hypothesis has a false observational consequence, then he may defeat him. "May" is a key word, because, as I have shown, the opponent can question both the premises and the implication link.

2. *Refutation due to inconsistency in the interlocutor's position,* that is,

$$((H \therefore \text{not-}T) \,\&\, T) \therefore \text{not-}H.$$

This is another effective way of winning a controversy. Suppose the opponent admits that T is a good theory. Then, if the proponent shows him that his hypothesis H is inconsistent with T, then the opponent is refuted.

Both the above schemes of victory are based on deductive arguments. But deductive arguments are not the only arguments available, and Aristotle himself had already taken inductive arguments into consideration. If the opponent denies a certain universal proposition, then the proponent may lead him to accept it (this is the sense of *epagein* and *epagogé*) by showing him that such a proposition holds good in, or is instantiated by, many particular cases,[25] and then, after so shifting the burden of proof, by challenging him to adduce a counterexample. If the opponent does not reply or is not able to find a negative case, the universal proposition is to be admitted.[26]

But other more sophisticated cases of inductive arguments exist. Consider the following situation. The opponent admits that H ∴ O, then in the course of the dialogue he also admits that p(O) = 1, for example, because O is already well known and other theories exist that explain or predict it. Then the proponent may try to refute him by using the following scheme taken from probabilistic logic:

3. *Refutation due to scarse or null increment of the initial probability of the interlocutor's hypothesis,* that is,

$$((H \therefore O) \,\&\, p(O) = 1) \therefore p(H, O) = p(H).$$

Another typical refutation scheme based on inductive arguments refers to a situation such as the following: the opponent says his own hypothesis is implied by a certain theory T, and during the discussion, he comes to admit that T is false, for example, because T ∴ O and not-O. The proponent can then show him that the credibility of his hypothesis is lower. Here, the probabilistic scheme goes like this:

4. *Refutation due to the explosion of the ground of the interlocutor's hypothesis,* that is,

$$((T \therefore H) \,\&\, p(T) = 0) \therefore p(H, \text{not-}T) < p(H).$$

Let us stop with such deductive and inductive refutation schemes as the above and consider the two main problems that arise regarding them. First, do deductive and inductive refutation schemes exhaust all possible ways of winning a dialectical victory? Second, under what conditions are refutation schemes efficient?

To answer affirmatively the first question means to classify all arguments as either deductive or inductive. This widespread view in textbooks of logic is disputable, however, first because such a classification is biased by the assumption there are only two logics, and second, because it is defective. Take an argument with the form ((H ∴ O) & O) ∴ H. Is it deductive or inductive? Or take an argument from authority such as the one used by Ingoli: is it elliptically deductive, elliptically inductive, or a legitimate argument of its own?

These questions can only be answered by taking into consideration the *context* in which an argument is put forward, in particular, the argumentative *field* to which it belongs (say, physics, biology, mathematics, etc.) and the *function* it performs in such a field (say, explaining phenomena, proving theorems, etc.). Thus, an argument with the form ((H ∴ O) & O) ∴ H is inductive in the field of empirical sciences with the function of confirming a hypothesis or suggesting it as worthy of consideration, but it would be considered deductive (and of course invalid) in the field of mathematics with the function of proving a theorem.

Once arguments are considered in their contexts, they can be grouped in such classes as explanations of states of affair, proofs of theorems, confirmations of hypotheses, predictions of events, suggestions of possible lines of research, and so on, and it may be difficult to establish which logic is proper to which class. Certain substantive controversies are sometimes about logical matters. It is a fact that different theories are often associated with different logics. For example, the controversy about Aristotelian science was not only about different explanations, but also about whether explanations of natural phenomena have to take the form of syllogistic deduction from metaphysical principles (Scholastics), induction from empirical data (Bacon), geometrical proof concerning an idealized model (Galileo), or mathematical derivation from principles (Descartes). The controversy about Darwin's theory was also about whether science has to use Baconian inferences or hypothetico-deductive inferences. The controversy between the Big Bang and the steady state theory of the universe also involved the question of whether cosmology is to be deductive or inductive.

As to the second main problem, conditions of efficiency, one might say that refutation schemes are efficient when the arguments they are based on are valid. This is clearly not enough. We have to consider that, when examined in a dialogue situation, an argument is quite different from those strings of exsanguine phrases or dead symbols formal logic deals with. An argument is a part of a living conversation between concrete interlocutors and the context of this conversation; that is, the field and function of the argument, as well as the specific factors holding in that situation and the commitments of the interlocutors, are relevant to its being a real argument and to its being efficient.

Let us take an example. Why do certain bodies, say, a piece of fir, float on water? The Aristotelians said:

Bodies with a preponderance of air float on water.
Fir has a preponderance of air.
∴ Fir floats on water.

Regarding this trivial syllogism, two things must be noted. First, in order for it to be a real argument and not just a string of phrases, its field and function are to be specified. Second, once this has been done and the argument has been classified, say, as an *explanatory argument,* in order for it to be evaluated, the requirements of the explanations are to be stated. Now, as it is an answer to a why-question, Galileo had no problem in considering this argument an explanation, but he had problems with considering it a good explanation because, given the requirements he set on explanations, he did not find the argument sound.

The general moral we can draw is that in order for an argument, or the refutation scheme based on it, to be efficient, that is, in order for it to bring about a dialectical victory, it has to be *convincing* or *sound.* That an argument is formally valid does not in itself guarantee that it is convincing. This is shown by the fact that there are valid arguments that are not convincing and invalid arguments that are convincing. The syllogism of the Aristotelians above is valid but it does not convince us today as it did not convince Galileo. In the same way, Ingoli's *petitio principii* we have met before is trivially valid (it has the form $p \therefore p$) but not convincing. In contrast, a circumstantial ad hominem argument may be convincing in many cases, although it is formally invalid.

Dialectics and Rhetoric

The next step is to examine the idea of a convincing or sound argument. Let me first recapitulate. I have started by putting

A wins a victory over B = A refutes B.

I have now shown that

A refutes B = A produces a convincing argument.

This means that I have started with dialectics (broadly speaking, the art of refuting) and ended up with rhetoric (broadly speaking, the art of convincing). But this is a point I must make clear, because, although it is becoming a fashionable topic, rhetoric is still shrouded in mystery.

Trevor Melia has usefully distinguished three senses of rhetoric: as the *act* of persuading, as the *analysis* of such acts, and as a *world-view.*[27] I shall use rhetoric in all these senses, but it is the first that is crucial and gives relish to the other two. I shall borrow it from chapter 1 of Book 1 of Aristotle's *Rhetoric,* although here Aristotle does not speak in terms of acts but of art or faculty, and he is not consistent with what he says later.

The problem with the first sense of rhetoric is what sort of act it is. According to what Aristotle says in chapter 1, rhetoric is not the art or faculty of mere persuading or convincing, but rather the art or faculty of using *modes* or *means* (*pisteis*)[28] of persuasion, more precisely, *technical* means. The proper technical means are logical (that is, arguments), because "the arousing of prejudice, pity, anger, and similar emotions has nothing to with the essential facts, but is merely a personal appeal to the man who is judging the case,"[29] and because "the modes of persuasion are the only true constituents of the art: everything else is merely accessory."[30] Moreover, the proper technical means are the same as dialectics or equivalent to them.[31]

As for the function of rhetoric, Aristotle writes that it "is not simply to succeed in persuading, but rather to discover the persuasive facts in each case."[32] He also writes that

> we must be able to employ persuasion, just as deduction can be employed, on opposite sides of a question, not in order that we may in employ it in both ways (for we must not make people believe what is wrong), but in order that we may see clearly what the facts are and, that, if another man argues unfairly, we on our part may be able to confute him.[33]

As I take it, this means that in order to refute someone, one has to convince him about the merit of the case under discussion by using arguments. Thus, according to my interpretation, rhetoric (1) uses arguments; (2) it leaves aside the speaker's personal character and the hearer's emotions, because they are accessory; (3) also leaves aside questions about style and arrangement of discourse, because they appeal to the man who is judging the case and do not affect the case itself; and (4) is instrumental to knowledge.[34] If, according to the first sense above, we now take rhetoric as the act or the practice of using convincing arguments, we may say that rhetoric is the set of all convincing arguments. But such an act or practice needs to be evaluated. This, in my view, is the function of dialectics.[35] According to a first definition, dialectics is then the *logic of convincing (or rhetorical) arguments.*

Let us now examine the concept of a convincing argument. Three preliminary questions arise: Convincing for whom? Convincing how? Convincing in what sense?

As to the first question, I have no doubt that an argument has to be convincing for the *given* community in which it is put forward. No argument can be absolutely convincing: as an argument is a string of a living dialogue between specific interlocutors who aim at changing their own specific belief systems, it is them that it has to convince. Nor can an argument be convincing for an ideal community or a universal audience. These are fictions invented by the Fathers, the secularized version of God's eye. Taken as a court, an ideal community is just another given community, a community of appeal.

As to the second question, is an argument convincing because it *does* convince or because it *ought* to convince, my answer is the latter. If, according to the sense of rhetoric above, the means of persuasion are argumentative, then if an argument is well construed, it ought to convince those it is addressed to, no matter how strong their personal resistance might be.

This also solves the third question: Is an argument convincing in a psychological or logical sense? Again, my answer is the latter. If an argument, in a given situation, is convincing, it is so because its conclusion is precisely the one everybody ought to accept in that situation.

But here the problem of the rules of inference crops up. What are these rules? What sort of logic is dialectics? As I have shown, refutation schemes are based on different sorts of rules. If B believes p and accepts r, than A can change B's belief if he shows that $p \therefore \neg r$, by using the rules of deductive logic. If B denies q (say, an empirical hypothesis), then A may try to convince him to change his belief if he proves that all the observational consequences of q are true, by using the rules of inductive logic. And so on. We can thus say that the rules of inference for convincing are the rules of formal logics, be they deductive, inductive, probabilistic, and so on. Notice that this is not to say that formal logics are by themselves enough to convince. Actually, formal logics do *not* convince, because they aim not at convincing but at inferring conclusions. This is to say that formal logics can be *used* to convince, that they are tools or instruments for convincing. In other words, formal logics may have a rhetorical function. Since dialectics is the logic of rhetorical arguments and since there is a plurality of logics, we have then a second definition of it: dialectics is the *logic of the rhetorical use of formal logics.*[36]

We are finally in the position to define a scientific convincing argument. We may say that a scientific argument, in a certain field of inquiry and for a certain purpose, is convincing if

1. its premises are accepted and the factors it relies on are admitted by the community, in that field and for that function;
2. the rules of inference are considered pertinent by the community; and
3. it is valid according to such rules.

This definition is not broad enough. As I have shown, there are certain arguments (such as ad hominem arguments, arguments from authority, etc.) that may be convincing although they violate or escape the rules of formal logics. There must, then, be other rules for appraising them. These cannot be but the (material) procedural constraints imposed on the debate. For example, an argument such as Ingoli's with the form $p \therefore p$ is bad (not convincing) because it violates the constraint of giving independent grounds to an answer (in that case, the answer to the question, Why *p?*). We have then to enlarge our definition of a convincing argument in order to also include these contraints in it. If we call "rules of debate" both the inference rules of formal logics and the material constraints ruling scientific debates, then we may say that a scientific argument, in a certain field of inquiry and for a certain function, is convincing if

1. its premises are accepted and the factors it relies on are admitted by the community, in that field and for that function;
2. the rules of debate are considered pertinent by the community; and
3. it is valid according to such rules.

Correspondingly, we have to enlarge our view of dialectics. If dialectics is the logic of all rhetorical arguments, then if we include material constraints into such a logic, we have a third definition that encompasses the previous two: dialectics is the *logic of debate.*

Science and Rhetoric

Galileo was engaged in a controversy against the Ptolemaics and won a victory. Newton was engaged in a controversy with Descartes and won a victory. And so on. The course of science is marked by a series of controversies between rival parties and subsequent victories of one party over another. Rhetorical analysis, in my sense, is the critical examination of the arguments put forward in these circumstancs. If I am right, all such victories depend on one party producing convincing arguments in the sense specified above. It is the job of the history-minded philosopher and the philosophy-oriented historian to carry on this analysis and to ascertain what arguments prevailed in what circumstances.

I am sure such an analysis will raise objections in several quarters. But I also think it solves certain problems about controversies better than that of our Fathers. First of all, it provides an explanation of why important controversies last a long time. Here the Fathers offer no other solution than psychological resistance stemming from prejudices, idols, opinions imbibed with mother's milk, as Bacon and Galileo put it. But there is more than this. Important controversies involve many of the factors of scientific dialectics, and when many factors are disputed at the same time, especially when values or assumptions are at stake, it is not easy and it may take a long time for one party to convince the other. Moreover, the material rules of debate are not as compelling as the formal ones. For example, a typical implicit constraint establishes that if one has the burden of proof and does not reply, that person can be considered refuted, but there are several ways and tricks to avoid that burden, such as silence or appeal to intuition awaiting better counterarguments.

Here psychological resistance plays a role. Although it does not matter—only arguments do—it is a rich source of devices for those who are in trouble in the course of a controversy. Einstein, to give a famous example, resorted to one such device (appeal to intimate conviction) when he wrote to Born:

> But I am convinced that someone will eventually come up with a theory whose objects, connected by laws, are not probabilities but considered facts, as used to be taken for granted until quite recently. I cannot, however, base this conviction on logical reasons, but can only produce my little finger as witness, that is, I offer no authority which would be able to command any kind of respect outside my hand.[37]

My rhetorical analysis also explains why controversies can hardly end up with a knockout. This is because persuasion involves degrees and it is a sort of accumulative process. We have also to consider that certain important premises and factors on which the rival parties in a controversy rely are often implicit, so arguments can hardly be immediately and clearly refuting.

When premises and factors are explicit, rhetorical arguments have the form of ad hominem arguments, in Locke's sense that they "press a man with consequences drawn from his own principles or concessions."[38] Although an interlocutor may withdraw his own concessions, this does not prevent the other interlocutor from defeating him.

Here is another advantage of rhetorical analysis. By focusing their attention on the logical relationships between cognitive claims and empirical evidence alone, the heirs of the Fathers came across the Duhem problem of crucial experiments. Within the framework of their logical analysis, there is no way of solving this problem except by relying on the scientist's personal sensibility or introducing tailor-made methodological rules. In the perspective of my rhetorical analysis, although there are no strictly logical crucial experiments, one can say there are *dialectically* crucial experiments. If, during a controversy, one party explicitly commits itself to a certain claim (say, the prediction of an outcome), then if that claim turns out to be false (say, the prediction does not happen), that party is refuted. Here, withdrawing the claim or adjusting the falsification, although logically possible, may be dialectically unfeasible.

Rhetorical analysis also has an advantage over content analysis. What about if two theories have the same observational consequences? Here content analysis faces serious problems because in this case the intrinsic content of the theories does not help to make an objective decision, which is then left to depend on personal idiosyncrasies or subjective interpretations of certain desiderata. Rhetorical analysis makes this decision dependent on the discussion between rival parties within a community. Such a discussion is open and its outcome is objective, in the sense that the force of the arguments supporting it is stringent for everybody.

Rhetorical analysis can also be used fruitfully to deal with the question of incommensurable theories. The obstacle we face in this case, that is, the problem of different references of the descriptive terms of the theories due to the different conceptual schemes they depend on, does not prevent a discussion between rival parties from taking place. As communication is possible even across such schemes, interlocutors may try to convince each other. It is enough for the parties to find a few shared premises to start with and then, by a sort of "rhetorical ascent," to go on step by step until one of them slowly yields while the other gets stronger and finally wins a victory.

Despite its advantages, rhetorical analysis is not without its cost. When the Fathers reacted to dialectics, they revived the argument with which Plato had stabbed the Sophists: What has rhetoric to do with reality and truth, which is the aim of science? People can be convinced that a donkey is a horse, but a donkey is not a horse.

Here is where rhetoric, as I take it, comes to be a world-view. By raising the Platonic objection, the Fathers proved to be realists. They thought that by mastering nature one can know it as it is—as Bacon put it, "[T]he fruits and works are as it were sponsors and sureties for the truth of philosophies"[39] or "that which has maximum utility in practice has minimum truth in knowledge."[40] But unless it is a stipulative definition of a philosophical stand, intervening successfully is no guarantee of representing correctly. Intervening successfully witnesses there is something out there to which we attach our concepts and categories; it is no evidence that this something is what our concepts and categories say it is. Put it differently, intervening successfully guarantees that our concepts *have* a reference not *what* this reference is.

Those of the Sons whose realistic faith has not been shaken by the earthquakes in the recent philosophy of science, logic, and meaning and nevertheless want to make room for rhetoric in science, take it as a device for the acceptance and better spread of scientific claims.[41] I admire their attachment to the view of the Fathers, but I cannot share their compromise. If there is no other accessible way of getting in touch with nature except by putting forward hypotheses, testing them against empirical evidence, *and* discussing them in front of a community, if there is no scientific knowledge apart from this interplay—in other words, if science is a game not with two players but with three—then rhetoric is no mere a device, but rather the tool of the very constitution of scientific knowledge. Where no discussions are engaged and no victories won, nature is dumb and science is blind.

Granted, reality is "out there" and the earth has rotated around the sun even before Galileo won his controversy against the Ptolemaics. But science does not deal with the earth rotating around the sun; it deals with the earth rotating around the sun being knowledge. Rhetoric is irrelevant to the former because the former does not depend on us, but rhetoric is constitutive of the latter because the latter is of our own making. Thus, we do not have to condemn scientific rhetoric as a contradiction, although two thousand years of Western philosophy have made it sound like this; we have rather to understand the structure of scientific rhetoric better.

A rhetorical world-view needs rhetoric twice, because it is a world-view and because it is rhetorical. But a rhetoric of rhetoric has no hope. At the very end, people take their decisions because they feel convinced.

Notes

1. Descartes (1628), p. 11.
2. Galilei (1624), p. 296.
3. Bacon (1620), p. 63.

4. Leibniz (1961), vol. 7, p. 202.
5. Ibid.
6. Rorty, (1982), p. 193.
7. Rorty (1989), p. 6.
8. Bacon, (1620), p. 42.
9. See Pera (1991a). I have given a more detailed account in Pera (1991b).
10. McMullin (1987) makes a distinction between "resolution" (when an agreement is reached on the ground of epistemic factors) and "closure" of a controversy (when nonepistemic factors are decisive). But if nonepistemic factors include such things as "the authority of the state, the loss of a research grant, or the laziness of a controversialist" (p. 78), the controversy either does not exist as a scientific controversy or is abandoned.
11. Aristotle, *Sophistical Refutations,* 2.
12. Aristotle, *Topics,* I, 10, 104a 6.
13. "A deduction is contentious if it starts from opinions that seem to be reputable, but are not really such". *Topics,* I, 1, 100b 23.
14. See Aristotle, *Sophistical Refutations,* 5, 167a 23ff: "For to refute is to contradict one and the same attribute . . . and to confute it from the propositions granted necessarily." See also *Topics,* VIII, 5 159b 18ff: "[T]he answerer should admit all views that seem to be the case and, of those that do not, all that are less implausible than the conclusion."
15. For this function, see De Pater (1965).
16. Ingoli (1616), p. 405.
17. Galilei (1624), p. 166.
18. Ingoli, p. 407.
19. Galilei, p. 178.
20. Ingoli, p. 404.
21. Galilee, p. 160.
22. Ingoli, pp. 406–407.
23. Aristotle, *Sophistical Refutations,* 1, 165a 4.
24. See ibid., 3.
25. See Aristotle, *Topics,* VIII, 1, 155b 34–35: "Thus if one desires to secure an admission that the knowledge of contraries is one, one should claim it not of contraries, but of opposites; for, if he grants this, one will then deduce that the knowledge of contraries is also the same, seeing that contraries are opposites; if he does not, one should secure the admission by induction, by formulating propositions to that effect in the case of particular contraries."
26. See ibid, VIII, 2, 157a 35–157b 1: "If one has made an induction on the strength of several cases and yet the answerer refuses to grant the universal proposition, then it is fair to demand his objection. But until one has oneself stated in what cases it is so, it is not fair to demand that he shall say in what cases it is not so; for one should make the induction first, and then demand the objection."
27. See Melia (1993).
28. See Aristotle, *Rhetoric,* 1355a 4: "the technical study of rhetoric is concerned with the modes of persuasion"; 1355b 27: "rhetoric may be defined as the faculty of observing in any given case the available means of persuasion."
29. Ibid., 1354a 17ff.
30. Ibid., 1354a 13.
31. Equivalent, because "the example is an induction, the enthymeme is a deduction." See ibid., 1356b 2.
32. Ibid., 1355b 10ff.
33. Ibid., 1355a 30ff.
34. As to (1), Aristotle (see n30 above) says that the proper rhetorical arguments are equivalent to those of dialectics. As to (2), he mentions, among the technical means of persuasion, also *ethos* and *pathos* in chapter 2 of Book 1 but he does not refer to them until Book 2. The same holds true for (3): style and arrangement are mentioned in chapter 2 of Book 1, but not used until Book 3. As to (4) I have shown that according to Aristotle rhetoric is useful "in order that we may see clearly what the facts are."

35. Here I depart from Aristotle's view about the relationships between dialectics and rhetoric. On this point he is ambiguous. He says neither that rhetoric includes dialectics, nor that dialectics includes rhetoric, but that rhetoric is a "counterpart," an "offshoot," a "branch" of dialectics, or "similar" to it. See *Rhetoric,* 1354a 1, 1356a 25ff. I also differ from Gross (1990), according to whom rhetoric is "the more general term that includes logic and dialectic" (p. 206). In my view, logic (or, better, formal logic*s,* for there are many) is the tool of rhetoric, because in order to convince one has to argue, hence to draw inferences and make use of rules of inference; whereas dialectics is the overall logic that regiment rhetoric and therefore, as I say later, the use of the pertinent logic. As I take style and arrangement as accessories and consider only arguments, I do not take rhetorical analysis as the "theory of the constitution of scientific texts" (p. 6). In my view, a rhetorical analysis of science or a rhetoric of science concerns the *logical structure* of scientific discourse, with respect to which style and arrangement are indifferent. I thus claim that the rhetorical or convincing force of, say, Galileo's *Dialogue* or Newton's *Principia* or *Opticks* is not affected by the former being written in a dialogical form and the latter in Euclidean style, although these elements might have been helpful for the reception of their views. If one takes "rhetorical" in the sense of "literary, or grammatical, or philological," as does Moss (1993, p. 23), I claim that rhetoric is not relevant to the cognitive status of science.

36. I am indebted for this to Barrotta (1992), § 16.

37. Einstein-Born (1971), p. 158.

38. See Locke (1692), vol. 2, p. 279. Hamblin (1970, pp. 161–162) has shed light on the Aristotelian origin of this sense of ad hominem.

39. Bacon (1620), I, § 73.

40. ibid., II, § 4.

41. See Moss (1993), postscript.

References

Aristotle, *Sophistical Refutations,* in *The Complete Works of Aristotle,* ed. J. Barnes, 2 vols., Bollingen Series LXXI, Princeton University Press, Princeton, N.J., 1984.

———, *Topics,* in *Complete Works.*

———, *Rhetoric,* in *Complete Works.*

Bacon, F. (1620), *Novum Organun,* in *The Works of Francis Bacon* (7 vols.), ed. J. Spedding, R. L. Ellis, and D. D. Heath, Longman, London 1860, vol. 1.

Barrotta, P. (1992), *Gli argomenti dell'economia,* Franco Angeli, Milano.

De Pater, W. A. (1965), *Les topiques d'Aristote et la dialectique platonicienne,* Editions St. Paul, Fribourg.

Descartes, R. (1628), *The Philosophical Writings of Descartes* (2 vols.), trans. J. Cottingham, R. Stoothoff, and D. Murdoch, Cambridge University Press, Cambridge 1985.

Einstein, A., and Born, M. (1971), *The Born-Einstein Letters,* trans. I. Born, Walker, New York.

Galilei, G. (1624), "Reply to Ingoli," trans. in M. Finocchiaro (ed.), *The Galileo Affair. A Documentary History,* University of California Press, Berkeley 1989, pp. 154–197.

Gross, A. G. (1990), *The Rhetoric of Science,* Harvard University Press, Cambridge, Mass.

Hamblin, C. L. (1970), *Fallacies,* Methuen, London.

Ingoli, F. (1616), *De situ et quiete terrae contra Copernici systemate disputatio,* in *Le opere di Galileo Galilei* (20 vols.), national edition, ed. A. Favaro, vol. 5, pp. 403–412.

Leibniz, G. W. (1961), *Philosophische Schriften* (7 vols.), trans. C. I. Gerhardt, G. Olms, Hildesheim.

Locke, J. (1692), *Essay on Human Understanding* (2 vols.), Dover, New York 1959.

McMullin, E. (1987), "Scientific Controversy and Its Termination," in H. Tristram Engelhardt, Jr., and A. Caplan (eds.), *Scientific Controversies,* Cambridge University Press, Cambridge 1987, pp. 49–91.

Melia, T. (1992), Essay review of P. Dear, *The Literary Structure of Scientific Argument: Historical Studies;* A. Gross, *The Rhetoric of Science;* G. Myers, *Writing Biology: Texts in the Social Construction of Scientific Rhetoric;* and L. Prelli, *A Rhetoric of Science: Inventing Scientific Discourse, Isis,* 83, 100–106.

Moss, Dietz J. (1993), *Novelties in the Heavens. Rhetoric and Scirence in the Copernican Controversy,* The University of Chicago Press, Chicago.

Pera, M. (1991a), "The Role and Value of Rhetoric in Science," in M. Pera and W. Shea (eds.), *Persuading Science: The Art of Scientific Rhetoric,* Watson Publishing, Canton, Mass, pp. 29–54.

———. (1991b), *Scienza e retorica,* Laterza, Roma-Bari.

Rorty, R. (1982), *Consequences of Pragmatism,* Harvester Press, Brighton.

———. (1989), *Contingency, Irony and Solidarity,* Cambridge University Press, Cambridge.

4

On the Cognitive Analysis of Scientific Controversies

RICHARD E. GRANDY

I want to begin by emphasizing that the definite article in my title should not be taken to imply that cognitive analyses are unique—there is no single view that could be called *the* view of cognitive science on scientific controversies or, I suspect, of anything else of interest (Giere 1992). In fact, one should be suspicious even of the claim that there are cognitive scientists, despite a number of reported sightings and self-reports. Very few people have been trained in a cognitive science program—most were trained as philosophers, psychologists, linguists, computer scientists, or some other field and then developed an interest and competence at cognitive science. There is even a serious and often heated debate about whether cognitive science(s) is singular or plural.

Perhaps less obvious, there is no single position in philosophy which is uniquely that of cognitive science(s). In this chapter I analyze three philosophers—Ron Giere, Miriam Solomon, and Paul Thagard, who explicitly describe themselves as taking a cognitive science approach to philosophical issues, and two others who incorporate significant elements from cognitive science though they may or may not identify themselves in that way—Philip Kitcher and Helen Longino.

One of the important ideas from cognitive psychology is that people organize their experience and attempt to explain and to predict events in the world in terms of models. There are a variety of terms for these—schema, frame, scripts—which differ in details that are important for the psychologists and researchers in artificial intelligence but not for us. The point is that people construct models for the purpose of understanding something. To understand the differences between sundry disciplines in approaching the analysis of science and the differences among cognitively oriented philosophers, one must consider both the kinds of models they deploy and the purposes they have.

My goal is to portray part of the range of positions associated with cognitive approaches to understanding science, beginning with some quasi-historical antecedents in logical empiricism, to point in the directions that I find most promising, and to sketch a map of some of the major obstacles and questions. The questions sometimes differ, depending on whether the writer is concerned with why there are scientific controversies, how they are resolved, why they are sometimes resolvable, or why they sometimes are persistent. And the acceptable explanations of any of these phenomena will depend on what the writer takes as sufficiently well-established background principles on which to rest an explanation.

To avoid some of the problems associated with generic descriptions of historical positions, instead of discussing the logical positivist or empiricist positions generically, I will discuss Rudolph Carnap's logical empiricist approach in the *Logical Foundations of Probability* (1950). Carnap's goal was to analyze the evaluation of theory with respect to evidence. Originally he intended to show that "all principles and theorems of inductive logic are analytic, and . . . hence the validity of inductive reasoning is not dependent upon any synthetic presuppositions" (Carnap 1950, p. v).

I want to make explicit the assumptions and expectations of Carnap's project. He was assuming, or at least expecting, that a unique metric of theory choice could be derived from a priori assumptions. I will call this the Theory Choice Algorithm Principle. He was assuming that the evidential basis would consist of statements that could be uncontroversially established with a probability of 1. I will call this the Observational Foundation Principle. Given these principles, it follows that if two scientists disagree about theory choice, then either

- they are using different evidential bases or
- (at least) one of them is not using the proper algorithm for theory evaluation.

Thus, if Carnap's original program could have been carried out, any scientific controversy must result from ignorance or inductive error; scientific controversies are an inefficiency in the progress of science. Furthermore, the scientific community, or at least the ideal one in which all scientists follow the algorithmic ideal, has no discernible function except to make the process of scientific development move faster by having more hands to make the tasks lighter. If Carnap's program does not succeed, then we must reevaluate the nature of scientific controversies.

Carnap also investigated a continuum of inductive methods indexed by a parameter weighing the relative value of old and new evidence, symbolized by lambda, and in 1963 reported that "[a]s far as we can judge the situation at the present time, it seems that an observer is free to choose any of the admissible values of lambda and thereby an inductive method. If we find that that X chooses a greater value of lambda than Y, then we recognize that X is more cautious than Y" (Carnap 1963, p. 75).

Once one has recognized that there is not a unique optimal algorithm, one might—though Carnap did not—ask whether there is anything interesting to be said from a normative point of view as to the structure of the scientific community. The answer depends, of course, on what one is trying to model. Given Carnap's assumptions about the nature of evidence and given no constraints of time or cost, one can simply let the community accumulate evidence until the theory evaluations converge sufficiently. However, if one wants to model theory evaluation by agents who have time and cost constraints, for example, who have finite lives and want to publish before they perish, then waiting for the observational cows to come home is not an acceptable answer.

Motivated by this concern, and also by the then prevalent criticism of Thomas Kuhn's work that it made theory choice a subjective matter of majority vote, in 1973 I wrote a still unpublished paper based on an idea from Shannon's (1964) approach to information theory. The engineering problem was how to improve signal detection given either imperfect detectors or noisy signal channels. He showed that with imperfect signal detectors, one could obtain an arbitrarily good improvement in the accuracy of signal detection by combining a sufficiently large number of the imperfect detectors and taking the result from the majority of the detectors. My analogy was to think of scientists as imperfect detectors of the signals sent by nature, perhaps under experimental questioning, regarding the best of a set of available theories.[1]

One of the aspects of this model that appeals to me is that it depends on only two assumptions, though these assumptions unfold into a host of interesting issues. One assumption is that the scientists/detectors are more likely to choose the better theory, or to detect the correct signal, than the alternatives. For the moment, let me only mention that for most purposes one can take "better theory" in a rather circumscribed way to mean the better theory from an available specified set for the purposes of fitting experimental results, making correct predictions, and cohering with other theories in the near future. This will depend the historical context, and it is not clear that Copernicus's was a better theory than Ptolemy's in 1550, though it certainly was by 1650. Moreover, one can be agnostic about metaphysical truth and realism—indeed, one could be either a realist or an anti-realist—and still adopt this assumption. If there are more than two theories in contention, then the assumption need only be that the better theory is more likely to be chosen than any individual competitor; one does not need the stronger assumption that the better theory is more likely to be chosen than the disjunction of the competitors.

The second assumption is that the scientists/detectors are independent of one another. Elaboration of this assumption is a complicated and subtle matter for which I can only give an approximate solution at this time. (See Maher 1993, for a related discussion.) The detectors cannot be objectively or statistically independent, for if each is more likely to detect the correct signal than not, then their conclusions will be correlated. The proper statement of the assumption is that their correlation is entirely due to their propensity to detect the correct signal. If the detectors are all made in the same factory, and thus are subject to the same biases and will produce the same result, true or false, then they are not independent in this sense. If the scientists were all trained by the same narrow-minded dissertation adviser and thus will produce the same result, true or false, regardless of the facts of the matter, then they are not independent in this sense.

This emphasis on the independence of judgment suggests an atomistic view of the scientists that is at odds with the emphasis I wish to place on the weighty role played by the social community of scientists. The process of being trained as a scientist in a particular specialty is simultaneously a process of making the trainee conform to the standard acceptable canons of the discipline, while (one hopes) preserving the freedom and independence of judgment that will enable the new specialist to explore and evaluate alternative approaches to anomalies in the discipline. The process of training ranges inclusively from training in specific experimental or calculational skills (depending on the subfield) to absorption of the cultural tales of the field (Traweek 1988).

This model of the scientific community is simple-minded in many ways, but it brings out an important point. The community must consist of, if not exclusively at least largely, inquirers who are responsive to information about the world—they must be rational inquirers as individuals. But there are also issues of optimality at the level of the community, and one can speak of rationality at the level of the structure of the community.

An oversimplified, overly quantitative illustration may help clarify the more general abstract points. Imagine five scientists/detectors, each with a probability of .6 of arriving at the better theory from a pair of theories. That is, the probability of any one scientist S choosing theory A is .6 if A is the better theory, and .4 otherwise, and similarly for theory B. Thus, relying on any single individual, we would have a .6 likelihood of ending with the better theory. But if we have five scientists and accept the majority opinion of the five, then the probability rises to .66. More interestingly and more complicatedly, if we require not a simple majority but a stronger majority, say four of five, then there is a .32 probability of choosing the better theory and .09 of choosing the worse theory, and a gap of .59 in those cases in which we have no decision. Extending the process a step further to unanimity among the five gives a .07 probability of getting the better theory, a .01 probability of being wrong, and a .92 chance of getting no decision.

The notion that one can assign an exact probability to the likelihood that a scientist will choose the better theory from a pair is of course pure philosophical fiction. But I think it is a fiction that is useful as a route to a more qualitative point. Without knowing the exact probabilities, if we believe that the members of a scientific community are more likely to choose the better than the worse theory, and that they are making independent judgments, then we know the general principles and trade-offs governing the selection of theories.

An alternative way of putting the main point is that there are many ways of improving reliability. One is to improve the reliability of individual scientists. This is, in effect, the strategy motivating the traditional "philosophical analysis of methodology" approach, but it presupposes that we can a priori determine the general structure of the universe, or that in the absence of that knowledge we can devise an optimal method of training scientists. It also assumes a high degree of effectiveness in science education, for which there is little evidence.

Two other approaches to improved reliability are possible, however, even with no improvement in the reliability of individuals. A larger community gives a higher potential reliability, though at the cost of educating and supporting that community. The setting of a higher plurality for consensus before closing discussion and moving on to the next item on the research agenda increases the probability of moving in the right direction but at the cost of making movement slower. This trade-off was already present in Carnap's discussion of alternative choices of a lambda value—the more cautious algorithm is slower to respond to new evidence, but nonetheless is less likely to be led on wild goose chases by short-term fluctuations.

One of the deep lessons that I believe has emerged from the study of problem-solving strategies, computational costs, and related questions is that most problem solving involves trade-offs. As soon as one considers the cost of computation or the opportunity costs of waiting longer for a more reliable answer, one has a conflict of constraints. Efficient problem solving always involves considerations of the utility of further investigation versus the costs of the resources expended. Folk wisdom, in English, recognizes this tension by enshrining the contradictory homilies "Look before you leap" and "He who hesitates is lost."

We have made the transition from thinking of the individual as the information processor, with the larger social structure being irrelevant, to thinking of the social structure as being the main unit of information processing, with the individuals being noninterchangeable elements in that structure. This attributes more significance to the social structure, and begins to emphasize the complexity of that structure.

One author who has recognized a role for the scientific community as a whole is Kitcher (see chapter 2). He argues that given the uncertainties of theory evaluation, it is important that there be a diversity of opinion on the correct research program to pursue. If everyone agreed on and pursued the most plausible direction, then potentially promising theories would be ignored, often at a considerable loss to the community. One could add to his point that a diversity of cognitive styles and willingness to take risks would be beneficial. Of course, in all of these matters, there is an essential tension between having sufficient communal agreement on some matters and leaving debate vigorous on others.

Kitcher has also discussed another relevant aspect in which my model requires desimplification. For illustrative purposes, I have treated each of the detectors/scientists as having one vote, for example, as making equal contributions to the final decision of the group. In fact, of course, various individuals have varying degrees of authority. If authority is rationally evaluated and earned, then it can add to the efficiency of the group; if it is arbitrary, then it can have a detrimental effect. Kitcher, in discussion, commented that it would not be rational to give authority on the basis of birthdate or nationality. While neither would be optimal, I think it worth observing that giving authority to those whose birthdays were on the fifteenth day of a month or, alternatively, in the first week of July would be preferable to a system that gave authority to one nationality. The loss of rationality would be less in the former case since the authority is (presumably) randomly distributed, and this is less likely to cause a biased decision.

A natural metaphysical question that arises at this point is whether one sees the scientific group or the individual as the basic unit of analysis and explanation. Two fairly representative but extreme positions are the following:

> What I propose . . . is a much more thoroughgoing contextualism than the one which urges us to remember that scientific inquiry occurs in a social context, or even that scientists are social actors whose interests drive their scientific work. What I urge is a contextualism which understands the cognitive processes of scientific inquiry not as opposed to the social, but as themselves social. This means that normativity, if it is possible at all, must be imposed on social processes and interactions, that is, that the rules or norms of justification that distinguish knowledge (or justified hypothesis-acceptance)

> from opinion must operate at the level of social as opposed to individual cognitive processes. (Longino 1992, p. 201)

Contrast that with the following position:

> The conclusion is simple. The most promising approach to a general theory of science is one that takes individual scientists as the basic units of analysis. It follows that we must look to the cognitive sciences for our most basic models, for it is these sciences that currently produce the best causal models of the cognitive activities of individual human agents. (Giere 1989, p. 8)

The view I am advocating accepts neither model, but sees the continuing dynamic interaction between group and individuals as critical. It is essential to see that although a group is in a sense constituted at a given time by a set of individuals, as a group changes over time members are attracted to the group or become part of it because of the properties of the group as a whole. The group and the perception of the group shape the cognitive behavior of those who join it. Moreover, epistemic evaluation seems appropriate for both individual and group processes, although the units and the measures of evaluation differ.

Scientific arguments, especially at times of controversy, do not come in neat packages except with rational or irrational reconstruction. For purposes of illustration, I will discuss some aspects of the controversy over continental drift. I assume that the general outline of the development of continental drift is familiar, but to recap briefly, Alfred Wegener proposed a detailed version of a theory of continental drift with elaborate supporting evidence in 1915. Wegener was not the first to propose continental movement as an explanation of various phenomena, nor was he the only one among his contemporaries, but his arguments were more elaborated and attracted greater attention. Although it attracted some adherents, it was a minority view until significant new evidence on seafloor spreading in the 1960s gave rise to a considerably different version of continental drift, which was fairly quickly accepted.

One of the convenient aspects of this controversy is that there was a major conference in Tulsa in 1926 whose proceedings provide a window to the controversy at that point with many perspectives. For example, one of the critics of Wegener at the conference gave the following abstract of his view:

> After considering the theory of continental drift with avowed impartiality, the author concludes by means of geophysical, geological and paleontologic reasoning that it should be rejected, because the original suggestion of the idea sprang from a similarity of form (coast lines of Africa and South America) which in itself constitutes no demonstration, because such a drift would have destroyed the similarity by faulting and because other contradictions destroy the necessary consequences of the hypothesis. (Willis 1928, p. 76)

The paper that follows hardly seems impartial despite the avowal, and the argument that Wegener's argument fails if his original inspiration can be shown faulty confuses the context of discovery with the context of argumentation. Yet the main argument has considerable force and is echoed in various forms by many of the contributors. Put briefly, it is difficult to reconcile the preservation of an exact fit between the eastern coast of South America and the western coast of Africa during an extended displacement while simultaneously appealing to the displacement as the cause of the creation of the Andes mountains on the west coast of South America.

On reading the Tulsa proceedings, it is clear that there is sharp disagreement about the coastline fit—Wegener maintains that the fit is excellent, while most of his critics (other than Willis) point out numerous discrepancies, and only van der Gracht, the organizer of the symposium, seems to take the view that a pretty good fit is all that is to be expected and itself constitutes significant evidence. One of the important points to be emphasized is that none of the evidence Wegener cites, and little of that of his critics, is first-hand observation. He relies on an enormous wealth of data from various disciplines. In fact, some of the resistance to his drift hypothesis seems to stem from the fact that he is urging an integrative approach at a time when many earth scientists saw separatist disciplinary specialization as the route to progress.

The geological community is even more dependent on shared community data than most other scientific communities, but this difference is only one of degree, not of kind. One of the complexities of the continental drift controversy was that it called upon and promised to unify data from relatively unrelated disciplines such as geophysics, geodesy, paleontology, paleobiology, paleoclimatology, and so on.

By way of illustrating the diversity, depth, and need for further work in cognitive approaches, I will compare two of the most well-articulated cognitive approaches, those of Giere (1988) and Thagard (1992), in their discussions of drift. Giere follows many other philosophers in seeing Wegener's theory as more plausible than the contractionist and fixist alternatives, and sees those who rejected Wegener's theory as influenced by less than rational considerations:

> That professional interests played an important role in the reaction to drift models is also borne out by a recent quantitative study. . . . In this study the most powerful predictor of attitudes toward mobilism was the number of publications—those who published more tended to be less favorable. Since most of the relevant publications over those years were within a stabilist framework, the number of publications is a plausible measure of investment in that approach. (Giere 1988, p. 239)

Thagard, on the other hand bases his analysis on the notion of explanatory coherence and a connectionist computer program to evaluate the coherence of sets of statements representing evidence and hypotheses. Applied to a set of statements that are intended to reconstruct Wegener's theory and evidence, the program concludes that the set is coherent; applied to a set of statements that are intended to reconstruct the theory and evidence of Wegener's opponent, the program concludes that set is coherent as well.

However, before concluding that Thagard has shown that Giere has unfairly maligned Wegener's opponents, we need to consider the reconstructed evidence. Among the evidential statements for Wegener's position are

> E1: The shape of the Atlantic coastlines match.
>
> E13: Strata on opposite sides of the Atlantic match.
>
> E17: Measurement of the position of Greenland indicates motion in excess of standard error.

Among those for the opponents are

> E8: The continents bordering the Atlantic can't be made to fit together consistently.
>
> E14: Rock formations on opposite sides of the Atlantic do not match.
>
> E18: Measurements made in 1926 show no movement of Greenland. (Thagard 1992, pp. 183–186)

It seems to me that what we have here is not only a theoretical controversy, but also a controversy of evidence. Without the next stage of analysis in which one considers the reasons why the two sides accepted different evidential bases, one cannot seriously adjudicate the issue of rationality.

And I am not certain one can ever make that adjudication in the traditional way. A number of writers have noted that in the drift controversy there were significant national divisions, especially that Americans tended to oppose drift. Prima facie, this seems to be evidence of irrational elements in the controversy, but when we consider the nature of the evidence and the fact that the evidence is literally very widespread and what is most familiar to individuals depends heavily on their geographical location, nationality may simply be an intermediate term that reflects the kind of evidence that is most familiar.

A wilder perspective comes from the work of Frank Sulloway (1996) on the importance of birth order in explaining individual positions in scientific controversies. In a large survey of almost three thousand scientific participants in twenty-eight controversies, Sulloway finds that birth order is a highly significant variable correlating with the position a scientist takes on a relevant controversy. Roughly, firstborns defend the status quo theory, and children born later in the family order support the innovative theory. This trend holds for the drift controversy among many others; Wegener, for example, is the second of two boys.

If this result is in fact correct, then we can see yet another dimension on which it is important for a scientific community to have variation. Too many firstborns in a discipline might tend to stifle innovation, while too few might lead to premature jettisoning of perfectly defensible theories. It seems likely that in different periods in various cultures there have been tendencies for firstborns to become, or in some cases not become, scientists in particular disciplines. If one had appropriate measures it would be of interest to compare the scientific developments across those cultures.

An author with an avowedly cognitive perspective whose views are fairly close but still significantly different from those I am advocating is Solomon (1994), who argues for the priority of the social over the individual and attempts to provide an analytic framework within which it is possible to distinguish a normatively appropriate consensus from an inappropriate one. She defines a consensus as normatively appropriate when it is the result of empirical success. Two disagreements between us may highlight the differences. First, she seems to take the notion of empirical success as nonmetaphorical agency: "Only when these so-called biasing factors operate on empirical successes in such a way that empirical successes select the theory on which consensus occurs is the consensus normative" (pp. 336–337). Furthermore, the metaphor of selection is used in various ways: "Consensus is thus normatively appropriate when the direction of consensus is selected by a theory with substantially superior empirical success" (p. 337).

Second, she seems to regard the consensus as the appropriate unit of evaluation, whereas I regard the individuals and community as the appropriate unit. In her terms, a community that is highly biased toward a theory, but where consensus would not quite coalesce without some empirical success, comes to a normatively appropriate consensus given very slight empirical success. And conversely, a community highly biased against a theory might eventually, after enormous irresistible empirical success, come to embrace the theory, and again this would be a normative consensus.

If I were confident that I could explicate, or that I understood, empirical success, I would instead offer as a normative judgment that a community is rational if, on balance, it would reach consensus on any of the available theories if one were to achieve significantly greater empirical success, and would remain without consensus otherwise. Notice that as in Solomon's model, a community of rational inquirers by my standard need not include any individuals who are totally free of biases, but the community must have a balance of biases and a reasonable degree of openness to evidence and alternatives.

A further set of points should be made about Thagard's approach. The first is that a theory is represented as a set of sentences, with no priority of hypotheses or evidence. The evaluative program gives no differential weights to the various elements, although it seems clear that the participants did. Essentially, this abstracts from the individuals, each of whom probably attached various weights to different hypotheses and (alleged) evidential statements, and presents a homogenized version that does not reflect the diversity that I have been arguing is essential to scientific development.

Second, the representation of the evidence in sentential form loses the significance of the maps and other images that appear throughout the discussions on all sides. Indeed, one could argue that the entire drift controversy begins and ends with a picture. The starting point for Wegener, and independently for many other drifters, is the seeming fit of the Atlantic coastlines. The final picture is the Eltanin-19 magnetic anomaly profile from a Pacific Ocean ridge, which is cited by most anti-drifters as the single piece of evidence that changed their minds (LeGrand 1990).

The third point is that even when one abstracts from individual differences in weighting and transposes the evidence into sentential form, the presentation of the sentences matters. The understanding of a scientific theory, especially by those opposed to it, is often highly influenced not only by their antecedent views but by the presentation, by the rhetoric of the debate. The sets of sentences I chose reflect this: Wegener said often that he first was led to the hypothesis by the coastline fit, and it bulks large in his early presentations (Wegener 1924, pp. 1–10). Over a period of time, as that evidence was questioned, or its force turned aside, Wegener leaned more heavily on the Greenland argument. Comparison of astronomical (lunar) observations over several decades gave an apparent (amazing) velocity of Greenland away from Europe of 20 meters per year.

In the fourth edition of his book, in 1929, this "proof" is given before other arguments (Wegener 1966, pp. 25–34). This turned out to be an unfortunate change, for when those measurements were shown to be unreliable (Nörland 1936) the drift hypothesis was even more tarnished. Among commentators, only Longwell (1944a,b,c) seems to have made the elementary, but relevant, logical point that although the geodetic evidence for drift had proven unsound, that kind of evidence might still be used to establish the hypothesis.

It is important to understand the differences between the model I am proposing and evolutionary models of scientific development. Although I certainly want to emphasize the role of the groups involved, there are significant differences. For example, although the socialization process is lengthy and its complexity should not be underestimated, there is clearly the possibility for a scientist to switch from one scientific research group to another in a way that a sparrow cannot become a robin. Perhaps more important, although ecological niches change and may at times become more demanding for most species, for example, because of drought, the niches that scientific theories are attempting to fit are consistently becoming more demanding in detail, precision, and variety. The empirical constraints on theories in a domain, with occasional deviations, become increasingly strict and so the theories are held to ever higher standards. Note that since these are empirical observational demands, the point holds regardless of whether one is an instrumentalist or some variety of realist.

In summary, the cognitive approach I have been advocating takes a wide and naturalistic view of cognition. Cognitive agents, including scientists, have differing computational abilities, differing representations of theory and evidence, different goals, different perceptions of the problems, constraints, and solutions. They also function within a very thick social structure that largely, though mostly implicitly, defines the acceptable theories, evidence, problems, and solutions. They also function in an environment that places increasingly stringent demands on the precision and scope of theories. Early mobilist and stabilist theories of continents had to contend with essentially no data concerning oceans floors, save for a few protruding islands. Cognitive analysis of our capacities, combined with the negative results from attempts to find unique inductive algorithms, and the intrinsic complexity of the world make it very easy to understand why there are controversies. What is required for a deeper understanding is to have a better analysis of how the random variations in representation and approach interact with the observational constraints to provide resolutions of controversy when they do.

Notes

I am indebted to members of the University of Wyoming philosophy department, to participants in the scientific controversies conference, and to Mark Cherry and Christopher Hitchcock for helpful commnts on earlier versions of the manuscript.

1. I subsequently learned from Arthur Kuflik that in the eighteenth century Condorcet had advanced the same mathematical argument for democracy!

References

Carnap, R. (1950), *Logical Foundations of Probability.* Chicago: University of Chicago Press.

———. (1963), "Carnap's intellectual autobiography," in P. A. Schilpp (ed.), *The Philosophy of Rudolf Carnap.* La Salle: Open Court.

Giere, R. (1988), *Explaining Science: A Cognitive Approach.* Chicago: University of Chicago Press.

———. (1989), "The units of analysis in science studies," in S. Fuller et al. (eds.), *The Cognitive Turn.* Dordrecht: Kluwer.

———. (ed.), (1992), *Cognitive Models of Science,* Minneapolis: University of Minnesota Press.

LeGrand, H. E. (1990), "Is one picture worth a thousand experiments?" in LeGrand (ed.), *Experimental Inquiries: Historical, Philosophical and Social Studies of Experimentation in Science.* Dordrecht: Kluwer, pp. 241–270.

Longino, H. (1992), "Essential tensions—Phase Two: Feminist, philosophical and social studies of science," in E. McMullin (ed.), *The Social Dimensions of Science.* Notre Dame: University of Notre Dame Press.

Longwell, C. R. (1944a), "Some thoughts on the evidence for continental drift," *American Journal of Science 242:* 218–231.

———. (1944b), "Further discussion of continental drift," *American Journal of Science 242:* 624.

———. (1944c), "The mobility of Greenland," *American Journal of Science 242:* 624.

Maher, P. (1993), *Betting on Theories,* Cambridge: Cambridge University Press.

Nörland, N. E. (1936), "Astronomical longitude and azimuth determination," *Monthly Notices of the Royal Astronomical Society 97:* 489–506.

Shannon, C., (1964), "The mathematical theory of communication," in *The Mathematical Theory of Communication,* C. Shannon and W. Weaver (eds.). Urbana: University of Illinois Press.

Solomon, M. (1994), "Social empiricism," *Nous 28*: 325–343.

Sulloway, F. J. (1996) *Born to Rebel.* New York: Random House.

Thagard, P. (1992), *Conceptual Revolutions.* Princeton, N.J.: Princeton University Press.

Traweek, S. (1988), *Beamtimes and Lifetimes: The World of High Energy Physicists.* Cambridge, Mass.: Harvard University Press.

Wegener, A. (1924), *The Origin of Oceans and Continents.* Trans. J. G. A. Skerl. Originally published as *Die Entstehung der Kontinente und Ozeane,* 3rd ed. (Braunschweig: Friedrich Vieweg und Sohn, 1922). New York: Dutton.

———. (1966), *The Origin of Oceans and Continents.* Trans. J. Biram. Originally published as *Die Entstehung der Kontinente und Ozeane,* 4th rev. ed. (Braunschweig: Friedrich Vieweg und Sohn, 1929). New York: Dover.

Willis, B. (1928), "Continental drift," in W. A. J. van Waterschoot van der Gracht (ed.), *Theories of Continental Drift: A Symposium.* Tulsa: American Association of Petroleum Engineers, pp. 76–82.

Part II

HISTORICAL AND CONTEMPORARY REFLECTIONS ON CONTROVERSIES

5

The Concept of the Individual and the Idea(l) of Method in Seventeenth-Century Natural Philosophy

PETER MACHAMER

The sixteenth and early seventeenth centuries saw the introduction of many changes in Europe. Indeed, in many ways it was the dawn of the modern world that is still with us. Not the least important of these was a change in a person's awareness of self, and a concomitant social belief in the individual as a the *locus* of power.[1] A person's conceptions of the relations of an individual human being to other persons, social institutions, and the natural world changed, as did the descriptions, metaphors, ideals, and values by which one conceived oneself. Epistemologically, socially, and economically, the individual was forced into the center of the new order of things.

The reasons behind this rise of individualism are many. It was the time, as Karl Marx (1845) said, of the break from feudalism to original capitalism; from, as Alexandre Koyre (1957) said, a closed world to an infinite universe; yet too, as Max Weber (1904–1905, p. 179) put it, from organic social organization in a fiscal monopolistic form to individualistic motives of rational legal acquisition by virtue of one's own ability and initiative. As empires gave way to nation states, and as relatively homogeneous elitist intellectual and theological traditions collapsed into narrowly defined sectarianisms and intellectual anarchy, people were forced into new beliefs and institutions, and back into themselves in order to justify and rationalize them.

This basic change in the way humans thought of themselves, and the concomitant changes in their interests, activities, and social structures, is one major component in the rise of modern science and was constitutive of this new human endeavor. In this chapter, I explore only one aspect of this broad, quite general claim about the rise of individualism and its relation of the scientific revolution. I shall try to sketch how a neo-Protagorean individualism relates to a new idea and ideal of scientific method (Machamer, 1992). Put succinctly, if somewhat cryptically, the rise of person-centered individualism necessitated a new concept of how knowledge could be reliably attained, how science or, better, natural philosophy should be carried on, and what was the nature of controversy and disputation.

The new method also implicitly contained certain paradoxical beliefs about the epistemological and social roles of individuals, which in their turn would give rise to a new problem—the objectivity of knowledge and science. How could ideas and practices that were first-person based ever become universal and objective? But this problem would only gradually emerge into seventeenth-century consciousness.

The Two *I*s of Individualism

Individualism has been stressed so far, but now a distinction needs to be made between two different though related aspects of this individualistic ego: the epistemic *I* and the entrepreneurial *I*. The first is well known to historians of philosophy. This is the role of the individual as the knowing subject. On this view, all knowledge ultimately comes through an individual human's senses. The second rises out of the social and economic changes of the times, and places the new natural philosophy squarely in a capitalistic context. This *I* demands credit for new discoveries and wishes to gain a superior leadership role. The entrepreneurial *I* wants to sell his new method to the world. Both senses of *I* are fundamental in the new idea of scientific method.

The Epistemic *I*

The new epistemology centers on an *I* who gains knowledge or comes to have ideas by interacting with the world (and sometimes God). It is this *I* that perceives and experiences the world and, reflexively, takes its own ideas as objects. It is identified as mind or intellect and, sometimes, in addition, as body (at least for some of its functions). This epistemic *I* (both in body and mind) is the locus of human reason, the basis of human emotions, the ascriber of human values, and the source for all human action. This description fits both (so-called) rationalists and empiricists of the seventeenth century. The epistemic *I* is characterized by processes and properties lying within an individual, yet they are universalizable and belong, at least potentially, to all individuals. Epistemology is thus democratized.

According to Francis Bacon (1620), "[I]t is necessary that a more perfect use and application of the human mind and intellect be introduced" (p. 42). In the *Novum Organon* (1620) Bacon draws attention to the sources of error—the idols, as he calls them. The emphasis of all of these—the idols of the tribe, the cave, the market, and the theater—is on the fact that people are misled because they do not examine broadly or systematically enough "*their own individual experiences of sense or discourse*" (p. 59, italics mine). They are fed, as Wittgenstein (1953) would much later say, on a one-sided diet of examples, and are too susceptible to persuasion by others. Here we see one form of Bacon's commitment the epistemological ego, for errors can be avoided by such attention to individual experiences. He tried to give more precise structure to the rules of this new method of invention, emphasizing that though it was hard to practice, it was easy to teach.

Galileo rationalizes and justifies the intellectual ability of the individual in a passage in his *Dialogues* (1632), which later was used against him in his trial. Late in Day 1, Galileo praises the genius of men like Michelangelo and goes on to claim that the genius of the mathematician doing philosophy can be perfect, just like God himself. God knows things extensively but

> taking man's understanding intensively, in so far as this term denotes understanding some proposition perfectly, I say that the human intellect does understand some of them perfectly, and thus in these has as much absolute certainty as Nature itself has. Of such are the mathematical sciences alone, that is geometry and arithmetic, in which the Divine intellect indeed knows infinitely more propositions, since it knows all. But with regard to those few which the human intellect does understand, I believe that its knowledge equals the Divine in objective certainty, for here it succeeds in understanding necessity. . . . (p. 103)

So it is that anyone who learns mathematics—and all people can—is able to attain objective knowledge.

Descartes combines the explicit emphasis on avoiding error and relying the first person as the way to truth, like Bacon, with the Galilean commitment to geometry as the mode of thinking with certitude. But Descartes, more that the others, stresses the need for justification and legitimation of the individualistic, first-person epistemological point of view. For Descartes, the foundation of knowledge itself is found in the *cogito* of every person's introspection. That foundation secures, in turn, the individual's knowledge stemming from experiences, experiments, and investigations of the world. Descartes' commitment to first-person-based knowledge in the *Meditations* (1640) is called the (subjective) way of ideas and is well known through many discussions in the history of philosophy.

Earlier in the *Discourse* (1638), Decartes gave the principle of democracy its due:

> Good sense is the best distributed thing in the world, for everyone thinks himself so well endowed with it that even those who are the hardest to please in everything else do not usually desire more of it than they possess. In this it is unlikely that everyone is mistaken. It indicates rather that the power of judging well and of distinguishing the true from the false—which we properly call "good sense" or "reason"—is naturally equal in all men, and consequently that the diversity of our opinions does not arise because some of us are more reasonable than others but solely because we direct our thoughts along different paths and do not attend to the same things. (p. 111)

Descartes' moral is, of course, is that all men ought to follow his method as laid out in the *Discourse,* for by doing so they will be enabled to avoid error and come to knowledge. In this he speaks just like Bacon. Descartes stresses all of these themes are again in the *Principles of Philosophy* (1644), where the epistemological individualism is, as it was in the *Meditations,* made the foundation for all knowledge.

Thomas Hobbes in *De Corpore* (*Elements of Philosophy,* 1655) democratically addresses the reader directly and with first person emphasis: "If you [the reader] will be a philosopher in good earnest, let your reason move upon the deep of your own cognitions and experience; those things that lie in confusion must be set asunder, distinguished, and everyone stamped with its own name, set in order . . ." (p. xii).

He becomes even more explicit in chapter 1, when he warms his reader to the common theme:

> Philosophy seems to me to be amongst men now, in the same manner as corn and wine are said to have been in ancient time. For from the beginning there were vines and ears of corn growing here and there in the fields; but no care was taken for the planting and sowing of them. Men lived therefore upon acorns; or if any were so bold as to venture upon eating of those unknown or doubtful fruits, they did it with danger to their health. In like manner every man brought Philosophy, that is, Natural Reason, into the world with him; for all men can reason to some degree, and concerning some things; but where there is need of a long series of reasons, there most men wander out of the way, and fall into error for want of method. (p. 1)

Here again lack of method leads to error, but lest you think individualism has been lost with Hobbes's systematic rationalization of philosophy in the service of development and acquisition of commodities, recall Hobbes's doctrine of thought and language.

For Hobbes, definitions are linguistic representations of things in order to signify to an individual's mind. Words are marks that cause a mind to have certain ideas, just as experiences so cause ideas. So all method, for Hobbes, comes down to careful use of language. Language is correlated with individual experiences of bodies in motion, so that these bodies—natural, human, and artificial—can be understood or clearly seen.

Robert Boyle, too, clearly writes in accord with a first-person epistemological mission. His works are replete with his recounting the experiments that he personally has constructed in order to gain knowledge about certain phenomena or to refute hypotheses that have been put forward by others. In all this the import is clear: a person must perform the requisite experiments in order to gain knowledge. So in his *Experiments and Notes about the Mechanical Origine or Production of Electricity* (1675), Boyle considers hypotheses by Kenelm Digby, Pierre Gassendi, and Descartes, and then begins his account of the phenomena he has observed all in order to show, as all hypotheses agree, that "Electrical attractions are not the Effects of a meer Quality, but a Substantial Emantion from the attracting Body" (p. 7).

The pedagogic and epistemological import of these exercises is explicitly put by Boyle when he urges all readers, in a democratic way, to consider the issues themselves:

> You will probably be the less dispos'd to believe, That Electrical Attractions must proceed from the Substantial Forms of the Attrahents, or that Chymical principle in them, if I acquaint you with some odd Trials. . . . And though, forebearing at present, to offer you my thoughts about the cause of these surprising Phaenoma, I propose it onely as a Probleme to your self and your curious Friends. . . . (p. 31)

The implication is clearly that he and his curious friends can solve it if they follow Boyle's method in these matters.

The Entrepreneurial *I*

There is second aspect of the ego for the new methodologists. This is the *I* as inventor, as creator. This *I* is the elitist genius who sees more clearly, more deeply, and more correctly what is true and what needs to be done. It is this *I* who is unique among human kind. This *I* by reporting discoveries and insights promotes himself and wishes credit as the originator of the new way of finding the truth. This is the *I* of priority disputes. This is the *I* who is contrasted with the they, the common herd—the *I* who wishes others to join him but who sees himself as the intellectual and social leader. The entrepreneurial *I* has both the sense of individual pride and self-congratulation and the sense of self-promotion and natural leadership.

One hallmark of this *I* is the increasing use of the first-person personal pronoun in the writings of the new methodologists. All of them indulge in first-person narratives, often accompanied by strong claims about their humility. One common rhetorical form in which the entrepreneurial *I* occurs is as a biographical tale about a narrator who is telling a story about himself or his experiences. The story is really meant as a parable about every person. Anyone can follow the same procedure or have the same experiences, if that person is properly trained in the method discovered by the narrator. Sometimes, the anti-authoritarian nature of the age comes through in such stories by way of assertions that people must freely choose to follow this path.

Francis Bacon (1620) writes:

> For my own part I have [braved all the dangers and unpleasantness]. For all those who before me have applied themselves to the invention of arts have but cast a glance or two upon facts and examples and experience, and straightway proceeded, as if invention were nothing more than an exercise of thought, to invoke their own spirits to give them oracles. I, on the contrary, dwelling purely and constantly among the facts of nature . . . have not sought . . . nor do I seek either to force or ensnare men's judgment, but I lead them to things themselves and the concordances of things, that they may see for themselves what they have, what they can dispute, what they can contribute to the common stock. (p. 14)

Yet Bacon's idea of the democratization of knowledge takes a paradoxical twist when he does not extend criticism of his own theory: "[T]hose that judge this doctrine of mine, I reject. . . . I cannot be asked to abide by the decision of a tribunal which is itself on trial" (p. 46).

For Galileo, the new ideal of the natural philosopher is an individual who is anti-authoritarian, extremely insightful and intelligent, and quite different from normal people. So at the very beginning, the dedication of *Dialogues* (1632) Galileo warms to this theme:

> [T]hough the difference between man and the other animals is enormous, yet one might say reasonably that it is little less than the difference among men themselves. What is the ratio of one to a thousand? Yet it is proverbial that one man is worth a thousand where a thousand are of less value than a single one. Such differences depend upon diverse mental abilities, and I reduce them to the difference between being or not being a philosopher; for philosophy, as the proper nutriment of those who can feed upon it, does in fact distinguish that single man from the common herd in a greater or less degree of merit according as his diet varies. (p. 3)

This is eloquent, but not Galileo at his best. Earlier in *Il Saggiatore* (1623) he had made a similar individualistic, iconoclastic point by contrasting the true philosopher (himself) with his opponent, called in that book Sarsi:

> Sarsi believes that all the hosts of good philosophers may be enclosed within walls of some sort. I believe, Sarsi, that they fly, and that they fly alone like eagles, and not like

starlings. It is true because eagles are scarce and little seen and less heard, whereas birds that fly in flocks fill the sky with shrieks and cries wherever they settle, and befoul the earth beneath them. But if true philosophers are like eagles, and not like the phoenix instead, Sig. Sarsi, the crowd of fools who know nothing is infinite; many are those who know little philosophy, few indeed, they who truly know some part of it, and only one knows, all, for that it God. To say plainly what I am trying to hint, and dealing with science as a method of demonstration and human reasoning capable of pursuit by mankind, I hold that the more it shall partake of perfection, the smaller number of conclusions it will promise to teach, and the fewer it will demonstrate; hence the less attractive it will be, and the smaller will be the number of followers. (p. 189)

Yet, despite this elitism, Galileo too holds that philosophical knowledge can be acquired by or awakened in any one who has an open mind and works hard. His use of Platonic anamnesis is democratic in claiming that knowledge can be elicited from any one (*Dialogues*, p. 12). Further, the structure of his dialogues exhibits this democratic commitment. It is in this mode that he constantly encourages Sagredo, the intelligent man of common sense, to learn what's right, and even holds out hope for Simplicio, the not-too-dogmatic Aristotelian.

It has not been well remarked in the literature that Descartes, even more than Bacon and Galileo, constantly reverts to the entrepreneurial use of the first-person pronoun "je" or "ego," or first-person singular verbs. And from this Ego, he easily generalizes to the *we*. The *we* as it occurs as early as in the *Regulae* (c. 1628) means "me and those like me" versus the ubiquitous *they,* who fall into error. Yet democratization is here too, for they could be bettered, if *they* would follow the Rules of Method as discovered by Descartes.

For example, Descartes, in typical seventeenth-century fashion, launches into a long biographical tale of how he personally arrived at his true method. He explains his dissatisfaction with his training, his admiration for the geometers, and how he discovered the need for a *mathesis universalis* (such as described and justified above). He becomes infatuated with his own story as he ends Rule Five of the *Regulae* (1628) by constructing a series of first-person claims designed to establish the importance, even to himself, of his own work: "I have resolved," "I have devoted," "I then I shall be able," "I embark," "I shall try," "I can readily recall," so that "when old age dims my memory I can readily recall it hereafter, if I need to by consulting this book [that I am now writing], and so that having disburdened my memory, I can henceforth devote my mind more freely to what remains" (pp. 9–20).

Descartes concludes his humble autobiographical narrative:

I consider myself very fortunate to have happened upon a certain path in my youth which led me to considerations and maxims from which I formed a method, whereby it seems to be, I can increase my knowledge gradually and raise it little by little to the highest point allowed by the mediocrity of my mind and the short duration of my life. (p. 112)

So he says his aim is not to teach, but only to reveal "how I have tried to direct my own [reason]." The implication in a publicly sold book that purports to teach method is that others ought to direct their reason similarly.

Even the modest Boyle, who often eschews any responsibility for giving causes, does decry those who do or watch experiments without having been properly philosophically trained (in his method):

> I consider too, that among those that are inclined to that philosophy, which I find I have been much imitated in calling Corpuscularian, there are many ingenious persons, especially among the nobility and gentry, who, having been first drawn to like this new way of philosophy by the sight of some experiments, which for their novelty or prettiness they were much pleased with, or for their strangeness they admired, have afterwards delighted themselves to make or see a variety of experiments, without having ever had the opportunity to be instructed in the rudiments or fundamentals of that philosophy whose pleasing or amazing productions have enamored them of it. (1667, p. 4)

Instruction is necessary in order to properly understand experiments, and Boyle is writing his tract in order to teach the fundamentals of the experimental philosophy to his protégé and anyone else who would learn. The entrepreneur is always a teacher and a leader.

The Paradox of Egos

The two aspects of the individual's ego, epistemic and entrepreneurial, are characterizable in terms of an individual's power. Both stress the role of the first person. However, the epistemic *I* strangely contrasts with the entrepreneurial *I*. The putative paradox is that it is the uniquely insightful entrepreneurial *I* who has found single-handedly the way to truth, the true method. But this method, or way to discover truth, must be teachable. Since all persons are equally epistemological subjects, all persons must be able to be taught, and can learn that which the entrepreneur has discovered. Only with such a democratic assumption can one claim that controversies and disputes can be resolved in virtue of proper training, for anyone who has been properly taught will see the force of the true method and so agree. So the unique entrepreneurial *I*, who alone is capable of teaching others what it knows, is by its epistemological nature equal to all other *I*s.

The paradox lies in claiming that the entrepreneurial has also has epistemological superiority. One might try to avoid this paradox by saying that the entrepreneurial *I* is only temporally first, or that the entrepreneur only accidentally arrived at the method. But neither of these comports well with claims to natural leadership or with the call that others should follow and be taught.

This universality of method also contrasts with theories that invoke a kind of esoteric wisdom available only to the chosen. Some alchemists and magicians held a version of an elitist theory by which only those properly initiated could become adept. Yet there, too, a kind of initiation is necessary in the new method for natural philosophy, for those who would learn must cast off error, be open-minded, work hard, and learn the principles of the true method. Dogmatists (those who believe in *other* methods) and those relying on textual authority will be unable to learn the method. By their bad training they are cut off from democratic community of true knowers.

Method

Throughout the seventeenth century there were calls for change. Most everyone connected with the rise of the new science felt the need to cast off the old ways so that new and better knowledge could be had. Ultimately changes were necessary so that people could prosper.

Francis Bacon in *The Great Instauration* (1607) criticizes the past work of others in natural philosophy, and sets out a general rationale for his own new plan:

> Men have neither the desire nor hope to encourage them to penetrate further. . . .
>
> For let a man look carefully into all that variety of books with which the arts and sciences abound, he will find everywhere endless repetitions of the same thing, varying in method of treatment, but not new in substance. . . .
>
> So that the state of learning as it now appears [is like that] represented in the old fable of Scylla, who had the head and face of a virgin, but her womb was hung round with barking monsters, from which she could not be delivered. For in like manner the sciences to which we are accustomed have certain general positions which are specious and flattering; but as soon as they come to particulars, which are the parts of generation, when they should produce fruit and works, then arise barking disputations, which are the end of the matter and all the issue they can yield. (pp. 13–14)

Bacon is stating a common feeling about the need for change. Change is needed to in order to free contemporaries from the old order. New methods of teaching and learning are needed to bring about a new philosophy, new knowledge, and the fruits therefrom.

Why did these seventeenth-century thinkers feel the need for a new science and a new method? The simple answer is that they had a vision of being in a new world, a vision that had science and individually based knowledge and power at its center. To develop in the new way, the individual was required to reject the old ways. The anti-Aristotelian slogan was to give up occult, hidden qualities and supplant them with what could be gleaned directly from the world. The anti-authoritarian cry of the modern age was to turn away from the books and to look to nature. But nature would not reveal her secrets easily: she had to be methodically approached and in some cases prodded.

Nature, the new world, was filled with the new and the novel. The natural philosopher no longer could be content with trying to carve nature at her joints, as Plato had held. The philosopher in this new world did not know what to expect from nature, and so had to take pains to discover what was there. Some brief background will recall some of the reasons why this situation came to be.

The world-traversing sea voyages of the late fifteenth and sixteenth centuries brought to Europe many new natural objects and marvelously different kinds of things and reports of experiences. Because of the new discoveries, botany became one of the fastest growing sciences in the Italian universities. It not only was part of the medical curriculum with its New-World, plant-based medicines, but also caused the introduction of public botanical gardens in Padua and Pisa in 1544. Students began to take field trips to examine their indigenous flora. These events had an important influence on the new idea of natural history that became so important to Francis Bacon. People even began to speculate on the novelties they would encounter upon reaching new planets. So Cyrano de Bergerac told of his voyage to the moon, while Kepler and later the Englishman Alexander Ross attempted to describe the beings that would be encountered on the other planets.

Moreover, it was not just nature that was novel. New nations, novel forms of government, new types of corporate entities, and new business methods were replacing old alliances and empires, aristocratic governments, and family- and guild-dominated trading practices. The new highly successful Dutch Republic broke off from the old Spanish empire. The divine right of kings was questioned by the writings of Jean Bodin and, dramatically, was laid to rest even for the commoners when the divine Charles I lost of his head. Banking houses like the Fugger's became the makers of emperors and entrepreneurs by bankrolling the wars of Europe and those who supplied the warmakers. The East India Trading Company even sold public shares in its trading ventures.

The content and structures of the world were new and were perceived to be new. The hallmark of experience was the unexpected. People speculated on how to understand the novel and sought new theories and practices to systematize the art of discovering new things, and to codify the resulting new knowledge.

Obviously the method by which this world of novel discoveries could be explored could not be dictated by traditional or a priori assumptions about the substance and character of what would be found, or about ideas concerning regularities in nature. A method was needed to give some sense about the reliability of inquiry or, minimally, to reassure that knowledge was possible and all was not chaos. A strategy was needed for approaching the novel, and for resolving any disputes and controversies that might arise.

If a method could not have substantive constraints, it must have procedural structure. The method too had to be codified in such form that it could be taught, which required that some consensus among people be reached. If such a teachable procedural method was not possible, then there could be no educational system and no common store of knowledge, and there would be complete individualistic epistemological relativism. In one form this led to skepticism, but all methodologists thought not only that knowledge was possible but also that each one had invented *the* method by which knowledge was to be acquired. Further, in an age of political chaos filled with violence and danger to all, intellectual as well as political anarchy and nihilism had to be avoided at all costs. Besides, one could not sell the results of skepticism.

Method generally means a way of proceeding. Structurally, methods can be based on substantive claims and distinctions or on procedural rules. I contend that the new scientific method of the seventeenth century was essentially and inevitably a procedural method designed to gain adherents to and train initiates in natural philosophy, and to halt disputes and controversies.

As inheritors of humanism, these new methodologists looked to the Greeks, where they found Aristotle's demonstrations, Plato's dialectic, Galen's theoretical empiricism, and the procedures of the geometers. But demonstrations had been roundly criticized by many over the centuries, among them Agrippa, Schegel, Catena, and Peter Ramus. Dialectic had been converted into dialogue, whose major goal was to teach people how to behave and speak in socially proper ways, and so could have little relevance for a method of natural philosophy. Even the new method of the sixteenth century was inadequate because it essentially was a method for communication and a teaching procedure (cf. Jardine, 1967; Ong, 1958). Thus, the Ramist tables simplified *extant* knowledge into a tractable order that could be learned by anyone wishing to know. But it was, in Father Ong's words, "a method of rhetorical rather than logical inspiration." It was a method for rhetorically presenting existing knowledge, not a method for discovery or for bringing structure to what was novel.

Even more important, the seventeenth-century iconoclasts had discarded nature's "joints" and her essential properties as scholastic trickery or occult qualities. The adherents of a new method all quickly became propositors of some version of a mechanical world picture. And so it became clear in very many ways that old methods would not work in this new world.

Despite the fact that until the mid-seventeenth-century Aristotelian forms of logic and the syllogism were still widely used both in and out of the universities, the use of the syllogism to a large extent was opposed and decried. From the ancient traditions, this left the geometers. In fact, the geometers, in one sense, had just been rediscovered in the late sixteenth century. There was much contemporary excitement about the *more geometrico* and about the new mathematics (and mechanics) that was being done.

The axiomatic method of the ancient geometers was both a substantive and a procedural method. The definitions were the substance. But the new method would not have analogs to points, lines, or planes. It would have to look for definitions that fit the world and define the world in new terms. Yet the new method had to be like geometry and be statable in rules. But, as noted, the ground or justification for these rules, since they could not be picked up directly from nature, had to have their origin from within the individual. There was nothing else a person could count on. Yet the necessity of geometrical demonstration and the certainty and conviction it produced seemed abstract enough to be adaptable to the new person-centered foundations.

Bacon (1620) contrasts previous philosophy and the intellectual sciences, with the mechanical arts (which were based on geometry) that have some life and are growing more perfect (p. 8). But even here, he says, very few men have searched, and even fewer have expanded knowledge. The reason is that they "have relaxed the severity of inquiry" and have not dwelt upon experience and the facts of nature as long as necessary, since even mechanics pursue experiments in "wandering inquiry, without any regular system of operation" (p. 11).

This emphasis on the mechanical as a model is one theme that will be held in common by even the most disputing claimants to the new method. Matter, motion, and sometimes mind as an additional substance, are the acceptable forms of nature for those building the new science.

The system that Bacon lays out in Book 2 of the *Novum Organon* is called the method of natural history, and its representational form is a series of tables of presentation. These tables present experiences, and their construction is what Bacon means by demonstration. To demonstrate something is to present it in these tabular forms.

In Bacon's system the individual philosopher must collect examples and instances of the phenomenon to be studied. The individual thinks about them, and arranges them in a first approximation of order in the form of the different tables. On Bacon's view it must be clear in the individual's experience when a given property is present or absent *in toto* or in degree. Anyone who pays attention, and is not misled by others, can recognize and record the properties and degrees of presence of properties correctly.

The next step is the induction, which is the construction by the philosopher of the tables of exclusion or rejection. This is followed by the Commencement of Interpretation, or the First Vintage. Recall that first vintages are made with immature grapes and so give thin, weak wine. So, Bacon says, take the shining or striking instances that are left after the rejections. These are where the property or form P has been found to correlate in each and all cases with another property. These are the paradigm cases. The philosopher must now move on to search for and sort out specific differences that further specify the two properties that have been found to be correlated. Bacon's example is that heat has been found to be correlated with motion in all instances. But it is clear that not all motions are heat, only certain types of motions, so the philosopher needs to differentiate those motions that are heat from those that are not. Bacon commits himself during this discussion to the view that all explanations of natural phenomena will ultimately have to deal with types of motion.

Bacon's method is essentially a set of procedures or rules for finding proper definitions. The two correlated forms or natures turn out to be instances where a species is correlated with a genus, and so now the differentiate must be sought by examining even more examples or experiences.

By contrast, Galileo is best seen as working within a tradition of Archimedian mechanics and the mixed sciences. I agree with William Wallace (chapter 6 in this volume) that Galileo very much had in mind the attempt to legitimize this science in terms of the traditional philosophy. He attempted to make acceptable the "lesser" practices of the mechanics by showing how they could be represented in philosophical terms acceptable to the Collegio Romano. I think this was the way Galileo sought to be considered as a philosopher, instead of just a mere mathematician and mechanician.

Throughout his life, though there were changes from *De Motu* (1590) through *Dialogo* (1632) to *Discoursi* (1638), Galileo believed that for something to be intelligible it had to be representable in terms of a mechanical model—of the balance, the inclined plane, and the pendulum, for example, the Archimedean simple machines. The mathematics for understanding nature was the proportional geometry of lines inscribed in or proscribed about circles. Experiences and experiments were just constructions that exemplified these machines, and set parameters as how they worked.

Here is one example, typical of Galileo, that combines these themes of great genius, democratic availability, and the use of a mechanical geometrical method to yield knowledge. In the Third Day of *Dialogues*, Galileo is talking about stellar observations, and the Tychonic objection that the stars are too far distant from us. He wants to answer this objection to Copernicanism by arguing that if the stars are stripped of their adventitious rays, their sparkle, and then the problem, would disappear. I do not now care about the adequacy, legitimacy, or truth of Galileo's argument; I care only about the way in which it proceeds and its paradigmatic Galilean character.

In *Sidereus Nuncius* (1610) as well a later in *Dialogues* (1632), Galileo had argued that the telescope stops these adventitious rays. So in *Dialogues* Sagredo reasonably objects that Galileo should not blame Tycho and these earlier thinkers for failing to draw conclusions about the adventitious irradiation because they did not have telescopes. Salviati answers that this would be true if "they could not have obtained the result without the telescope . . . but one could do it without, and *I* have done so, and here is the method I have used." He hangs a rope between himself and the star being observed (Vega) and finds the distance at which the cord just hides the body of the star. This blocks the rays, and so allows the calculation of the visual angle of just the star's body (by the table of chords) (pp. 361–362).

I submit this is a typical Galilean experiment. It is a mechanical construction, but also it puts the experiment into a context claiming that he (as a genius) found it, even though anyone else could have found it in principle. Further, it can be replicated by anyone now that he has revealed it and, in principle, could have been done before.

The use of experiment and a commitment to mechanical principles as the form of explanation can also be noticed in Day 4, where Galileo tries to give a mechanical explanation of the actions of the lodestone (p. 409). Or, again, even more clearly when he explains the two events that a theory of the tides must account for, the monthly and annual periods, by first showing how a clock works and then showing how the principle of the pendulum applies to these cases (p. 450). The demonstration uses the principle of the chord applied, as in *Discorsi,* to the inclined plane (p. 451).

The method of the true philosopher, for Galileo as for Bacon, involves showing how mechanical motions serve to explain the phenomenon. Galileo further requires that a person actually have the experience of these motions by constructing a machine or mechanical device and by representing the motions geometrically. The device serves both to help discover the proper mechanism and to validate the existence and operation in nature of such a type of machine.

True method for Descartes, as for Bacon and Galileo, is contrasted with what most mortals in fact do. In Rule 4 of the *Regulae:*

> So blind is the curiosity with which mortals are possessed that they often direct their minds down untrodden paths, in the groundless hope that they will chance upon what they are seeking, rather like someone who is consumed with such a senseless desire to discover treasure that he continually roams the streets to see if he can find any that a passerby might have dropped. This is how almost every chemist, most geometers, and many philosophers pursue their research. . . . [Method contrasts with this haphazard proceeding.] By "a method" I mean reliable rules which are easy to apply, and such that if one follows them exactly, one will never take what is false to be true or fruitlessly expend one's mental efforts, but will gradually and constantly increase one's knowledge till one arrives at a true understanding of everything within one's capacity. (pp. 15–11)

But Descartes, too, resolved "to seek no knowledge other than that which could be found in myself or else in the great book of the world" (p. 115), and he delighted in mathematics, but at first "thought it was only of service in the mechanical arts" (p. 114). Later he, too, found out that mathematics, geometry, was to be used to understand the book The World (which he does in Le Monde, where his model was optics).

To understand method, says Hobbes, it is necessary to understand that philosophy is the figuring out, by true ratiocination, of causes from effects, and effects from causes. So method "is the shortest way of finding out effects by their known causes or of causes by their known effects." This is the science of causes or *dioti.* All other science, *oti,* is either perception by sense, or the imagination or memory remaining after perception (1655, 1.6.1, pp. 65–66).

The ways of method, for Hobbes, are composition and division, or synthetic and analytic. What this method comes down to, whether in invention or teaching, in all the sciences, is finding the right definitions, which are universal equations in which the parts are clear. For Hobbes, the sciences, after definitions,

> proceed in the same manner by which [things were] found out; namely, that in the first place those things to be demonstrated which immediately succeed universal definitions (prima philosophia). Next those things that may be demonstrated by simple motion (in

> which geometry consist). After geometry, such things as may be taught or showed by manifest action, that is, by thrusting from, or pulling towards. And after these the motion or mutation of the invisible parts of things, and the doctrine of sense and imagination (physics); and of the internal passions, especially those of men, in which are comprehended the grounds of civil duties or civil philosophy. (p. 87)

Demonstrations

As noted above, all the methodologists rejected the Aristotelian syllogism as the proper way to present method. The syllogism could not be used to discover new things. Nor since it was held to be tied to Scholastic metaphysics could the syllogism be counted on to handle novelties or new discoveries. But some method of demonstration was still required. Knowledge, in order to be taught, had to be codified in ways that would allow people to be trained.

If we look closely at the seventeenth-century methodologists, we find something quite remarkable in their concept of demonstration. The common model for rational representation is described in terms of what is easily visible or what can be clearly and distinctly seen. As Hobbes (1655) put it, "Demonstration was understood by them for that sort of ratiocination that placed the thing they were to prove, as it were before men's eyes" (p. 86). The representations of knowledge are always spatial displays. Very often the preferred form is taken to be proportional geometry. But tables laying out agreements and exclusions, definitions in terms of subjects being included in predicates, and even pictorial diagrams also fall into the spatial modes of representations. Causal, and thus explanatory, relations are conceived in terms of spatial, often mechanical models and metaphors.

Necessity attends to these representations because they can be seen to be true by anyone who properly attends to them. Spatial relation as primary mode of understanding lends itself well to an ontology of body and motion, for these are easily picturable. This, despite subtle differences, was the ontology of the all new methodologists (even, I would say, including Kepler). The seventeenth century thought in spatial terms; this mode of understanding and representation they took to be the prototype of the intelligible.

Experience, and derivatively the experiences that come from manipulative experiments, is crucial. Experience is essential, for it is only in experience that an individual, cognizing subject can interact with the world. This is one aspect of the common theme about learning from the book of nature. More specifically, experience and the experiences attendant upon one's own actions play a part in making geometry clear through the process of geometrical construction. Action also is part of the goal of method, and it lies in practices and the making of products that result in more commodious living.

The system of operation or method that is needed cannot be logic, for, according to Bacon, history shows that the pursuit of logic left to its own course is not to be trusted, for logic is too weak for the disease, for example, inadequate to sort out the confusions in natural philosophy: logic, "though it may be very properly applied to civil business and to those arts which rest on discourse and opinion is not nearly subtle enough to deal with nature." We need, says Bacon, demonstrations, but not of the usual syllogistic type. We do not need "that method of discovery and proof according to which the most general principles are first established, and then the intermediate axioms are tried and proved by them, [for this method] is the parent of error and the curse of all science" (1620, p. 67). For Bacon "the best demonstration by far is experience" (p. 67) but experiences must be obtained in an orderly and systematic manner. Note that experience for Bacon is an action, something to be done by an individual. The proper experiences do not just happen to a person. This is the haphazard method that he criticized.

Galileo's mechanical and common sense experiments are also described in such perceptual terms. A man of genius, a philosopher like Salviati or his academician friend, takes up a phenomenon to be explained or, as in *Dialogues* a controversy to be resolved. If the interlocutors do not follow the method of the philosopher, they will reach false conclusions like that illustrated in the story that Galileo tells in Day 2 about the famous doctor who carries out an anatomical dissection aimed at quieting a controversy between the Galenists and the Peripatetics. The traditional philosopher observing this finally remarks: "You have made me see this matter so plainly and palpably that if Aristotle's text were not contrary to it, stating clearly that the nerves originate in the heart, I should be forced to admit it to be true" (1632, p. 110).

Here is a theme of the new method that we have seen before in different ways. Demonstration (especially in anatomy) epistemologically (as well as etymologically) consists in laying a phenomenon before oneself and others. This "laying out" exhibits the structure of the phenomenon, exhibits its true nature. What is laid out provides an experience for those seeing it. It carries informational certainly that causes assent.

For Descartes, this is the point of analysis, the reduction to simples. Analysis lays out the parts, the substructures. For Bacon, the laying out was done by constructing the tables that organized and analyzed (interpreted) experience. The tables spatially represented presence, absence, and difference. For Galileo, the exhibition or laying out that commanded assent was the reduction and representation of the phenomenon in a mechanical model, and the geometrical description of how the machine worked. Descartes makes this point in a more abstract terms in Rule 14:

> Accordingly, in all reasoning it is only by means of comparison that we attain an exact knowledge of the truth. This is the way syllogisms work. But as we have frequently insisted, the syllogistic forms are of no help in grasping the truth of things. . . . [C]omparisons are said to be simple and straight forward only when the things sought and the initial data participate equally in [a common nature]. The chief part of human endeavor is simply to reduce these proportions to the point where an equality between what we are seeking and what we already know is clearly visible. (1628, p. 57)

Equations lay out parts of things, so this just repeats the general point about demonstration. But equations that make the known and the unknown "visible" are not wholly metaphorical. Equations provide a visual experience that carries its informational certainly with it. Descartes becomes clearly Galilean as he continues:

> Nothing can be reduced to such an equality except what admits of differences of degree, and everything covered by the term magnitude. So we understand that all we have to deal with here are magnitudes in general. (pp. 57–58)

Controversies, Disputes, and the Paradox of Method

It was thought that following *the* method (whomsoever's it was) would put an end to all controversies. A good method would not let controversies arise.

Of course, many controversies arose. In the seventeenth century controversies were carried on about the cosmologies of Ptolemy, Tycho Brahe, and Copernicus. The existence of a vacuum and the nature of air were new, exhausting topics for dispute. Disputes about the nature of heat, light, and magnetism attracted many and became incendiary topics. And alternative world-views confronted each other about the nature of matter and the causal forms of its efficacy. In short, substantive debates about the nature of the world flourished. Ancillary, but no less heated, controversies centered on how different forms of religion could be reconciled with the new science, how natural philosophy ought to be incorporated into the curriculum, and what are the basic categories needed in order to explain natural phenomena.

Yet, the doctrines of method all held that disputes or controversies are due to ignorance or perversity. Controversies are stupid and accomplish nothing. Only those who cannot reason properly will find it necessary to dispute. Obviously, as noted, the ideal of universality and consensus contrasts starkly with the increasing number of disputes that engage these scientific entrepreneurs, and with the entrepreneurial claims of each that he alone has found the true method.

Bacon holds that disputations and controversy only can occur when there is no proper method. Proper method for Descartes, too, is used to resolved disputes by finding the truth:

> But whenever two persons make opposite judgments about the same thing, it is certain that at least one of them is mistaken, and neither, it seems, has knowledge. For if the reasoning of one of them were certain and evident, he would be able to lay it before the other in such a way as eventually to convince his intellect as well. (c. 1628, 2, p. 11)

For Hobbes lack of a proper method will lead people to "dispute and wrangle like men who are not well in their wits" (1635, p. 2). Definitions for Hobbes remove any equivocation and represent to the mind a univocal, universal notion of the thing defined. But they are not subject to dispute:

> For when a master instructing his scholar; if the scholar understand all parts of the thing defined, which are resolved in the definition, and yet will not will not admit the definition, there needs by no further controversy betwixt them, it being all one as if he refused to be taught. But if he understand nothing, then certainly the definition is faulty; for the nature of a definition consists in this, that it exhibit a clear idea of the thing defined; and principles are either known by themselves or they are not principles. (1.6.15, p. 84)

The paradox of method (analogous to the paradox of ego) was that each candidate for the title of the true method was unique and claimed to be able to solve all the problems of natural philosophy. Yet each was different. So there were inevitable disputes about which was the real true method despite the fact that each held that true method would not allow for disputes or controversy. So accordingly, controversy was not really possible, yet there raged throughout the century both methodological and substantive controversies. The sixteenth-century disputes about method are part of and concurrent with the substantive disputes about the sciences, the proper system of knowledge, and what should be the form of the curriculum.

Despite the existence of these disputes and debates about the substance of natural philosophy, and even about what was its proper form, I find that by the early seventeenth century the question of method had in effect been resolved. From 1607 onward there were some disputes about the empirical character and theoretical nature of natural philosophy, about how exactly the geometrical model ought to be employed, and about the exact role of experiment and experience, but these are minor in contrast with the common elements that I have tried to sketch above.

Professionalism and the Goals of Natural Philosophy

The new method was necessary in order to establish science as a legitimate new activity. The entrepreneurs wanted to have natural philosophy considered as an important new profession with its own status and rewards, and also to encourage others to become professionals or at least to acknowledge the values and pleasures of science as a new form of leisure activity. Science first gets institutionalized as a new kind of club, where the members do not talk about politics or religion or tell dirty jokes, but rather tell tales about their novel experiences, review theoretical claims, and do experiments.

Bacon's remarks are indicative of the new sensibility of his age, reflecting the need for the professionalization of natural philosophy. His account of the reasons why natural philosophy has not made progress includes the following:

> [N]atural philosophy, even among those who have attended to it, has scarcely ever possessed, especially in later times, a disengaged and whole man (unless it were some monk studying in his cell, or some gentleman in a country home), but that it has been merely a passage and bridge to something else. And so this great mother of the sciences has with strange indignity been degraded to the offices of a servant, having to attend on the business of medicine or mathematics. (1620, p. 77)

Finally, it remains to speak about the goals that people held for the new method. Since the method should have guaranteed certainty, and since its application should have ended all disputes, it would seem that no additional reason for adopting a new method would be needed. But perhaps these methodologists felt just a little uneasy about their epistemological claims.

The ultimate goal for introducing this new and improved method, for Bacon in the *Novum Organon,* can be seen if one only "considers what are the true ends of knowledge. . . . [one should] seek it not for the pleasure of mind, or for contention, or for superiority over others, or for profit, or fame, or power, or any of these inferior things, but for the benefit and use of life, and that they perfect and govern it in charity" (p. 15). The benefit and use of life contrast with the entrepreneurial spirit displayed in Bacon's style of writing. His supplications to Elizabeth and later James I to aid him in his setting up his institute for philosophical studies show his true institutional goals.

Bacon said his method ultimately was to bring humans power as well as understanding. Galileo's ideal was the artisan who could control things, and much of his payoff was in the money that he hoped would come from principles of navigation, and better ballistic principles for cannon shots. Descartes also talks of the use to which the fruits of method are put. In his self-deprecating fashion he held:

> I have never made much of the products of my own mind, and so long as the only fruits I gathered from the method I use were my own satisfaction regarding certain difficulties in the speculative sciences, or else my attempts to govern my own conduct by the principles I learned from it, I did not think I was obliged to write anything about it. [But as I developed knowledge particularly in physics and its principles] they opened my eyes to the possibility of gaining knowledge which would be very useful in life, and of discovering a practical philosophy which might replace speculative philosophy taught in the schools. . . . This is desirable not only for the invention of innumerable devices which would facilitate our enjoyment of the fruits of the earth and all the goods we find there, but also, and most importantly, for the maintenance of health. (1638, part 6, pp. 142–143)

Characteristically, Hobbes is most explicit in his conception of what the end or scope of philosophy ought to be: it should provide for

> use to our benefit of effects formerly seen; or that, by application of bodies to one another, we may produce the like effects of those we conceive in our mind . . . for the commodity of human life. . . . (1655, 1.6, p. 7)

> The end of knowledge is power. . . . The utility of philosophy is, especially, of natural philosophy and geometry [is] best understood by reckoning up the chief commodities of which mankind is capable. . . . The greatest commodities of mankind are the arts; namely, of measuring matter and motion; of moving ponderous bodies; of architecture; of navigation; of making instruments for all uses; of calculating the celestial motions for aspects of the stars, and parts of time; of geography &c; These benefits are enjoyed by almost all the people of Europe, by most of those of Asia, and by some in Africa; but the Americans, and they that live near the Poles, do totally want them. But why? have not all men one kind of soul, and the same faculties of mind? What then makes this difference except philosophy? Philosophy; therefore, is the cause of all these benefits. But the utility of moral and civil philosophy is to be estimated not so much by the commodities we have by knowing these sciences, as by the calamities we receive from not knowing them . . . war, and especially civil war. (pp. 7–8)

Conclusion

As a brief conclusion, perhaps it is worth noting some things that Bacon, Galileo, Descartes, Hobbes, and Boyle all share:

- Each wanted power, to be the leader, to design *the* new system of science, knowledge, and method.
- Each wanted to command all other natural philosophers in a collective endeavor, and wanted acknowledgment by them as a leader.
- Each held a democratic commitment that all men could learn the new method, and yet that he was the only one who could have invented it.
- Each thought motion and matter or body was somehow the key to the new science.
- Each was impressed by the mechanical sciences, mechanical devices, and machines.

- Each stated the goal of the new philosophy and method in terms of truth, but ultimately justified or validated truth claims by appeal to practical benefits.
- Only one of them held a university position, and that for a brief while.

Notes

I owe thanks to the members of the Vico Equenze colloquium (June 1993) who commented helpfully on previous drafts, and to members of my seminar in Pittsburgh, particularly Chris Jones. Thanks also to Donna Kline, who offered invaluable advice.

1. This is not to say that no forms of individualism, epistemological or entrepreneurial, existed before this time. Indeed, Morris (1987) argues that individualism started with the twelveth-century renaissance in the areas of religion and social relations. While this is true and there certainly were precursor currents, the full-blown modern form of centering all responsibilities, rights, and hopes on the individual alone seems only to be codified during the seventeenth century.

References

Bacon, Francis (1607), *The Great Instauration,* reprinted in Fulton H. Anderson, (ed.), *The New Organon and Related Writings,* Indianapolis: Bobbs-Merrill 1960.

——— (1620), *Novum Organon,* in *The New Organon.*

Boyle, Robert (1667), *The Origin of Forms and Qualities According to the Corpuscular Philosophy,* 2nd ed., Oxford. Reprinted in M. A. Stewart (ed.), *Selected Papers of Robert Boyle,* Indianapolis: Hackett 1991.

——— (1675), *Experiments and Notes about the Mechanical Origine or Production of Electricity,* London. Reprinted in Boyle, *Electricity and Magnetism,* Oxford: Old Ashmolean Reprints VII, 1927.

Descartes, Rene (c. 1628), *Regulae* (*Rules for the Direction of the Mind*). Reprinted in John Gottingham, Robert Stoothoff, and Dugald Murdoch (eds.), *The Philosophical Writings of Descartes* (2 vols.), vol. 1, Cambridge: Cambridge University Press 1985.

——— (1638), *Discourse on Method,* in *Philosophical Writings.*

——— (1640), *Meditations,* in *Philosophical Writings.*

——— (1644), *Principles of Philosophy,* in *Philosophical Writings.*

Galilei, Galileo (1632), *Dialogues on the Two Chief World Systems,* trans. S. Drake, Berkeley: University of California Press, 1967.

——— (1610) *Sidereas Nuncius* (*The Starry Messenger*), in S. Drake, *Discoveries, and Opinions of Galileo,* Philadelphia: Anchor Books (1957)

——— (1623), *Il Saggiatore* (*The Assayer*). Reprinted in *The Controversy of the Comets of 1618*, trans. S. Drake and C. D. O'Malley, Philadelphia: University of Pennsylvania Press 1960.

——— (1638), *Discorsi* (*Discourses on the Two New Sciences*), trans. S. Drake, Madison: University of Wisconsin Press.

Hobbes, Thomas (1655), *Elements of Philosophy, The First Section Concerning Body.* Reprint. W. Molesworth ed. (1839), Aalen Germany: Scientia P. 166, 187, 188, 194 1962.

Jardine, Lisa (1967), *Francis Bacon,* Cambridge: Cambridge University Press 1967.

Koyre, Alexandre (1958), *From the Closed World to the Infinite Universe,* New York: Harper and Row.

Machamer, Peter (1992), "The person-centered rhetoric," in M. Pera and W. Shed (eds.), *Persuading Science*, Canton, Ma.: Science History Publishers.

Marx, Karl, and Frederick Engels (1845), *The German Ideology.* Reprinted in Karl Marx and Frederick Engels, *Collected Works* (25 vols.), vol. 5, New York: International Publishers 1976.

Morris, Colin (1987), *The Discovery of the Individual 1050—1200,* Toronto: University of Toronto Press.

Ong, Walter (1958), *Ramus: Method and the Decay of Dialogue,* Cambridge, Mass.: Harvard University Press 1958.

Weber, Max (1904–1905), *The Protestant Ethic and the Spirit of Capitalism,* trans. Talcott Parsons, New York: Scribner's 1958.

Wittgenstein, Ludwig (1953) *Philosophical Investigations*, trans. G.E.M. Anscombe, New York: MacMillan.

6

Dialectics, Experiments, and Mathematics in Galileo

WILLIAM A. WALLACE

Perhaps it is not out of place, in a book devoted to scientific controversies, to begin this chapter with a long-standing controversy over Galileo and his methodology. This is not so much a scientific controversy as it is one relating to the history of science, but it is instructive for the light it can shed on how scientific controversies are ultimately resolved. I refer to the *Methodenstreit* initiated by Ernst Cassirer (1922), developed by John Herman Randall, Jr. (1940, 1976), and contested ever since by a host of writers including Neil Gilbert (1963), William Edwards (1983), Adriano Carugo and Alistair Crombie (1983), and myself (Wallace, 1984). The point of the controversy is whether or not Galileo was influenced by the Paduan Aristotelians, in particular, Jacopo Zabarella, when developing the new sciences for which he is justly famous. My contention now is that the lengthy controversy has finally been resolved: Galileo indeed *was* influenced by Zabarella, but in a novel way not foreseen by any of the protagonists before Edwards and I published the Latin text of his *Tractatio de demonstratione* (Galilei 1988; Berti 1991). The *Tractatio* occupies a major part of MS Gal. 27, entitled *Dialletica,* which had been sitting for many years in the Biblioteca Nazionale Centrale in Florence. It was known to Antonio Favaro, the editor of the National Edition of Galileo's works, but he did not think it worth transcribing for his edition, and so it has remained unknown to scholars for over four centuries (GG9: 273–282).[1] The designation written on the manuscript, "*Dialettica,*" explains the "Dialectics" in my title.

The path leading from Zabarella to Galileo surely was not easy to foresee. Zabarella's teaching came, not by way of an anticlerical Averroist interpretation of Aristotle, as Cassirer and Randall had maintained, but via Jesuit professors of the Collegio Romano who read Aristotle with the eyes of Aquinas rather than those of Averroes

(Wallace 1988). And this was especially fortuitous in that it enabled Galileo to import, within the general framework provided by Aristotle's *Posterior Analytics,* both experimental and mathematical techniques that would become the hallmark of his future scientific work.[2]

Galileo's treatise on demonstration was appropriated with only slight modifications from a complete course on logic and methodology offered at the Collegio Romano by the Jesuit Paolo Della Valle in 1588 (Wallace 1992b, pp. 27–40). The last question of the treatise is devoted to the *regressus demonstrativus* or demonstrative regress, the distinctive methodology of the Paduan Aristotelians. Della Valle had in turn appropriated the teaching from the logic text of Zabarella, the main proponent of the method at the University of Padua. Della Valle did so through the intermediary of another Jesuit, Giovanni Lorini, and this serves to explain a few terminological changes along the line, but the doctrine remained the same nonetheless (Wallace 1988, pp. 143–145). Thus, there can be no doubt that Galileo understood Paduan methodology. The bulk of this chapter is devoted to showing how he employed what it taught, its *logica docens,* in his scientific writings, and how this teaching enabled him to make several discoveries that were very much controverted in the early seventeenth century. These related not only to the novelties in the heavens but also to the new science of motion that occupied him to the end of his life.

The Demonstrative Regress

As Galileo presents the teaching, it involves two demonstrations, one "of the fact" and the other "of the reasoned fact" (Galilei 1988, pp. 108–113; Wallace 1992b, pp. 180–184). Galileo refers to these demonstrations as "progressions" and notes that they are separated by an intermediate stage. The first progression argues from effect to cause, and the second argues in the reverse direction, thus "regressing" from cause to effect. For the process to work, the demonstration of the fact must come first, and the effect must initially be more known than the cause, though in the end the two must be seen as convertible. The intermediate stage effects the transition from it to the second demonstration. The transition itself involves time and work, for testing when experimentation is needed and for computation where mathematics is involved, so that the causal connection can be made clear and precise. The result is then seen in the second progression, when the cause, having been grasped "formally" or precisely as it *is* the cause, and indeed the unique cause in view of the convertibility condition, is shown to be necessarily connected with the effect. Only at this stage is knowledge that is strictly scientific attained, for then one knows the reasoned fact, the proper cause of the effect that is being investigated. The entire process, known as the demonstrative *regressus,* may be schematized as follows:

First progression: from effect to cause—the cause is materially suspected but not yet recognized formally as the cause. This generally presupposes that the effect is more known to the senses than the cause and that it awakens interest or curiosity, thus serving as the starting point of the investigation. At the end of this progression the cause comes to be suspected as plausible, for example, known "materially," as really existing, thus as the terminus of a "demonstration of the fact" (Latin *demonstratio quia,* Greek *to hoti*), but known only in a material way and not yet as necessarily connected with the effect.

Intermediate stage: the work of the intellect, testing to see if this is a cause convertible with the effect, eliminating other possibilities. This usually requires a period of time, during which the work is that of the mind (*negotiatio intellectus*), not the senses, although sensible experience plays an important and essential part. Basic to this stage is a *mentale ipsius causae examen* (literally "a mental examination of the cause itself"), where the Latin *examen* corresponds to the Greek *peira,* a term that is the root for the Latin *periculum* (meaning test) or *experimentum* (meaning experiment or experience). The main task is thus one of testing, for example, investigating and eliminating other possibilities, and so seeing the cause as required wherever the effect is present. At the end of this period, the cause is grasped "formally" by the mind, that is, precisely as it is the cause, and the unique cause, of the particular effect.

Second progression: from the cause, recognized "formally" as the cause, to its proper effects. At this stage the necessary connection between cause and effect is grasped. The cause is seen as ontologically prior to the effect and thus as more knowable in itself, even though the effect is more apparent to the senses. The cause is also seen to explain the effect, for example, to give a proper reason why the phenomenon appears the way it does. On this account the second progression constitutes a "demonstration of the reasoned fact" (Latin *demonstratio propter quid,* Greek *to dioti*).

In Zabarella's account the intermediate stage, the work of the intellect or the *examen mentale* with its testing procedures (note the Greek *peira,* source of the Latin *periculum*), carries a heavy burden (Zabarella 1597, p. 486; Olivieri 1978, pp. 164–166). Charles Schmitt (1969) has made a detailed study of Zabarella's use of the term *periculum* or experiment as compared with Galileo's use of the same in his early writings. Surprisingly, Zabarella turns out to be more the empiricist than Galileo. Furthermore, in his 1597 commentary on the *Posterior Analytics,* Zabarella identifies the precise point at which Aristotle himself employs the regress (cols. 836–840). This is in his study of the heavenly bodies, where Aristotle reasoned to the facts that the moon is a sphere and that the planets are closer to the earth than the fixed stars (*Posterior Analytics* I.13, 78a31–b12). Both of these demonstrations pertain to the "mixed science" of astronomy, which uses mathematics to explain the phenomena of the heavens (Lennox 1986; Wallace 1992a, pp. 107–111). On both counts, then, experiment and mathematics, the Paduan regress was open to innovation on precisely the points that would be exploited by Galileo.

Early Experiments with Motion

To turn now to how Galileo made use of this teaching, his *logica utens,* I examine first his preliminary studies of motion at Pisa in 1590, shortly after he had written the *Treatise on Demonstration,* then turn to his discoveries with the telescope, and after that examine his advanced studies of motion at Padua. My concern is to show how in all these cases Galileo employed the demonstrative regress, though in varying ways as dictated by the subject matter. Galileo does not identify the regress as such in his writings, mentioning the "demonstrative progression" (*progressione dimostrativa*) only once, in his 1612 analysis of floating bodies (GG4:67.23). This is not unusual, for the Latin *regressus* has no counterpart in Greek, and Aristotle himself did not use the term. Nor did Aristotle identify syllogisms in his scientific treatises, but this is no sign that he failed to employ them.

Perhaps Galileo's greatest innovation in the study of motion was his use of the inclined plane to slow the descent of bodies under the influence of gravity. The basic insight behind this experiment is found in chapter 14 of Galileo's early *De motu,* "On motion" (GG2:296–302; Drabkin and Drake 1960, pp. 63–69). The *De motu* was composed in 1590, the year after the *Treatise on Demonstration,* and is now conserved at Florence in MS Gal. 71. If the weight of a body can be decreased by positioning it on an incline, thought Galileo, its velocity down the incline will be proportionally slowed. The demonstration he offers is geometrical, but it invokes several suppositions and on this account may be seen as a demonstration *ex suppositione.* If these suppositions are granted, the conclusion follows directly: the ratio of speeds down the incline will be as the length of the incline to its vertical height, because the weight of the body varies in precisely that proportion. His reasoning process here, arranged in the form of the demonstrative regress, is as follows (from Wallace 1992a, pp. 251–255):

First progression:

Effect: Heavy bodies descend along planes inclined to the horizontal more swiftly the greater the angle of inclination.

Cause: Their heaviness on the incline increases with the angle of inclination.

Intermediate stage: Geometrical analysis shows that the ratio of the force required to overcome weight on an incline to that required to overcome weight vertically is as the ratio of the vertical height to the oblique distance along the incline (GG1:298).

Suppositions: (1) that heavy bodies move downward by reason of their weight (*gravitas*), and thus their speed of fall is directly proportional to their weights (GG1:262).

Again: (2) that there is no accidental resistance (*nulla existente accidentali resistentia*) occasioned by the roughness of the moving body or of the inclined plane, or by the shape of the body; that the plane is, so to speak, incorporeal, or at least that it is very carefully smoothed and perfectly hard; and that the moving body is perfectly smooth and of a perfectly spherical shape (GG1:298–299).

Further: (3) under such conditions, that any given body can be moved on a plane parallel to the horizon by a force smaller than any given force whatever (GG1:299–300).

Second progression:

Cause: The weight of a heavy body on an incline is to its vertical weight as its vertical height is to the length of the incline.

Effect: The ratio of its speeds down the incline will be as the ratio of the length of the incline to its vertical height (GG1:298).

Galileo uses the term *periculum* for test or experiment five times in this treatise (Schmitt 1969, pp. 114–123). One occurrence is in connection with his first supposition in the schema, the Aristotelian principle that speed of fall (*V*) is directly proportional to the falling body's weight (*W*). Galileo says that if one performs the *periculum* or experiment these ratios will not actually be observed, and he attributes the discrepancy to "accidental causes" (GG1:273; Wallace 1983). Another place is in connection with the third supposition. Here Galileo states that one should not be surprised if a *periculum* or experiment does not verify this, for two reasons: external impediments prevent it, and a plane surface cannot be parallel to the horizon because the earth's surface is spherical (GG1:301). But if these difficulties can be overcome, the proof will be valid on the basis of the second supposition.

The mathematical argument that forms the basis of Galileo's proof is shown in Figure 6.1. Here is the familiar circle used to analyze the lever in Aristotle's *Mechanical Problems,* a work not known until the Renaissance. Orthogonal lines, BF ⊥ HG and BL ⊥ NO, make evident that the inclined plane actually obeys the law of the lever, and thus a body's weight on the incline (*W*), and consequently its speed downward (V), is diminished in the ratios BK/BF and BM/BL, as indicated in the figure.

Another example of the demonstrative regress occurs in chapter 19 of the *De motu* treatise, where Galileo uses it against Aristotle to explain why bodies increase their speed, or accelerate, during fall (GG2:315–323; Drabkin and Drake 1960, pp. 85–94). The argument may be diagrammed as follows:

First progression:

Effect: There is an observable increase in the speed of natural falling motion toward the end of the motion.

Cause: The falling body is less heavy at the beginning of its motion than it is at its end.

Intermediate stage: Supposition: heavy bodies move downward by reason of their weight (*gravitas*), and thus their speed of fall is directly proportional to their weights (GG1:262). The explanations offered by Aristotle and others invoke only accidental causes and do not arrive at the essential cause of the acceleration (GG1:317). That is, the weight of the body does not increase as it approaches its proper place; the body is not pushed by the medium rushing in behind it to fill the void created by its motion, since it is only accidental that it moves in a plenum; nor does the body encounter less resistance by having to separate fewer parts of the medium as it approaches the end of its motion (GG1:316–317).

Rather, the natural and intrinsic weight (*naturalis et intrinseca gravitas*) of the body remains constant. Thus, it is necessary to find some external force (*vis extrinseca*) that lightens the body at the beginning of its fall. This can only be the impelling force (*virtus impellens*) or lightness that sustains the body before it begins to fall and continually diminishes throughout its fall.

Such an impelling force is found not only when bodies are thrown upward before their descent, but also in cases where natural fall is not preceded by such a forced motion (GG1:318–320).

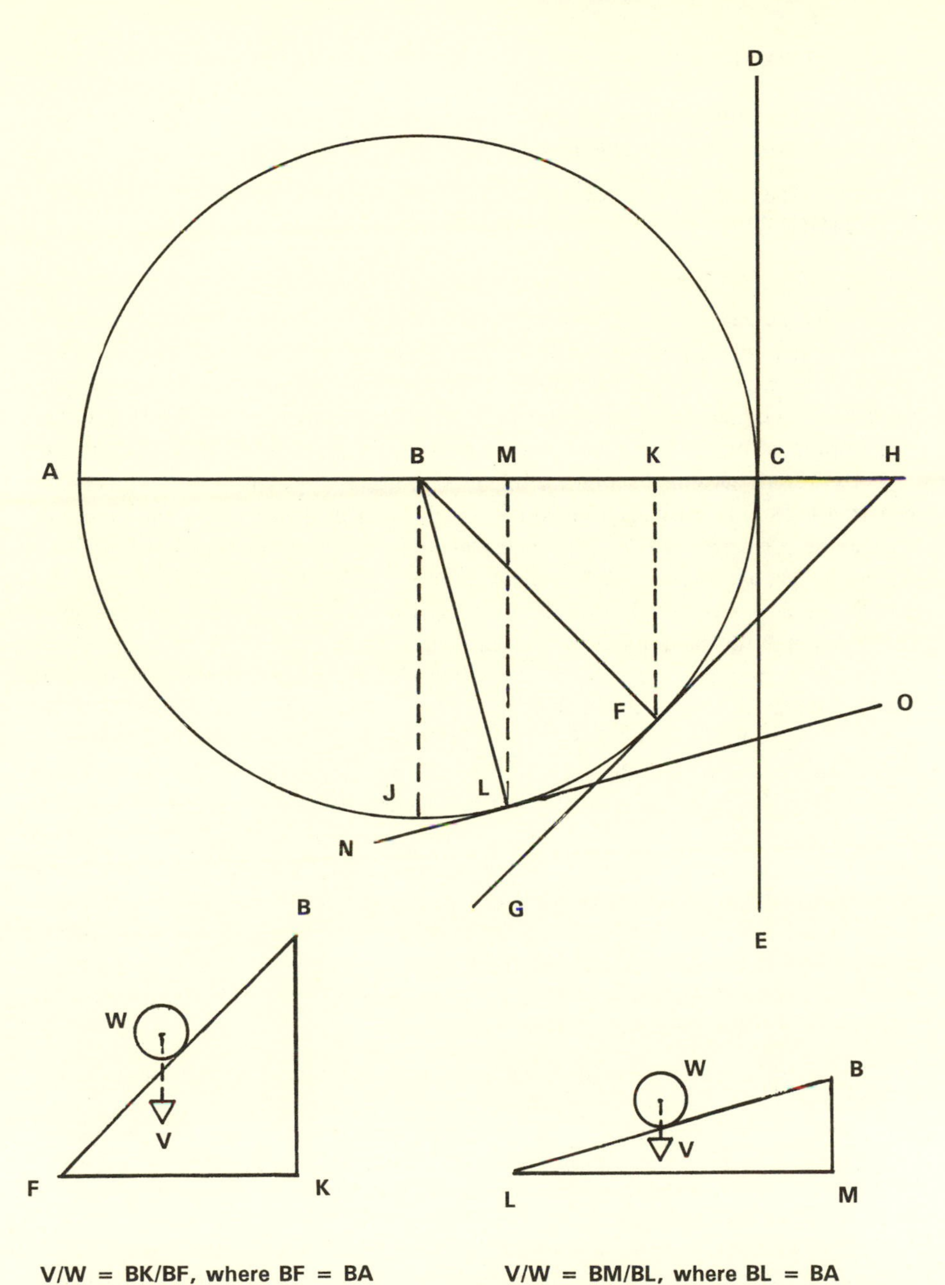

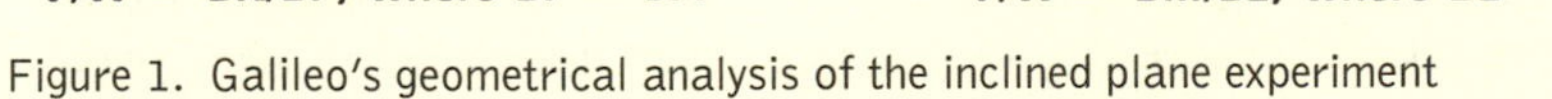

Figure 1. Galileo's geometrical analysis of the inclined plane experiment

Second progression:

> Cause: The effective weight of the body continually increases as this impelling force weakens and acts less against the body's essential weight.
>
> Effect: The body moves faster and faster throughout its fall from beginning to end (GG1:319).

As opposed to Aristotle's cause, Galileo feels that he has discovered the *vera causa* of velocity increase, namely, the decrease in the body's weight at the beginning of its fall. Note here the supposition on which the argument is based: Aristotle's dynamic law, $V \propto W$, which Galileo has already admitted cannot be verified by *periculum* or experiment. Galileo bases his explanation on an upwardly directed *impetus* or "levity" impressed on the body that is self-expending with time. At the summit of the object's upward motion, that *impetus* imparted to it exactly balances the body's natural *gravitas* and so the body comes to vertical rest. Then, as the *impetus* continues to decrease, the body gets heavier and heavier and so it increases its speed of fall, finally reaching what we would call a "terminal velocity" (GG1:317–320). This final state is not observed, it was thought in Galileo's time, for in the ordinary case bodies do not fall far enough for the terminal state to be achieved.

A final *periculum* or experiment to which Galileo makes reference in the early manuscript on motion occurs in chapter 22 of *De motu,* where he speaks of dropping objects from "a high tower" (GG2:333–337; Drabkin and Drake 1960, pp. 106–110). Since Galileo was at Pisa when he wrote this, he probably performed these tests from the Leaning Tower, as his student Vincenzo Viviani later reported (GG19:606.210–218). One of Galileo's professors at Pisa, Girolamo Borro, had taught that, when two equal bodies of lead and wood are thrown simultaneously from a window, the lighter body invariably reaches the ground before the heavier one. Borro maintained that he had established this by a public *experimentum* (Borro 1575, p. 215). Galileo contested this result. While conceding that the lighter body moves more swiftly at the beginning of its motion, Galileo argued that the heavier body quickly overtakes it and reaches the ground far ahead. Thus he wrote:

> [Borro holds] that air is heavy in its own region, from which it follows that things which have more air are heavier in the region of air—and this is also Aristotle's opinion. Thus a wooden sphere, for example, since it has more air in it than a leaden one, has three heavy elements, air, water, and earth; while the leaden one, since it has less air in it, has, as it were, only two heavy elements: the result is that the wooden sphere falls [in air] more swiftly than the leaden. . . .
>
> But experience [*experientia*] shows the opposite. For it is true that wood moves more swiftly than lead in the beginning of its motion; but a little later the motion of the lead is so accelerated that it leaves the wood behind it. And if they are both let fall from a high tower, the lead moves far out in front. This is something I have often tested [*De hoc saepe periculum feci*]. Therefore we must try to find a firmer cause on the basis of firmer hypotheses. (GG1:333–334)

Galileo then argued that the lighter body cannot conserve its upward impetus as well as the heavy body, and thus it falls quickly at first, but the heavy body soon overcomes its upward impetus and so catches up with, and then passes, the lighter body.

As Thomas Settle (1983) has shown, if Galileo did test this from the Leaning Tower, there is a simple explanation of the phenomenon he observed. The tower contains seven stories that are reached from an interior stairway. If one leans somewhat to drop the objects with hands extended over the ledge, and does so at successively higher stories, one finds that at the lowest stories the wood reaches the ground before the lead, whereas at the higher stories the lead arrives well ahead of the wood. The experiment actually has been duplicated and confirms Galileo's finding. Settle and others speculate that the heavier object induces arm fatigue in one holding it out over the ledge, and this causes the holder to have a slower release or to pull up on the heavier object, thus delaying its initial fall.

All of the foregoing materials pertain to the 1590 period. Galileo had wanted to publish the treatise on motion, but he had doubts about the "true causes" he had proposed in it because of his failure to obtain experimental confirmation of his results. He kept the manuscript in his possession, nonetheless, and when he finally did discover the correct law of falling bodies, as I shall show later, he inserted a draft of his discovery among the folios of the manuscript, thus signaling its role in the discovery process (Fredette 1972; Camerota 1992).

Novelties in the Heavens

We move now to the next period of experimental activity, mainly at Padua and roughly from 1604 to 1612, during which Galileo also made his important discoveries with the telescope. Here again he imported mathematical techniques into the demonstrative regress and thus was working in the "mixed" or "middle science" (*scientia media*) tradition. For the work with the telescope with which he revealed his "novelties in the heavens," Galileo relied mainly on projective geometry, whereas with his more advanced studies of the kinematics of motion he had to investigate properties of conic sections, particularly those of the parabola. Since the uses of projective geometry are simpler, I begin with them and later consider the more complicated experiments of his kinematical researches, even though this reverses the chronological order of their performance.

Precisely how the regress works in astronomy may be seen from a study of Galileo's treatise on the sphere, the *Trattato della sfera ovvero Cosmografia,* which he composed at Padua around 1602. The context is his explanation in the *Trattato* of the aspects and phases of the moon and the ways these vary with the moon's synoptic and sidereal periods (GG2:251–253). These phenomena depend only on relative positions within the earth-moon and earth-sun systems and do not require commitment to either geocentrism or heliocentrism, being equally well explained in either. Basic to the explanation is the conviction that these aspects and phases are effects (*effetti*) for which it is possible to assign the cause (*la causa;* GG2:250). Among the causes Galileo enumerates are that the moon is spherical in shape, that it is not luminous by nature but receives its light from the sun, and that the orientations of the two with respect to the earth are what cause the various aspects and the places and times of their appearances. The argument follows closely the paradigm provided by Aristotle in *Posterior Analytics* (I.13) to show that the moon is a sphere. It involves only one supposition, that light travels in straight lines, and this is what governs the intermediate stage. This allows one to use projective geometry to establish the convertibility condition, namely, that *only* external illumination falling on a shape that is spherical will cause the moon to exhibit the phases it does at precise positions and times observable from the earth. The reasoning may be summarized as follows (from Wallace 1992a, pp. 194–197):

First progression:

Effect: The moon's aspects and phases.

Cause: Its spherical shape, illumined by the sun, at various positions and times.

Intermediate stage: The moon is not luminous by nature; it is externally illumined by the sun, and it is observed from many different angles; *only* a shape that is spherical and this illumination will, under these circumstances, exhibit the aspects and phases it does at precise positions and times observable from the earth. The precise phenomena can be calculated from the supposition (*ex suppositione*) that light travels in straight lines, using theorems proved in projective geometry.

Second progression:

Cause: The moon's spherical shape, illumined by the sun, at various positions and times.

Effect: The moon's aspects and phases, calculated using the laws of geometrical optics.

When Galileo made his exciting discoveries with the telescope in 1609–1610 this same paradigm was ready at hand for further exploitation. Others before him had constructed telescopes, and some had even looked at the heavens with them, but none would formulate the "necessary demonstrations" Galileo would propose on the basis of his observations. We know that between November 30 and December 18 of 1609 Galileo studied the moon with his new instrument and made no fewer than eight drawings of the appearances he observed. On January 7, 1610, he wrote to Antonio de'Medici in Florence that, from the data he had obtained, "sane reasoning cannot conclude otherwise" than that the moon's surface contains mountains and valleys similar to, but larger than, those spread over the surface of the earth (GG10:273). Thus, within about a month, by his own account, Galileo had demonstrated to his personal satisfaction that there are mountains on the moon.

The regress that supports this reasoning may be schematized as follows (from Wallace 1992a, pp. 198–201):

First progression:

Effect: Sharply defined spots on illuminated parts of the moon's surface, an irregular line at the terminator, with points of light emerging in the dark parts.

Cause: The surface of the moon is rough and uneven, with bulges and depressions (GG3.1:62–63).

Intermediate stage: Dark part of spots have their side toward the sun; shadows diminish as the sun climbs higher; points of light in the dark area gradually increase in brightness and size, finally connect with the dark area; "we are driven to conclude by necessity" that *only* prominences and depressions can explain the appearances "for certain and beyond doubt" (GG3.1:64–69).

Second progression:

Cause: Changing illumination from the sun's rays on mountains of calculable height rising from the moon's surface.

Effect: All of the observed appearances (GG3.1:69–70).

Here again there are the two progressions, the first *quia* from effect to cause, the second *propter quid* from cause to effect, with the intermediate stage establishing the convertibility condition and thus the connection between the two. The implied supposition, not indicated here, is the same as that underlying the *Trattato della sfera* demonstrations, namely, that light travels in straight lines. Those who see this and carefully observe the phenomena, wrote Galileo, "are driven to conclude by necessity" that *only* prominences and depressions on the moon's surface can explain its appearances "for certain and beyond doubt" (GG3.1.64–69; Galileo 1989, pp. 39–48).

On the very evening in which Galileo wrote to Antonio de'Medici that he had demonstrated the existence of mountains on the moon, he noted a further strange phenomenon, namely, that the planet Jupiter was "accompanied by three fixed stars" (GG10:277). That was on January 7, 1610. The next night Galileo turned his telescope on the heavens again, hoping to see that Jupiter had moved to the west of these stars, as Ptolemaic computations then predicted (GG3.1.80). To his surprise this time he found the planet to be east of them. His attempt to resolve that anomaly led him to a program of observing Jupiter and its strange companions whenever he could over a two-month period. By January 11 he had concluded that they were not fixed stars that could be used to determine the motion of Jupiter, but rather were small bodies, never observed before, that were moving along with Jupiter and indeed were actually circling it. "I therefore arrived at the conclusion, entirely beyond doubt [*omnique procul dubio*]," he wrote, "that in the heavens there are three stars wandering about Jupiter like Venus and Mercury around the sun" (GG3.1.81). On January 13 he saw a fourth object for the first time, and by the 15th he had convinced himself that it was doing the same (GG3.1.82). So within a week of his curiosity having been aroused by the anomaly, he had completed the demonstrative regress and had convinced himself that Jupiter has four satellites revolving about it, as it made its own majestic revolution around the center of the universe (GG3.1.80–95; Galileo 1989, pp. 64–84).

The reasoning process Galileo employed in this discovery may be outlined as follows (from Wallace 1992a, pp. 201–203):

First progression:

Effect: Four little stars accompany Jupiter, always in a straight line with it, and move along the line with respect to each other and to Jupiter.

Cause: The stars are planets of Jupiter, circling around it at various periods and distances from it.

Intermediate stage: Sixty-five observations between January 7 and March 2, analyzing in detail their variations in position, how they separate off from Jupiter and each other and merge with them in successive observations; inference to the *only* possible motion that explains these details; concluding "no one can doubt" (*nemini dubium esse potest*) that they complete revolutions around Jupiter in the plane of the ecliptic, each at a fixed radius and with its characteristic time of revolution (GG3.1.94).

Second progression:

Cause: Four satellites of Jupiter always accompany it, in direct and retrograde motion, with their own distances from it and periods of revolution (GG4:210), as it revolves around the center in twelve years.

Effect: Seen on edge, the satellites produce the appearance of four points of light, moving back and forth on a line with the planet and parallel to the ecliptic.

The basic supposition is again that light travels in straight lines. Over and above that, of course, one has to know enough projective geometry to recognize that satellites circling around an equatorial plane, when seen on edge, would appear to be moving back and forth on a line parallel to the planet's equator and along the elliptic. Galileo quickly saw the convertibility of the geometry involved, going from the straight-line motion he actually observed to the circular motion that alone could cause it, and then regressing from the cause back to the effects he had so carefully observed.

Space does not permit me to exhibit the completely analogous reasoning process by which Galileo, in December of 1610, having by then observed the phases of Venus, could demonstrate that the planet is in orbit around the sun. The geometry in this case is considerably more complex than that required to complete the intermediate stages outlined in the previous examples of the regress. But when one understands the geometry involved, it is a simple matter to understand why, when seen from the earth, Venus exhibits the phases it does and its changes in size and appearance. One can also see why there is no possibility that Venus could be rotating around the earth, but *must* be orbiting the sun (Wallace 1992a, pp. 203–207). It is this demonstration, along with the previous ones, that Galileo clearly had in mind when in 1615, in his famous *Letter to the Grand Duchess Christina,* he wrote so glowingly about his "necessary demonstrations based on sensible experience." It is perhaps significant that he uses this expression or its equivalent over forty times in that much-quoted letter (Moss 1986).

The "Tabletop" Experiments

These astronomical discoveries, of course, are truly wonderful demonstrations, and one can readily understand why, as their significance was grasped, they brought Galileo almost immediate fame throughout Europe. And yet in the final analysis they are not as important as the series of experiments on motion and falling bodies he performed at Padua immediately prior to the telescopic discoveries in the years 1604–1609. In these tests, known as the "table-top" experiments, Galileo used an inclined plane placed on the edge of a table to establish (1) the correct speed law, that velocity is proportional not to the distance of fall, as he earlier thought, but to the square root of distance; (2) the correct distance law, that distance of fall is proportional to the square of the time of fall; and (3) that the path a body follows when projected horizontally at uniform velocity and then allowed to fall under the influence of gravity is a semi-parabola. All of these results were established by Galileo through the use of the demonstrative regress, as I will now explain.

Around 1602, while in correspondence with Guidobaldo del Monte, Galileo experimented with the pendulum as an alternative to the inclined plane, because, although the bob of the pendulum moves along the arc of a circle rather than a chord, it eliminates the surface friction always present on the plane (Naylor 1974). Galileo had already rejected the Aristotelian dynamic law, that speed of fall is uniform and simply proportional to weight. In 1604 he wrote to Paolo Sarpi stating that speed increases with distance of fall, and from this principle he was trying to deduce various properties of falling motion (GG10:115–116). Shortly after that he apparently initiated experiments with an inclined plane situated on the top of a table with its base at or near the table's edge, thus allowing a ball to roll down the incline and then drop freely to the floor. These experiments were totally unknown until about 1972, when Stillman Drake uncovered folios in MS Gal. 72 that gave evidence of them (Drake 1973, 1978). Since then they have been analyzed in detail and duplicated by Drake, Ronald Naylor (1976, 1980, 1990), and David Hill (1979, 1986, 1988). Collectively their results show that Galileo was engaged in a serious research program in the first decade of the seventeenth century, achieving an experimental accuracy within three percent when testing his calculated results.

As shown in figure 6.2, this program made use of four different, but connected, types of experiment, probably made in the progression as shown from top to bottom. The first type, shown in figure 6.2a, was designed to ascertain the correct speed law, to show that distance of horizontal projection after various distances of roll down the incline do *not* vary as the distance of roll but rather as the square root of that distance.[3] With this knowledge in hand, Galileo then began to work on defining the characteristics of the curves that result when the angle of inclination is varied. This is the second type of experiment, fairly complex, shown in figure 6.2b. The curves shown there approach more and more a semi-parabolic form the smaller the angle of incline, suggesting that a straight horizontal projection might yield the semi-parabola.[4] The problem then became one of achieving such projection while at the same time having a way to vary and measure the ball's velocity on leaving the tabletop. Galileo's solution is shown in figure 6.2c, illustrating his design of different deflectors to produce

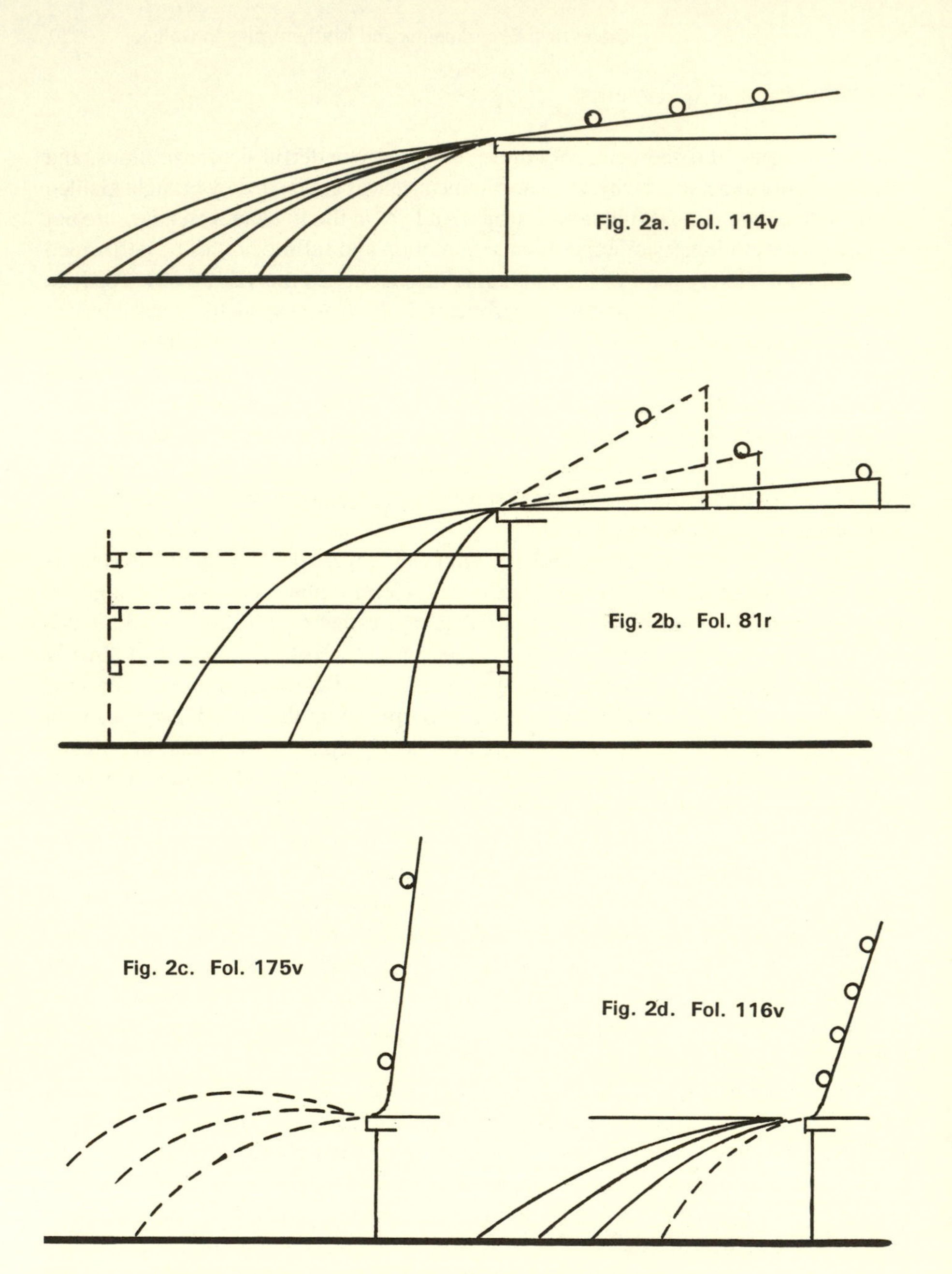

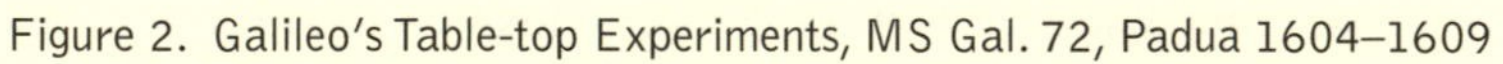
Figure 2. Galileo's Table-top Experiments, MS Gal. 72, Padua 1604–1609

a variety of curves,[5] and in figure 6.2d, the one he finally used to achieve horizontal projection, along with the series of semi-parabolic curves he eventually produced.[6]

The key result that emerges from these experiments is that the speed of bodies in free fall, instantiated by balls that are no longer on the incline but have left it and are falling naturally, varies directly as their time of fall. From this principle, explicitly stated at the beginning of the Third Day of the discourses of the *Two New Sciences,* Galileo derives most of the propositions he presents in the Third and Fourth Days of that work (Wallace 1992a, pp. 287–289). His reasoning in establishing that principle may be abbreviated as follows:

First progression:

Effect: The various properties of heavy bodies moving with a motion that is naturally accelerated.

Cause: Their falling at a speed is directly proportional to their time of fall.

Intermediate stage: This is proved kinematically, because only a falling speed directly proportional to the time of fall can produce distances that satisfy the odd-number rule and the times-squared rule in vertical fall, the double-distance rule when the vertical speed is converted to horizontal speed, and the semi-parabolic path when free fall occurs after the vertical speed has been converted to horizontal speed—by geometrical demonstration, from the supposition (*ex suppositione*) that all impediments such as friction, the resistance of the medium, and all other accidental factors have been removed.

It is also argued from physical considerations: for nature itself (*instituti ipsiusmet naturae*) causes the falling motion of a heavy body, which is a natural motion, to increase in the simplest way: by adding equal increments to the speed in equal intervals of time. It is also argued from disproof of the simplest alternative, since speed does not increase directly with the distance of fall but rather with the square root of that distance.

It is confirmed experimentally, for physical experiments (*naturalia experimenta*) show that all these metrical properties are verified within degrees of accuracy that allow for slight departures owing to impediments and accidental causes (GG2:261, 8:197).

Second progression:

Cause: A heavy body that is naturally accelerated in free fall at a speed that is directly proportional to its time of fall from rest.

Effect: Metrical properties described by the odd-number, times-squared, and double-distance rules and by paths of semi-parabolic projection.

As can be seen here, the regress is employed once again to arrive at the true cause, what becomes for Galileo the definition of naturally accelerated motion. The first progression is *a posteriori,* from effect to cause, and the second *a priori,* from cause to effect. The intermediate stage, the work of the intellect, carries the burden of proof, as heretofore. Actually its wording as shown here follows closely Galileo's Latin text in his draft of this passage, the *De motu accelerato* fragment now bound in MS Gal. 71 (GG2:226), where Galileo inserted it after writing it out. It also appears in the *Two New Sciences,* and with almost identical wording (GG8:198).

The demonstration here, like the earlier ones, is explicitly made *ex suppositione,* that is, on the supposition that all impediments to the falling motion, such as friction, resistance of the medium, and accidental factors, have been removed. The proof is based partly on the elimination of the simplest alternative, that speed of fall is based on distance of fall, as Galileo himself had first thought. But the direct proof is experimental. Note the reference to "physical experiment*s*," pointedly in the plural. The reference is not to the simple inclined-plane experiment described in the *Two New Sciences,* as it has commonly been taken, but to the whole gamut of experiments, tabletop included, performed at Padua before the discoveries with the telescope. Note further that Galileo no longer identifies the weight of the falling body as the cause of its fall, as in his early formulations. Now he is interested solely in the kinematic factors that bear on the quantitative aspects of naturally accelerated motion. As for the ultimate physical cause of the fall, he identifies this simply as "nature," the ultimate explanatory principle in Aristotelian physics. So he himself is working unambiguously in the tradition of a mathematical physics, a "mixed" or "middle science." What he proposes to do for dynamics is what Archimedes has done for statics, that is, provide a hitherto unknown science of local motion based on mathematics and not on physical principles alone (Wallace 1984, pp. 272–276; 1992a, pp. 270–273, 285–293).

Mathematical Physics

Galileo, like Newton after him, thus regarded himself as a mathematical physicist. It is interesting to see where he conceived his own genius to lie in working out the demonstrations he offered, proofs that would characterize his "new science." In my view his ability lay in knowing how to pose suppositions that permit experiments to be made, and then verifying, in the experiments themselves, that the suppositions hold up within the degree of accuracy required to justify them. Galileo himself hints at this in a passage of the Second Day of the *Dialogue on the Two World Systems* of 1632, where Simplicio and Salviati are discussing whether a sphere touches a plane at a point (GG7: 233–234). Here Simplicio is arguing, in effect, that mathematics cannot be used in physics because abstract spheres are not the same as material spheres and this difference vitiates Galileo's calculations. That is not Aristotle's argument, though it was used by Peripatetics of Galileo's day. Aristotle himself acknowledged the validity of "mixed sciences," those that use mathematics to establish conclusions in physics, and indeed used them, as I have shown, to exemplify the demonstrative regress. So Galileo's reply to Simplicio's argument is authentically Aristotelian, being based on the materials in the logical treatises of his MS Gal. 27 (Wallace 1981; 1992a, pp. 139–149):

> I readily grant you all these things, but they are beside the point. . . . By your own statement, spheres and planes are either not to be found in the world, or if they found they are spoiled upon being used for this effect. It would therefore have been better for you

> to grant the conclusion conditionally [*condizionatamente*]; that is, for you to have said that if there were given a material sphere and plane that were so perfect and remained so, they would touch one another in a single point, and then to have denied that such were to be had. . . .
>
> Whenever you apply a material sphere to a material plane in the concrete, you apply a sphere which is not perfect to a plane which is not perfect, and you say that these do not touch each other in one point. But I tell you that even in the abstract, an immaterial sphere which is not a perfect sphere can touch an immaterial plane which is not perfectly flat in not one point, but over a part of its surface, so that what happens in the concrete up to this point happens in the same way in the abstract. . . .
>
> Do you know what does happen, Simplicio? Just as the computer who wants his calculations to deal with sugar, silk, and wool must discount the boxes, bales, and other packings, so the geometrical physicist [*filosofo geometra,* Galileo's term for the mathematical physicist of his day], when he wants to recognize in the concrete the effects which he has demonstrated in the abstract, must deduct the impediments of the matter [*gli impedimenti della materia*], and, if he is able to do so, I assure you that his results are in no less agreement than arithmetical computations. *The errors, then, lie not in the abstractness or concreteness, not in geometry or in physics, but in the calculator who does not know how to make an accurate calculation.* Hence if you had a perfect sphere and a perfect plane, even though they were material, you would have no doubt that they touched in one point. . . . (GG7:232–234)

As Galileo explains, one must understand the differences between abstract spheres and material spheres, and know how to "deduct the impediments of the matter," if one is to make accurate calculations in physics. To do so the mathematical physicist must proceed "conditionally" in his discipline. Galileo's term is *condizionatamente,* which has the same meaning as *ex conditione* or *ex suppositione.* The condition or supposition is that the scientist recognize these impediments and devise experiments that can circumvent them. If he can do so, his results will be true within the limits he has set for himself on the basis of his own suppositions. The calculator who is unable to do this, who does not understand the mathematics or who is unable to perform experiments of this type, will always get erroneous results. But then the fault is not in the mathematical physics but rather in the experimenter who lacks the expertise to verify his insights.

Galileo the Controversialist

Although Galileo has a well deserved reputation as a polemicist and controversialist, his scientific writings to this point were not essentially polemical. His early work on motion at Pisa did develop out of disputations at the university in which Galileo upheld a progressive view, and undoubtedly he was concerned that his arguments would be rejected by conservative Aristotelians, including his teacher Francesco Buonamici (GG1:398.4–19, 412.19–22). Still, his ideas mirrored in some ways those already advanced by Giovanni Battista Benedetti and others (Wallace 1987). The important point is that they were never published, and this alone would explain why they did not spark any controversies. His discoveries with the telescope, of course, touched on a very controversial subject, the nature of the heavens, and yet his manner of reporting them in the *Sidereus Nuncius* gave his adversaries little ground for rejecting them. The main problem they posed was their factual status, for those not having access to a telescope with sufficient magnification and resolving power would be tempted to dismiss the phenomena he reported as optical illusions. Albert Van Helden has shown that astronomers who might have been expected to reject them on philosophical grounds, such as the Jesuit professors at the Collegio Romano, actually verified Galileo's findings as soon as they had constructed a good telescope themselves (Galilei 1989, pp. 110–112). This situation seems quite representative, for despite some early opposition from irresponsible authors such as Martin Horky, Francesco Sizzi, and Giulio Cesare Lagalla, Galileo's findings were soon accepted without argument by astronomers throughout Europe.

A similar situation obtains with regard to the tabletop experiments performed by Galileo at Padua in the first decade of the seventeenth century. Like the materials present in MSS Gal. 27, 46, and 71, none of the findings recorded in MS Gal. 72 was known in Galileo's day. They stimulate controversy in our day over how they are to be interpreted, but that is properly a problem for historians and not for scientists—unless one is to consider the possibility mentioned by Gideon Freudenthal of a scientist having a controversy with himself. In that event the older Galileo was arguing with the younger, and what he was contesting was simply the unproved suppositions on which his earlier "demonstrations" had been based.

Shortly after 1610, however, Galileo himself became deeply involved in controversy, and unfortunately this state continued more or less uninterruptedly until the end of his life. Many of these controversies were more theological than they were scientific, being concerned with how the scriptures were to be interpreted and what latitude should be allowed to those who departed from the traditional teachings of the Church. As to the scientific problems with which he had then to deal, most of these were not solvable with the information available to Galileo or anyone else, and so were not amenable to the use of the demonstrative regress. Much of Galileo's part in them, however, can be understood in terms of various adaptations he seems to have made when applying regressive methods to situations where certitude could not be attained and one had to resort to probable reasoning. Three cases that illustrate these adaptations, which invoke a type of dialectical (as opposed to demonstrative) regress, turn out to be representative of his future work. All three took place in the second decade of the seventeenth century, shortly after his remarkable success with the tabletop experiments and his discoveries with the telescope. The first two involved actual controversies: Galileo's dispute with Ludovico delle Colombe at Florence in 1612 over the true cause of flotation, and his prolonged debate with Christopher Scheiner in 1618 over the nature of sunspots. The third controversy occurred in the interval between the other two and was not itself a controversy, although it was to give rise to one a decade and a half later. This was his proposal to Cardinal Orsini in 1616 sketching his argument from the tides to prove the earth's motion, which was to have disastrous consequences for Galileo when he reformulated it in the *Dialogue* of 1632.

Galileo's preferred technique in controversy was to set up two mutually exclusive or dichotomous explanations for a particular phenomenon and then devise various observational or experimental texts that would serve to eliminate the one and thus leave the other. When coupled with geometrical methods of proof, this technique lends itself to a *reductio ad impossibile* for one of the alternatives and thus supplies indirect proof for the other. The dichotomy itself functions as a *suppositio* in the proof, and is particularly effective if it is proposed by, or is acceptable to, the other party to the controversy.

In the dispute with Colombe the supposition was that a body's motion downward in a medium was caused either by the shape of the body (Colombe's alternative) or by the weight of the body in the medium in which it is placed (Galileo's alternative). The argument Galileo proposed in support of his side is based on hydrostatic principles, properly applied through geometrical analysis, to make clear the proper cause of flotation. The conclusion to which it came is that the true, intrinsic, and proper cause of flotation and submergence, excluding mediate and accidental causes, is the weight of a body relative to that of the medium. That is to say, it is not the body's absolute weight or *gravitas* that determines whether the body will float or not, but rather its *propria gravitas,* its weight in the medium in which it is immersed, considering that the body is buoyed up by a force equal to the weight of the fluid it is able to displace. For Galileo, this alone explains why one body will float in a medium and others will not, how much will protrude above the surface when it does, and how a medium can support a weight heavier than itself (GG4:79).

To meet Colombe's counterarguments, Galileo admits that a body's shape may affect the speed of its motion through a medium, but this is not the proper cause of its motion. This can be demonstrated by experimenting with a mass of wax molded into various shapes; its position in the medium is determined by its weight and not by any particular shape it is made to assume. The special case of a thin plate of ebony floating on water can then be explained by an accidental cause. Here Galileo formulates the ingenious proposal that the volume of air enclosed by ridges and below the water's surface, when joined to the unwetted top surface of the plate, adds to the plate's buoyancy and so causes it to float (GG4:107–111). Thus he sidesteps the problem of surface tension, focusing on an equilibrium situation in which the causes that might produce motion cancel out and a simple volumetric solution can be provided using geometrical principles (Wallace 1992a, pp. 276–278).

The dispute with Scheiner over sunspots lent itself to the same technique. According to Scheiner the observed appearances of the spots may be explained in one of two ways: either as spots moving on or near the sun's surface or as spots rotating in a celestial sphere outside the sun, presumably "stars" or planets. Anxious to preserve the sun's unalterability and incorruptibility as a heavenly body, Scheiner opted for the second alternative, leaving Galileo the opportunity to exploit the first. This he did by subscribing to Scheiner's dichotomy and then attacking the latter's position, again through the use of geometrical analysis—this time using the principles of optics rather than those of hydrostatics.

Geometrical optics, Galileo states, provides necessary demonstrations that the spots are not outside the sun but are contiguous with its surface. In particular, the spots appear thinner when near the edge of the sun than when close to its center; the distances they travel increase as they approach the center and decrease as they recede toward the edge; and they separate more and more as they approach the center—for one who knows *perspettiva,* "a clear argument [*manifesto argomento*] that the sun is a globe and that the spots are close to the sun's surface" (GG5:119; Shea 1972, pp. 55–57).

Furthermore, close observation shows that the appearances of the spots are not those of stars (*stelle*); they more resemble clouds (*nugole*) that form and dissolve and so change size and shape. Thus, Galileo observes, it is not certain that the same spots return after a complete revolution, nor is it certain that the sun itself rotates on its axis, although it appears to do so (GG5:133).

It is interesting to note that, although Galileo claims to incorporate "necessary demonstrations" (*dimostrazioni necessarie*) in his overall argument, and so is successful in negating Scheiner's position, he himself advances only probable opinion as to what the spots ultimately might be. Thus his conclusions may be summarized as follows: the spots are definitely not stars or planets rotating in their own celestial orbits around the sun somewhere between it and earth; it is probable that they are clouds in a medium surrounding the sun's surface; and it is more probable that the sun itself rotates and carries this medium and its clouds along with it than that these have an independent circular motion around the sun (Wallace 1992a, pp. 207–211).

The third case I examine does not invoke a dichotomy as does the first two but employs causal argument to assign degrees of probability to various possible explanations for a given phenomenon, along lines already seen in the dispute with Scheiner. In this and similar cases Galileo's various causal maxims assume importance, namely, that there is only one true and primary cause for any one effect; that effects similar in kind must be reducible to a single true and primary cause; that there is a fixed and constant connection between cause and effect, so that any alteration in the one will be accompanied by a fixed and constant alteration in the other, and so on (Mertz 1980; Wallace 1983, pp. 612, 622). In this case the effect to be explained is the ebb and flow of the tides in the various oceans and seas on the earth's surface, which Galileo suspects might be connected with the motion of the earth. In his *Letter to the Grand Duchess Christina* of 1615 Galileo had made reference to "physical effects whose causes perhaps cannot be determined in any other way" (GG5:311) without indicating precisely what he had in mind. Apparently he discussed this with a young friend, Alessandro Orsini, who had just been made a cardinal and who asked Galileo to write out his argument. Galileo did so on January 8, 1616, in a letter now entitled *Discourse on the Tides* (GG5:377–395).

Galileo begins by noting that sensory appearances show that the tides involve a true local motion in the sea, and thus to find their cause one must investigate the various ways motion can be imparted to water. He further notes the complexity of tidal phenomena, and on this account will see if any of the possible movers can reasonably be assigned as the primary cause. To this he then proposes to add secondary or concomitant causes to account for the diversity of the tides' movements. Since the motion of the container can often explain the motion of the fluid it contains, Galileo speculates that "the cause of the tides could reside in some motion of the basins containing the seawater," thus focusing on the motion of the terrestrial globe as "more probable" than any other cause previously assigned (GG5:381). On this basis he takes the motion of the earth hypothetically (*ex hypothesi*) and, from its two motions, one of annual revolution around the sun, the other of diurnal rotation on its axis, explains how it might function as a primary cause of the back-and-forth motion of the water on its surface. This cause will obviously not be enough to account for the particular details of tidal phenomena, and so to it he adds additional causes, among them the gravity of seawater, the length and depth of the basin in which it is contained, the frequency of its oscillations, and the ways these might be coordinated with the movement of various parts of the earth.

Galileo concludes on the note that with this explanation he is able to harmonize the earth's motion and the tides, "taking the former as the cause of the latter, and the latter as a sign of and an argument for the former" (GG5:393). His expression here clearly signals the use of the demonstrative *regressus,* despite the fact that the argument he is proposing is just as clearly not a demonstration. To take account of both features, we propose to modify our earlier formulation of the *regressus* to accommodate it to probable argument. The revised form is the "dialectical regress" to which reference has already been made. When applied to Galileo's early statement of the tidal argument the *regressus* may be seen to proceed as follows (from Wallace 1992a, pp. 211–216):

Possible cause: from an effect to one or more hypothetical causes that might be sufficient to produce it.

Effect: The ebb and flow of the tides in various oceans and seas on the earth's surface.

Possible Cause: Primarily a twofold motion of the earth, secondarily by auxiliary factors.

Dialectical inquiry: use of probable reasoning and correlations to specify in detail the causal factors that produce the effect. The motion of a container can explain the motion of water within it; the diurnal and annual motions of the earth produce unequal motions at different parts of the earth's surface; the oscillations set up in bodies of water by these unequal motions vary in period depending on the lengths and depths of the sea basins. These unequal motions also are of two types and have two components, one vertical, seen mainly at the extremity of the basins, the other horizontal, seen mainly at their middle; in very large seas differential factors further operate to produce more movement in some parts than in others.

Tidal periods of twelve hours are produced by the primary cause; those of six, four, three, and two hours are produced additionally by various combinations of secondary causes. The motion of the moon is a fictitious cause that has nothing to do with tidal motions (GG5:381–393).

Probable cause: from one or more causes now regarded as probable to the actual production of the effect.

Probable Cause: Twofold motion of the earth, acting on bodies of water of different shapes and sizes.

Effect: An ebb and flow of tides at characteristic periods in the respective basins.

Note that there is no air of controversy in this initial presentation of the tidal argument. It was written, as already observed, after Galileo's letter to Christina. It was written also *after* Cardinal Bellarmine's letter to Foscarini (and Galileo) warning against using the earth's motion, without offering demonstrative proof, to question the Church's traditional interpretation of Scripture, and *before* the Church's decree against teaching or defending Copernicanism, which was dated March 5, 1616. Thus it reflects Galileo's thought on the tidal proof at a relatively tranquil period in his life—well before he got embroiled in the bitter controversies over scriptural interpretation that would lead ultimately to his trial and condemnation by the Church in 1633.

The subsequent history of the "Galileo Affair" has been rehearsed so many times that it does not require repetition here (Finocchiaro 1989). Suffice it to mention that Galileo was a skilled controversialist, and he did not fail to use all of the means of dialectics and rhetoric to argue the case for the earth's motion (Moss 1983, 1986, 1993; Finocchiaro 1980). On the other hand, his expressed intention was only to make that case "persuasible" (*persuasibile,* GG7:30.22), and not once did he ever claim to have demonstrated the earth's motion as an epistemic conclusion. Unfortunately, this fact seems often overlooked in the vast literature that now surrounds the infamous affair.

Regarding the various views of scientific controversies proposed at this conference, most can be seen as verified in one way or another in the work of Galileo. Some analyses apply more readily to recent science than they do to that of the early modern period, and for this reason, as for reasons of brevity, I shall be selective in my comments. Following Philip Kitcher's taxonomy (see chapter 1 in this volume), I would say that Galileo subscribed to a rationalistic model of controversy wherein the issues argued would have testable consequences and where epistemic closure would be sought through necessary arguments or ones that engender highly probable conclusions. In his dialectics Galileo was particularly expert at blocking his opponents' lines of escape. In those cases where he could offer demonstrations on the model of Aristotle's *Posterior Analytics,* my impression is that he won over his adversaries rather quickly considering the novelty of the conclusions to which he had come. Similarly, employing Aristides Baltas's taxonomy (see chapter 2 in this volume), Galileo again followed a rationalistic model. His "constitutive assumptions" were generally unexpressed but they are clearly those of Aristotelian and commonsense realism combined with those of Euclidean geometry. His "controversial assumptions," on the other hand, were explicitly recognized and were generally formulated as suppositions (*suppositiones*) that, in his physics, would have argumentative force similar to that of postulates or petitions in classical mathematics.

Marcello Pera's analysis (see chapter 3 in this volume) comes closer to Galileo's timeframe and thus is more applicable to my account. Galileo's epistemology was still premodern, and his discussion was dialectical, largely internal and epistemic. He definitely proposed some of the premises of his arguments as propositions to be agreed upon, and these again he labeled "suppositions." His logic was clearly both a posteriori (or inductive) and a priori (or deductive), each constituting a different phase or progression in the demonstrative regress, and it aimed for both formal and content validity. Rhetoric and dialectic were frequently intermingled in Galileo's discourse, particularly in his prolonged crusade in support of the Copernican opinion, although Pera's view of the respective spheres of rhetoric and dialectic is somewhat at variance with my own (Wallace 1992a, pp. 128–130).

A similar appraisal might be made of Peter Machamer's account (chapter 6), which agrees in most particulars with the analysis presented in this chapter. Additional points of agreement may be noted in two of the remaining contributions. With regard to Gideon Freudenthal's analysis (chapter 7), all of Galileo's discoveries had cognitive content, no previously available solutions were at hand, and vital cosmic and religious interests were at stake. Strictly speaking, no new system of physics was yet in question, since Galileo's methods mainly involved adjustments within an Aristotelian-Archimedean-Euclidean framework. And respecting Mauritio Mamiani's canons (see chapter 8), disputes were generally at the levels of new observations or experiments, all involving matters of fact. There was little theory in the modern sense, and any unobservables that might have been involved were entities or measurements that previously had escaped observation in ordinary sense experience.

Finally, when the chapters are considered as a whole it becomes apparent that, as a controversialist, Galileo the scientist had few equals. But the main lesson I have been urging in this chapter is somewhat different. Put simply, the best way to end a scientific controversy is to produce a convincing demonstration. I readily grant that such demonstrations are difficult to come by when one is working at the frontiers of knowledge. But Galileo faced that situation, too, and he succeeded where all others since Aristotle had failed. That is why we rightly honor him as the Father of Modern Science.[7]

Notes

1. Antonio Favaro, ed., *Le Opere di Galileo Galilei,* 20 vols. in 21 (Florence, 1890–1909, repr. 1968), 9: 273–282 [cited hereafter as GG].

2. Averroes was a rationalist in his understanding of Aristotle and saw no role for mathematics in the study of nature. Aquinas, on the other hand, was more the empiricist, as was his teacher Albertus Magnus. He also developed Aristotle's teaching on the mixed sciences, which he called "middle sciences" or *scientiae mediae* because they are intermediate between mathematics and physics, and used them as a model for the science of revealed theology. See Aquinas's *Summa theologiae,* Part I, Quest. 1, Art. 2.

3. This result is based on Hill's (1988) interpretation of a diagram found on fol. 114v of MS Gal. 72 (pp. 658–659). Galileo lists a series of numbers for different lengths of horizontal projection along the floor, namely, 253, 337, 395, 451, 495, 534, and 573. Hill, in attempting to duplicate Galileo's figures, has found that increasingly longer lengths of roll down an inclined plane inclined at an angle of 12° to the table top, with the lengths standing in the ratio of 1:2:3:4:5:6:7, will yield Galileo's figures approximately. If one takes the starting length of roll at 400, the successive lengths will be 800, 1200, 1600, 2000, 2400, and 2800. Taking the square root of the middle figure in this sequence, 1600, and fitting it to the middle figure in the horizontal projections, 451, one obtains a sequence very similar to Galileo's, namely, 226, 319, 390, 451, 505, 552, and 596. This would seem to confirm that the distance of horizontal projection, which is a measure of the ball's velocity on leaving the incline, is as the square root of the length of roll down the incline. Arguing *a pari,* this would seem to show that velocity of fall is *not* proportional to distance of fall, as Galileo had conjectured in his letter to Sarpi, but rather is proportional to the square root of that distance.

4. The diagram here is based on a figure drawn by Galileo on fol. 81r of MS Gal. 72. This has been analyzed in various ways by Naylor and Hill; the explanation here follows Hill's interpretation. On the figure, Galileo has written numerals for all the horizontal intervals at the different vertical levels, all of which are reproduced by Hill (1988, pp. 647–648). According to Hill's calculations, the curves approach a parabolic form the farther they extend away from the table. Hill speculates that they were generated by rolling balls down inclines of various angles of inclination, with the balls then being allowed to drop through different vertical distances,

either to the floor or to a board set at some intermediate height between the floor and the table top. Hill identifies four different heights and three different angles of inclination used in the experiment. As the angle of inclination decreases, the curves approach semi-parabolic form. This would seem to suggest that horizontal projection after a roll (which cannot be achieved with this experimental setup) would yield the sought-after parabolic form.

5. This diagram is sketched by Galileo on fol. 175v of MS Gal. 72. It is reproduced by Naylor (1980, p. 558).

6. This is the famous diagram on fol. 116v of MS Gal. 72, which has been subjected to many analyses since Drake (1973) first called attention to it. On the folio Galileo lists the height of the table, 828 units, and also various heights of fall down an incline, namely, 300, 600, 800, 828, and 1000 units. Along the horizontal at the level of the floor he then records measurements of horizontal projection, writing the figures 800, 1172, 1328, 1340, and 1500. For the last four figures he then provides a second set of figures, namely, 1131, 1306, 1330, and 1460, presumably his calculations of what the distances should be if the 800 figure is taken as the baseline and one is attempting to show that successive heights of fall are in the same ratio as the squares of the distances of horizontal projection. Should this relationship be verified experimentally, one would have proof that the velocity of fall is directly proportional to the time of fall, the principle Galileo would use for his theorems on naturally accelerated motion in the *Two New Sciences*. The proof is sketched in Wallace (1981a, pp. 154–156).

7. Since this essay was first presented at Vico Equenze Colloquium in June of 1993, I have further stressed the historical importance of the demonstrative regress appropriated by Galileo from the Paduan Aristotelians via the Roman Jesuits. My more fully developed thesis is that the use of suppositions within the intermediate stage of the regress, particularly suppositions relating to experimentation and mathematical reasoning, is still relevant in the present day for understanding how controversies have arisen, and then been resolved, in the development of modern science. For the detailed development of this thesis, see Wallace (1996), especially chs. 8–10.

References

Berti, E. (1991), "La theoria aristotelica della dimostrazione nella "Tractatio" omonima di Galilei," in M. Ciliberto and C. Vasoli (eds.), *Filosofia e cultura,* per Eugenio Garin. Rome: Editori Riuniti, 327–350.

Borro, G. (1576), *De motu gravium et levium.* Florence: Georgius Marescottus.

Camerota, N. (1992), *Gli Scritti De motu antiquiora di Galileo Galilei: Il Ms Gal. 71, Un'analisi storico-critica.* Cagliari: CUEC Editori.

Carugo, A., and Crombie, A. C. (1983), "The Jesuits and Galileo's Ideas of Science and of Nature," *Annali dell'Istituto e Museo di Storia della Scienza di Firenze,* 8.2:3–68.

Cassirer, E. (1922), *Das Erkenntnisproblem in der Philosophie und Wissenschaft der neueren Zeit* (2 vols.), vol.1. Berlin.

Drabkin, I. E., and Drake, S. (1960), *On Motion and on Mechanics.* Madison: University of Wisconsin Press.

Drake, S. (1973), "Galileo's Experimental Confirmation of Horizontal Inertia: Unpublished Manuscripts," *Isis,* 64:291–305.

——— (1978), *Galileo at Work: His Scientific Autobiography.* Chicago: University of Chicago Press.

Edwards, W. F. (1983), "Paduan Aristotelianism and the Origin of Modern Theories of Method," in L. Olivieri (ed.), *Aristotelismo Veneto e Scienza Moderna* (2 vols.), 1:206–220 Padua: Editrice Antenore.

Favaro, A. (ed.). (1968), *Le Opere di Galileo Galilei* (20 vols. in 21) 1890–1909. Florence: G. Barbèra.

Finocchiaro, M. A. (1980), *Galileo and the Art of Reasoning.* Boston Studies in the Philosophy of Science, 61. Boston: Reidel.

——— (1989), *The Galileo Affair: A Documentary History.* Berkeley: University of California Press.

Fredette, R. (1972), "Galileo's *De motu antiquiora*," *Physis* 14: 321–348.

Galilei, G. (1988), *Tractatio de praecognitionibus et praecognitis* and *Tractatio de demonstratione,* W. F. Edwards and W. A. Wallace (eds.). Padua: Editrice Antenore.

——— (1989), *Sidereus Nuncius* (*The Sidereal Messenger*). 1610. A. Van Helden (trans.). Chicago: University of Chicago Press.

Gilbert, N. (1963), "Galileo and the School of Padua," *Journal of the History of Philosophy,* 1:223–231.

Hill, D. K. (1979), "A Note on a Galilean Worksheet," *Isis,* 70:269–271.

——— (1986), "Galileo's Work on 116v: A New Analysis," *Isis,* 77:283–291.

——— (1988), "Dissecting Trajectories: Galileo's Early Experiments on Projectile Motion and the Law of Fall," *Isis* 79:646–668.

Lennox, J. G. (1986), "Aristotle, Galileo, and 'Mixed Sciences'," in W. A. Wallace (ed.), *Reinterpreting Galileo.* Washington, D.C.: The Catholic University of America Press, 29–51.

Mertz, D. W. (1980), "On Galileo's Method of Causal Proportionality," *Studies in History and Philosophy of Science,* 11:229–242.

Moss, J. D. (1983), "Galileo's *Letter to Christina:* Some Rhetorical Considerations," *Renaissance Quarterly,* 36:547–576.

——— (1986), "The Rhetoric of Proof in Galileo's Writings on the Copernican System," in W. A. Wallace (ed.), *Reinterpreting Galileo.* Washington, D.C.: The Catholic University of America Press, 179–204.

——— (1993), *Novelties in the Heavens: Rhetoric and Science in the Copernican Controversy.* Chicago: University of Chicago Press.

Naylor, R. (1974), "The Evolution of an Experiment: Guidobaldo del Monte and Galileo's *Discorsi* Demonstration of the Parabolic Trajectory," *Physis* 16:323–346.

——— (1976), "The Search for the Parabolic Trajectory," *Annals of Science* 33:153–172.

——— (1980), "Galileo's Theory of Projectile Motion," *Isis* 71:550–570.

——— (1990), "Galileo's Method of Analysis and Synthesis," *Isis* 81:695–707.

Olivieri, L. (1978), "Galileo Galilei e la tradizione aristotelica," *Verifiche* 7:147–166.

Randall, J. H., Jr. (1940), "The Development of Scientific Method in the School of Padua," *Journal of the History of Ideas,* 1:177–206.

——— (1976), "Paduan Aristotelianism Reconsidered," in E. P. Mahoney (ed.), *Philosophy and Humanism: Renaissance Essays in Honor of Paul Oskar Kristeller.* New York: Columbia University Press, 275–282.

Schmitt, C. B. (1969), "Experience and Experiment: a Comparison of Zabarella's View with Galileo's in *De motu,*" *Studies in the Renaissance* 16:80–138.

Settle, T. B. (1983), "Galileo and Early Experimentation," in R. Aris et al. (eds.), *Springs of Scientific Creativity: Essays on Founders of Modern Science.* Minneapolis: University of Minnesota Press, 3–20.

Shea, W. R. (1972), *Galileo's Intellectual Revolution: Middle Period, 1610–1632.* New York: Science History Publications.

Wallace, W. A., (1981a), *Prelude to Galileo: Essays on Medieval and Sixteenth-Century Sources of Galileo's Thought.* Boston Studies in the Philosophy of Science, 62. Dordrecht: Reidel.

——— (1981b), "Aristotle and Galileo: The Uses of *Hupothesis* (*Suppositio*) in Scientific Reasoning," in D. J. O'Meara (ed.), *Studies in Aristotle.* Washington, D.C.: The Catholic University of America Press, 47–77.

——— (1983), "The Problem of Causality in Galileo's Science," *Review of Metaphysics* 36:607–632.

——— (1984), *Galileo and His Sources: The Heritage of the Collegio Romano in Galileo's Science.* Princeton, N.J.: Princeton University Press.

——— ed. (1986), *Reinterpreting Galileo.* Studies in Philosophy and the History of Philosophy, 15. Washington, D.C.: The Catholic University of America Press.

——— (1987), "Science and Philosophy at the Collegio Romano in the Time of Benedetti," in *Atti Convegno Internazionale di Studio "Giovan Battista Benedetti e il suo tempo."* Venice: Istituto Veneto di Scienze, Lettere ed Arti, 113–126.

——— (1988), "Randall *Redivivus:* Galileo and the Paduan Aristotelians," *Journal of the History of Ideas,* 49:133–149.

——— (1992a), *Galileo's Logic of Discovery and Proof. The Background, Content, and Use of His Appropriated Treatises on Aristotle's* Posterior Analytics. Boston Studies in the Philosophy of Science, 137. Dordrecht: Kluwer.

——— (1992b), *Galileo's Logical Treatises. A Translation, with Notes and Commentary, of His Appropriated Latin Questions on Aristotle's* Posterior Analytics. Boston Studies in the Philosophy of Science, 138. Dordrecht: Kluwer.

——— (1995), "Circularityand the Demonstrative *Regressus*: From Pietro d'Abano to Galileo Galilei," *Vivarium* 33:76–97.

——— (1996), *The Modeling of Nature: Philosophy of Science and Philosophy of Nature in Synthesis.* Washington, D.C.: The Catholic University of America Press.

Zabarella, J. (1597), *Opera logica.* Cologne: Zetzner.

7

A Rational Controversy over Compounding Forces

GIDEON FREUDENTHAL

In this chapter, I attempt first to clarify the concept of "scientific controversy" in general, and then apply it to one historical case." The interest in scientific controversies as a phenomenon sui generis is relatively new. Of course, the fact that controversies arise in science was not just recently discovered. However, in a traditional interpretation of science, controversies *should not* have arisen. Conceiving of science as proceeding from a firm foundation of indubitable facts on the one hand, and according to evident logical principles on the other, there was only one path to follow, and this led toward truth. Disagreement resulted from *error* of one side (or both). "Legitimate scientific controversies" seemed to be an oxymoron.

In more recent approaches, science is conceived of as a complex social system. Institutions and the involved economic and political aspects, instrumental practices, power, and rhetoric come to the fore, rather than theory and justification. Theories now appear entrenched in practice, and since no clear-cut decision on individual truth claims is possible, it is widely accepted that all knowledge results from a (more or less controversial) "negotiation" both among practitioners of one group and also across the borders of such groups. However, the fact that science is a messy business does not recommend that the means of its study should be blunt and its historical narrative undifferentiated; on the contrary I propose therefore an analytical approach to science, especially to scientific controversies, that clearly distinguishes between the means of study and its object. The means should be analytical and differentiated; the phenomenon studied should be recognized and presented in its complexity. Both demands are compatible in science itself, and there is no reason why they should not also be compatible in science studies. I introduce an analytical concept of "controversy" below.

The case study presented will not fulfill all the demands mentioned thus far. The case studied—a disagreement between Honoré Fabri and John Wallis over the compounding of forces in the mid seventeenth century—was but one episode in a much more comprehensive controversy involving many scholars belonging to different "schools" and ranging over many topics. At stake were the concept of "force" in mechanics, the notion of a new kind of magnitude ("vectors") that seemed incompatible with the previously accepted notions, the role of conservation principles in science, among others. All these are not discussed here. Rather, I concentrate exclusively on Fabri's criticism of John Wallis's conception of compounding forces and on Wallis's answer to this criticism. I hence abstract from all noncognitive issues involved and from all overarching cognitive issues in which these two moves in the overall controversy were embedded.

What Is a Controversy?

"Controversy" is not a well-defined technical term: there are many different kinds of disagreement (in science as well). I'll use "scientific controversy" to refer specifically to a persistent antagonistic discussion over a disagreement concerning a substantial scientific issue that is not resolvable by standard means of the discipline involved. Thus, a discussion over a disagreement between scientists is not necessarily also a scientific controversy. All discussions that are not over scientific issues do not fall under "scientific controversy," nor do all "moves" within proper scientific controversies. A "controversy" has to be over a cognitive content and conducted with arguments; a *scientific* controversy must have a specific scientific content. The borders of "scientific" should be just as precise or vague as those of "science" in the same context.[1] Moreover, the discussion has to be persistent and antagonistic in order to count as a controversy: a disagreement resolved after a single exchange will not qualify, nor will a disagreement that prompts the discussants to begin a common research in order to resolve their disagreement. Moreover, a discussion will not qualify as a controversy either if it recognizably could have been resolved by the standard ways of the discipline involved, or if it could not have been resolved at all. In both cases there is no justification to conduct an ongoing discussion.[2]

Of course, many actual scientific controversies involve aspects that do not support their classification as proper controversies. But the definition above is not intended to characterize any historically known controversy, nor does it imply that such proper controversies exist. Rather, it refers to one component of some controversies, a component, however, that seems sufficiently important to justify its separate consideration. It is also true that analytically differentiated scientific controversies often overlap, such that the resolution of one and the beginning of the other are not noticed at all. The continuity of the historical event is quite independent from the distinctions introduced here.[3] However, these complications are irrelevant to the point of principle made above. They merely complicate the necessary interpretation. Such cases should be clearly distinguished from those in which a controversy in one area (say, science) indeed *cognitively* depends on differences of opinion in another domain (say, metaphysics or politics). In these cases, it is the controversy proper as defined above that necessitates the consideration of other domains that function as presuppositions of the controversy under scrutiny.[4]

Not every discussion over a scientific matter of fact turns into a proper controversy. Within an existing scientific community and under normal circumstances (for example, as long as techniques of measurement, etc., are not cast in doubt), a discussion over the height of Mount Everest would not qualify as a controversy. Such a question is to be settled by establishing "facts" or retrieving the relevant information that is readily available. To justify the conduct of a controversy, there must not be a readily available solution for the disagreement underlying the controversy.

But in fact, the requirements are farther reaching: if there is or seems to be a way to resolve the disagreement by some available information, or if there exists a recognized method to generate this information, be it by experiment or inference, then the discussion should not count as a controversy. A disagreement becomes a controversy because it cannot be resolved by the standard ways of the profession involved, for otherwise the conduct of a controversy (instead of generating the knowledge in question) does not make sense. Usually, therefore, a controversy is over what counts in the relevant community as an interpretation rather than over the truth of matters of fact or the validity of an inference; within a given frame of reference, differences of opinion on these issues can be cleared up and mere mistakes corrected.

Moreover, the conduct of the controversy must appear as a reasonable means to achieve progress in the matter. Questions that cannot be settled may prompt endless discussions, but these would not qualify as proper "controversies." This consideration can well distinguish between a scientific "controversy" and a religious controversy. Religion is of course the locus classicus of so-called controversies, but these are rather the classic example of endless disputes that can never be resolved.

A controversy exists when each side attempts to demonstrate the adequacy of its position by showing its capability in explaining the cases adduced as exemplifications, by demonstrating its fertility in explaining new cases, or by proving consistent with neighboring explanations, basic principles, and so on. On the other hand, it is also a standard technique in "controversies" to demonstrate the inadequacy of the rival conception in all these respects. "Critique" is an essential component of a "controversy."[5] Each side attempts primarily to prove its superiority over its rival, be it by demonstrating the adequacy of one's own position or the inadequacy of the rival's. Victory—relatively greater adequacy—is the immediate goal; objective adequacy can be the result thereof.

A controversy is possible because the competing views are embedded in the same general conceptual scheme, and the very same reason makes the controversy imperative. If two different views belonged to different conceptual schemes, then it would not only be difficult (if at all possible) to communicate, but the different views could not be incompatible. Controversies are called for because the adversaries adopt the same general conceptual system albeit in mutually exclusive interpretations, and it is the sharing of the same general system that renders them possible.[6]

However, it seems that the resolution of a controversy rarely simply proves one side right, the other wrong. Rather, a genuine resolution of a controversy is achieved within a reformed conceptual system that supersedes the system in which the controversy arose and yet showing a relatively greater influence of one of the positions over the other. This is the reason why controversies may prove productive for the theory or discipline involved.[7] This characteristic of the resolution of a controversy implies that the rival positions can be seen as two different interpretations of the same basic conceptual scheme and not as different conceptual schemes.

If the conditions above are satisfied, that is, the controversy is over a scientific issue that cannot be resolved by standard procedures, then the resolution can indeed consist in scientific development, either conceptual or in instrumentation and so on. A new conceptual system is said to have conceptually developed if it can reconstruct the previous positions but cannot be reconstructed within the previous system.[8]

A last and difficult problem remains. I have characterized "scientific controversy" on the basis of epistemic issues and independently from the social setting involved. It seems that this approach implies the commitment to some kind of Platonism such that the mere existence of incompatible views suffices for a controversy to exist and that the real exchange of pro and con arguments is not required. Would, then, my unresolvable disagreement with an Aristotelian thesis count as a controversy, even though Aristotle cannot answer the criticism and point to weaknesses of my alternative position? A first answer to this question has already been given above by the requirement that a controversy be persistent. But there is also a point of principle attached to this issue.

It seems to me that here, too, the analytical procedure and the real state of affairs are confused, only in the reverse. The fact that all cognitive activities are performed by living people in a social context should not lead to reducing science to a social activity analyzed exclusively with sociological methods, nor should the application of logical methods entail the conclusion that the existence of a scientific controversy does not depend on social agents. Hence, in reference to the example given, my analysis of a controversy should proceed from the fact of the existence of an antagonistic, persistent discussion, thus clarifying (as much as possible) what part of the discussion forms a scientific controversy. The existence of the controversy as a social phenomenon is thus presuppose by this analysis.

The case study below displays a pattern that does not look at all like a controversy, but rather like the imagined case of my disagreement with Aristotle. It nevertheless displays a quite typical pattern of controversies: instead of an ongoing exchange between two parties, we find only the criticism of the views of B by C. However, the views of B are similar to those of B1, B2, and so on, proposed at the same time, and are expressed in criticism of A's views (also shared by A1, A2, etc.). What appears here as a simple criticism of B by C is in fact one thread of a complex controversy. In accord with the view advocated here that the logical analysis of a controversy may abstract from its social components, this episode can by all means be studied as a controversy.

To sum up, "scientific controversy" refers to an persistent antagonistic discussion over a disagreement concerning a substantial scientific issue that is not resolvable by standard means of the discipline involved. To count as a "scientific controversy" the discussion should not be reducible either to a theoretical disagreement or to an antagonistic discussion. Both characteristics have to apply together, and there must not be a resolution in sight.

How Are Rational Controversies Possible?

The requirements discussed thus far may seem merely normative, but they actually explain some important features of controversies. Consider the dynamics of controversies and the recurrent complaint over "misunderstandings." Misunderstandings are often the symptom of a disagreement not yet uncovered. They show that the meaning attached to propositions or terms is different. Of course, participants often complain over alleged misunderstandings as a tactical move, but partially sharing conceptual systems *must* generate misunderstandings. And trying to reach the root of misunderstandings initiates an exploration of the conceptual system involved. This is but another aspect of the conduct of a proper controversy. Whereas an antagonistic discussion may often consist in the repetition of the same claims on both sides, such a dispute will not count as a proper controversy, in which arguments must be adduced. A disagreement that arises between different interpretations of the same general conceptual system prompts the attempt to uncover its sources. Since the participants share the same conceptual system, the controversy often takes the form of an exploration of lineage of premises supporting the originally contested issue. We therefore encounter in controversies the tendency to move "deeper," to foundations (methods, principles, etc.).

In conducting a controversy, the opponents search the source of their disagreement. The conduct of a controversy hence consists in an exploration of the conceptual system involved, often in its elaboration, because there is no reason to presuppose that the entire system to which a system belongs exists in actu. This does not merely mean that the individual is not necessarily in command of the entire system, but that the system is not yet necessarily fully developed—both in its extension and in its presuppositions and implications—and exactly this may happen in the course of a controversy.

The nature of the ongoing discussion also shows why a controversy is both similar to and different from a competition between rival scientific theories. A controversy is not "a normal scientific dispute about which of two competing views was best equipped to deal with the empirical data" (Papineau 1977, p. 144, referring to the "*vis-viva* controversy"). The rivalry between competing theories presupposes the existence of two distinct theories such that both can be tested by the empirical data; a "controversy" is typically over the question of which view is the "correct" interpretation of a common theory. This crucial difference is conspicuous in the form in which the disagreement is negotiated: the ongoing discussion and the specific dynamics of controversies (the exploration of lineage of premises supporting the originally contested issue) as well as their style (proliferation of "misunderstandings," etc.). However, a controversy may end in a competition between theories. This is the case when the controversy brings to pinpoints the principal points of disagreement and also the fact that the latter cannot be resolved; a separate exploration of the different possibilities and the test of which view is "best equipped to deal with the empirical data" can then ensue.

Note finally that the notion of "controversy" as developed above points to the fact that different meanings are an essential feature of controversies. Such partial differences of meaning within a shared general conceptual system explain both the possibility of communication and its obstacles, both the possibility of the controversy and its necessity. The controversy is possible because it arises from a disagreement over a specific issue, and this testifies to a common ground that can also serve as such in the controversy. The controversy is necessary because the existence of mutually exclusive positions within the same conceptual system is intolerable in science. Finally, the considerations above also explain the abundance of complaints over "misunderstandings" in controversies.

The following presentation of the controversy between John Wallis and Honoré Fabri should exemplify the general characterization of scientific controversies sketched above.[9]

The Wallis-Fabri Controversy over the Compounding of "Impetuses"

In the end of the second part of his *Specimen Dynamicum* (1695), Leibniz discusses different questions concerning the theorem of compounding of motions; the last sentences of this discussion (and of the essay) read as follows:

> It is also understood from these matters that the composition of motions or the resolution of one motion into two or any number whatever can safely be used, even though, according to Wallis [*apud Wallisium*], one ingenious man [*vir ingeniosus*]) has raised plausible doubts. For the matter certainly deserves to be proved and cannot be assumed to be known in itself, as many have done. (Leibniz 1969, p. 450; 1849 6:254)

A short time afterward, in his review of Wallis's *Opera Mathematica* (1695), Leibniz mentions again the critique of this "vir ingeniosus"; this time the critiquer is mentioned by name: Honoré Fabri.[10] In the following, I will endeavor to show that Wallis and Fabri conduct a typical controversy, the first (but not final) resolution of which was later achieved by Leibniz.

The Traditional Problems of Compounding Forces

The topic of the controversy between Wallis and Fabri is the theorem of the compounding of forces as it was introduced into early modern physics. The theorem first presupposes that "force" is a directed magnitude, a vector of sorts, and can therefore be represented by a directed straight line. The length and the direction of the line represent the magnitude of the force and its direction, respectively. The theorem states that the effect of two simultaneously applied forces will be equivalent to the effect of a single force such that if two sides of a parallelogram represent the component forces, then the diagonal of the parallelogram represents the resultant force. The theorem is therefore also known as the "parallelogram rule."

Now, a related theorem was traditionally known: the parallelogram rule for velocities. It states that velocities of a body may be compounded according to the parallelogram rule to yield the resultant velocity. If the time in which these motions are performed is equal, then the rule presupposes little but elementary geometry. This theorem was known in antiquity and can be found, for example, in the pseudo-Aristotelian *Mechanica.* However, when forces are involved the issue no longer concerns geometry but causality. Once this was realized, there were good reasons to doubt whether the parallelogram rule is also valid for forces. The principal doubt concerned the "principle of independence" or "superposition" stating that the simultaneous action of two forces is equivalent to their successive action, hence that the effect of a force is independent of the simultaneous application of other forces.

Traditionally, forces that neither coincide in their direction nor are orthogonal to each other were conceived as "contrary" and "opposed." Now, being "opposed" meant that they "struggled" with one another and that the stronger prevailed and determined the ensuing motion. Hence, no compound motion existed, because one force always prevailed over the other. A simultaneous action of two forces was understood to result in subsequent effects. Thus, for example, the trajectory of a projectile shot obliquely upward was conceived as consisting of two subsequent distinct motions: one in a straight line in the direction of the shot, and an ensuing vertical fall downward. The physical impossibility of compound motion was also reflected in language to form a conceptual inconsistency—at the time one would speak of a "contradiction." Thus, saying that two different forces simultaneously act on the same body was understood to imply that the body "has" two motions, hence simultaneously moves in two different directions—and this certainly is inconsistent (although not evidently a contradiction).

Referring to the special albeit most important case of oblique projection, Niccolò Tartaglia succinctly formulated the received view:

> Proposition V
> No uniformly heavy body can go through any interval of time or of space with mixed natural and violent motion.

The argument provided refers to the essential properties of natural and violent motions, respectively: the projectile "would go increasing its speed according to its share in the natural motion"[11] and similarly that it would diminish in speed according to its participation in the violent motion[12] "which would be an absurd thing, for such a body would have to be increasing and diminishing in speed at the same time" (Tartaglia 1537; Drake and Drabkin 1969, p. 80).[13]

Recall, however, that the problem does not end with Tartaglia or with "preclassical mechanics" in general. Galileo himself restricted his endeavors to compounding vertical and horizontal, for example, orthogonal motions, which traditionally were conceived as indifferent rather than opposed to each other (see Damerow et al. 1991, ch. 3). Centuries later, in Bertrand Russell's work we meet with late traces of the original problem, and a few years ago the same problem was prominent in discussions over realism (Russell 1900, p. 50; 1903, pp. 451, 449; Cartwright 1983). In its later version, however, the problem belongs exclusively to philosophy and is not discussed any more in physics.

After this short sketch of the background, I now turn to the controversy between Wallis and Fabri.

Fabri's Criticism of the Compounding of Forces and Motions

Fabri's criticism of the compounding of motions does not call into doubt the results achieved by the application of the parallelogram rule. Aa early as 1646, in his *Tractatus physicus de motu locali,* Fabri derives the correct law of reflection by applying the parallelogram rule within the conceptual system of his physics (Fabri 1646, p. 243). The same is true for the case of the motion of a body simultaneously acted upon by two forces. Fabri does not at all doubt that the parallelogram rule is valid also in this case. The critique developed by Fabri is not directed at the parallelogram rule; rather, it is directed at the interpretation of the same as referring to the compounding of "impetuses" or "motions."

The Inconsistency of the Parallelogram rule with the Rules of Addition. As said above, Fabri rejects the notion of compounding "motions" or "impetuses" (which are proportional to the speeds) because the sum of the (scalar) impetuses (or speeds) is not equal to the resultant impetus (or speed) produced. Fabri's main concern is hence to avoid the antinomy implicit in the notion that the impetus is, on the one hand, the cause of motion and proportional to speed, while on the other hand, the sum of two (scalar) impetuses is not proportional to the velocity produced. In short, $a^2 + b^2 = c^2$; $a + b \neq c$.

Fabri's own interpretation of the parallelogram rule shows him to basically be a Cartesian: impetus is proportional to speed and indifferent to direction. Every actual motion is determined toward some direction; the determination of a motion is a property of the same and is of a magnitude proportional to the speed. Determinations may be compounded and resolved according to the parallelogram rule (there is no law stating the conservation of "determination" in interaction) and the diagonal represents the resultant determination (and, of course, the speed, too). Since the impetus that is proportional to the speed (and determination) of the resultant is in all cases with the exception of coinciding forces smaller than the sum of the component (scalar) impetuses, there arises a seeming antinomy. This is resolved by introducing the proposition that the difference between the sum of the impetuses of the components and the impetus of the resultant will be destroyed:

> Indeed when there are two impetuses determined along different lines but not diametrically opposed, they conflict according to the different degrees of opposition. . . . Therefore since the whole impetus does not have the whole motion, which this double determination hinders, some degrees are destroyed, lest they be in vain [*frustra*]. (Fabri 1646, p. 70; see also Fabri 1669, pp. 194–195, 198)

It does not come as a surprise that Fabri is able to calculate the correct results on the basis of this conception. If the impetus has to adjust according to the results of compounding determinations, and if determinations are directed velocities, then the actual results must be the same as if Fabri compounded (vectorial) velocities (at least as long as only one body is involved and mass does not have to be considered).

This result gives rise to the criticism that at least one of the concepts involved, probably "impetus," is superfluous. "Determination" (or "momentum") and "velocity" would have sufficed. The question therefore arises as to why Fabri does not give up altogether the concept of a nonconserved impetus, which is indifferent to direction, when his argument rests exclusively on the concept of determination. The reason is, I believe, that the physical concept "determination" referred, just as its logical origin, to a "further specification" of a concept (and ontologically, to a specifying property of an entity), in the present case, of speed or motion. A *directed* speed is first of all a speed, the direction of which is specified by its determination. A determination without an underlying speed would be much the same as the Cheshire cat's smile shining in the dark when the cat itself has faded away.[14] Thus, the application of the parallelogram rule to dynamics leads Fabri to the concept "determination" as a vector quantity, and to a concept of "impetus" as a scalar quantity. Loosely speaking, we can say that the motivation behind these two concepts is satisfied by our concepts of "momentum" and "kinetic energy." The assertion that both entities are conserved and the application of both to determine the outcome of physical interactions were possible only after Leibniz substituted for "impetus" or "motion" the concept of *vis viva*, which is proportional not to velocity but to the square of velocity.

The Reality of the Components. Having shown that compounding impetuses according to the parallelogram rule is incompatible with the notion of the conservation of the (scalar) quantity of impetus, Fabri ventures to show that maintaining the reality of the components would violate the parallelogram rule. The argument seems to rest on a blatant mistake, but is quite sound, as I argue below.

Suppose two equal impetuses are applied to a body in opposite directions; the body will remain in rest, and according to Fabri, the impetuses must be destroyed lest they be in vain. Fabri now attempts to prove that according to the alternative interpretation, namely, that the impetuses are not destroyed but compounded, the parallelogram rule is violated (fig. 7.1). (According to the latter interpretation, the body will of course be at rest as well.)

> Let there be two equal impetuses in the ball C and let them be preserved; it is clear that the ball will remain in rest. Let the impetuses be as AC, DC and let a third one as BC be

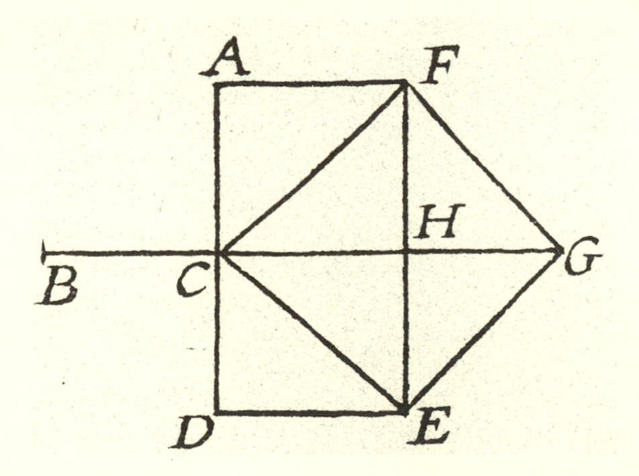

Figure 7.1 Fabri, 1669, Figure 72.

> impressed; out of BC, DC there will arise CF, and out of BC, AC there will arise CE; and out of CF, CE there will arise CG. Hence the mixed motion will be CG, whereas the simple is CH. (Fabri 1669, p. 202)

Now, the argument seems to rest on a blatant mistake, since according to our concepts Fabri is simply putting the third applied impetus (BC) twice into account. And this indeed is Fabri's criticism: you maintain that impetuses can be compounded and are real, then the absurd consequence follows that the presence of two equal and opposite impetuses would double the velocity produced by the third.

Fabri's contention that the compounding of forces is a theorem on causal relations and not "addition" (a mathematical operation) has two intimately connected consequences: first, that imparting motion to a body does not mean that an impetus has been transferred from the mover onto the projectile, and second, that the impetuses of the mover and of the projectile, which are not identical, are therefore also not necessarily equal.[15]

Resolving Arising Difficulties. Fabri's conception of the destruction of superfluous impetus and of compounding of determinations meets with difficulties when the collision of bodies in motion (and therefore with impetuses) is considered. Suppose that two equal elastic bodies A and B move in the same line in opposite directions. The bodies collide and rebound with their original speeds in opposite directions. If the impetus of A produces in B a new impetus and a new determination in the opposite direction and vice versa, then each body must have in the instant of impact (and therefore, in the absence of a subsequent interaction, later as well) two impetuses and two determinations: the original and the new ones, and the speeds must either be double the original ones or cancel out each other and equal zero. Both solutions are inadmissable.

Fabri resolves the difficulty in the following way. In the moment of impact each body really has two impetuses (which are indifferent to direction) but three determinations: The original one, and two in the opposite direction. The first of these is produced by the resistance of the other body independent of the body's motion (just as the surface produces such a determination in the case of reflection), and the other is produced together with the new impetus by the impetus of the other body. The result of the compounding of these determinations $[1 + (-1) + (-1)]$ is hence one determination in the reverse direction accompanying two impetuses, but since one impetus is then in vain (frustra), it will immediately be destroyed (Fabri 1669, p. 197) Fabri is thus able to explain why the bodies will rebound with speeds equal to their original ones without maintaining that the original impetuses are conserved.

Finally, Fabri also criticizes the notion of the resolution of a motion into components. First, a simple motion cannot be considered compounded: it is simple. Second, there is no justification to prefer an orthogonal resolution to any other. The (Cartesian) justification adduced in favor of the thesis criticized is that the surface "resists only the perpendicular motion and not an oblique one as well, which is very false" (Fabri 1669, p. 195). And indeed, the argument is false if "motion" is a scalar magnitude and if "resistance" is conceived as preventing the continuation of the motion. And since

Fabri does not see any physical justification for the standard orthogonal resolution of "motion," he concludes that the resolution is completely arbitrary. Naturally, Fabri too uses this very same resolution, but he resolves determinations rather than motions. And Fabri believes that his concept of "determination," which is defined as directed magnitude, provides a conceptual justification for this specific resolution, a justification that cannot be given for "motions" with no direction. The justification refers to the traditional (Aristotelian) concept of "opposition," but opposition of directions is not conceived as dichotomic. "Determinations" can be rather "more or less" opposed, and their combination is not impossible but produces a third "intermediate" entity.[16]

Wallis's Answer to Fabri

The Context of Wallis's Discussion with Fabri. Wallis discusses problems related to the compounding of motions in several contexts. First, Wallis introduces in chapter 10 of his *Mechanica* ("De Motibus Compositis, Acceleratis, Reterdatis, & Projectorum") the addition and subtraction of motions along a line, either in the same or in opposite directions, respectively. He then deals with motions that are not on the same line, and it is here that the parallelogram rule is indiscriminately introduced for velocities and forces. Wallis states that the resultant is proportional to the diagonal of the parallelogram the sides of which are proportional to the components:

> When a mobile receives two directed impetuses; say from its position along two right lines which make an angle; and when the speeds are uniform and in the same proportion to each other as the lengths of the sides of the parallelogram; the mobile will traverse the diagonal of the parallelogram with a speed which is in the same proportion to the given ones as the diagonal to the sides of the parallelogram. (ch. 10, prop. 6; Wallis 1695, p. 998)

Wallis not only supposes here that impetuses and speeds are proportional to each other, but also tacitly assumes that forces are indifferent to each other. More important, even, is the fact that Wallis speaks of "directed impetus." Having thus a concept of a directed magnitude producing speed and proportional to it, he does not need an additional concept such as Fabri's "determination." In this and in the following paragraphs he generalizes the parallelogram rule to motions that are not in the same plane, to accelerated and retarded motions, and so on, and does not mention Fabri in this context.

It is only when Wallis discusses reflection in chapter 13 ("De Elatere, & Restitione seu Reflexione") that he mentions Fabri's critique and justifies his own conception. Wallis explains the rebounding of a body hitting an obstacle with reference to "elasticity"; his presentation is very clear: "When a heavy body in motion impinges on a firm obstacle, and when one or both of the bodies are elastic, it will reflect or rebound with the same speed [*celeritas*] with which it approached and in the same right line" (ch. 13, prop. 1; Wallis 1695, p. 1018). Wallis makes three points to elucidate this proposition:

> I say first, that this pressure will last continuously until the elastic force, which is contrary to the compression, is equipollent to the compressing force. . . .
>
> This compressing force is double the momentum of the impinging body. This is so because the force transferred by the impinging heavy body is equal to its momentum

> (namely, when its motion is completely exhausted), say, mrPC. But because of the equal resistance of the obstacle (which confers to the compression an equal force although in the opposite direction) namely another mrPC . . . the total compressive force is, therefore, 2mrPC.
>
> . . . and one half or one of the mrPC is spent on the obstacle . . . and the other mrPC on the pressing [*advectum*] body. Therefore, since it is by the same force, the body A will deflect with the same speed with which it came. . . .
>
> I say finally that the body will rebound in the same straight line. Since the spring was bent inward by the directly impinging body and since it is by the very same elastic force that the body will restitute its former form, per. def. 1 of the same [chapter] (and therefore will return the same way it approached), the force will impart to the rebounding body the same direction which it received from him, albeit in the contrary sense. And therefore the body will return via the same straight line. (Wallis 1695, p. 1019)

Thus Wallis shares in fact some principal considerations of Fabri: Wallis, too, accepts the notion that in reflection a new cause must be introduced to account for the new (reverse) motion, but since he conceives "impetus" as a directed force, he does not need the concept of "determination" and applies the parallelogram rule to the impetuses themselves. It is in the context of the resolution of impetus into components that Wallis criticizes Fabri.

Wallis's Resolution of Momentum. Wallis discusses the resolution of the "motion" in the context of "reflection":

> Proposition II
> When a heavy body in motion impinges obliquely on a firm obstacle; and let one of both or both be elastic; it will reflect with the same speed (& and in the same plane) in such a way that the angle of reflection will be equal to the angle of incidence. (Wallis 1695, p. 1021)

The proof rests on proposition 6, chapter 10, that is, on resolving the oblique motion into two, the one parallel to the surface of the obstacle, the other perpendicular to it. Since the parallel component meets with no resistance, it will remain unchanged; the perpendicular component will produce an elastic force, which will make the body rebound in the same line and with the same speed. These two new components can be represented by two sides of a parallelogram, and its diagonal will represent the resultant, that is, the motion of the rebounding body. It is then easy to show that the angle of incidence equals the angle of reflection.

In the scholium to this proposition, Wallis answers Fabri's objections concerning first the fact that a simple motion (that is, along a straight line) cannot be considered as a compounded one, and second that since the resultant can be the diagonal of infinitely many parallelograms, the proposed resolution (for example, into a parallel and a perpendicular component) is arbitrary.

> If a beginner (or somebody who is superior to a beginner) asked, why I affirm without justification that this most simple motion . . . is composed of two? Or else, if I want it compounded, and since it can be compounded with equal justification in thousand other ways, why I claim without justification that it is compounded in this way neglecting all others (neither proving that it is composed nor that it is not composed in any other way)?

> I answer that no motion can be so simple that it cannot be resolved into many components. And if I preferred this way to others, I did so according to my right, to apply that composition (since I could have applied any) which is most convenient for the present purpose. (Wallis 1695, p. 1022)

Wallis then proceeds to give examples of how the sum 12 can be composed in different ways out of integers and likewise out of fractions, and concludes by mentioning Fabri as having raised the doubts that force him to give these clarifications (Wallis 1695, p. 1023).

The Shibolet of the Controversy: Wallis's Statical Dynamics. It is quite clear that Wallis does not understand Fabri's objections since for him "force," "momentum," and "motion" are directed magnitudes, and since he has no principle of conservation of "motion" or "momentum," he also has no scruples about applying the parallelogram rule to "momentum." The fact that the (scalar) sum of the components is not contained in the resultant is not even mentioned by him. Nor does Wallis seem to have been disturbed by the logical problems concerning the notion of a compound motion, that is, by the question of how the reality of the component causes can be affirmed without any observable component effects.

However, Wallis provides an interesting clue as to the background of his understanding of compounding forces. In his discussion of Fabri's objections, Wallis gives examples of two legitimate resolutions into components: first, the possible resolutions of the number 12, and second, the momentum mrPC. The examples given for the possible compositions of the number 12 out of integers are $3 \cdot 4$, $2 \cdot 6$, and $1 \cdot 12$; Wallis now argues that if 12 should be resolved and the divisor 2 is given, then there is only one correct answer, namely, that 12 is composed of $2 \cdot 6$. Wallis hastens to add that he does not maintain that the other possible compositions were false but that they are of no use in the present task (*quod ad praesens negotium sint inutiles*; Wallis 1695, p. 1022). An analogous justification is given for the resolution of a momentum: if the momentum is mrPC [where "P" stands for weight (pondus) and "C" for speed (*celeritas*)] and the weight is given as 2mP, then it necessarily follows that the velocity is 1/2rC.

Note that this nonarbitrary resolution of the "momentum" mrPC into components does not refer at all to its resolution into independent momenta (each of which has the dimension PC) but rather into the "components" weight and velocity! It is therefore clear that the numerical and the momentum example are instances of *multiplication*, not of *addition*. In the case of momentum it is not the existence of two independent momenta that is assumed but that of weight and velocity.

Now, both methods are mathematically equivalent. The same result is obtained whether the parallelogram rule is applied and all momenta add up (PC + PC), or first the center of gravity is determined by pure geometrical methods (and its value calculated) and then the momentum is found by multiplication with the arithmetical mean of the speeds of all the points $[(\sum P \cdot \sum C)/n]$. These procedures are equivalent, but only because the assumption of the independence of the forces from each other is in fact valid. Wallis avoids the whole issue of the independence of the components by applying without further ado a method that is only valid because the independence of the components does indeed hold, that is, by presupposing it.

Wallis's disregard for the presupposition underlying his concept of compounding forces is enhanced by his understanding of "momentum." For Wallis it means both our present-day *momentum* and also the "moment of force" or torque. The latter concept was at the center of statics and was considered evident. In Wallis's words:

> I call the momentum of whatever magnitude (in relation to the axis of conversion or what stands for it) this one which arises out of the magnitude and its distance (or out of the magnitude and the distance of the center of gravity) from that axis. And therefore, when the magnitude and the distance are given, the momentum is given, or by the momentum and one of the other magnitudes, the remaining is given. That is what you call the method of momenta. (Wallis's letter to Leibniz, July 30, 1697; Leibniz 1849, IV:33; Wallis 1695, p. 682)

While here Wallis was introducing *weight* and *distance* from the axis as the components of momentum, he also explicitly defines "momentum" as the product of *weight* and *speed* albeit in the context of statics:

> And this is the foundation of all machines for facilitating motion. For in whatever ratio the weight is increased, the speed is diminished in the same ratio; whence it is that the product of the weight and the speed for any moving force is the same. (Wallis's letter to Oldenburg, November 15, 1668; Oldenburg 1965, V:168)

Thus, Wallis uses "momentum" (mrPC) for two concepts: the product of weight and distance from the fulcrum, and the product of weight and velocity. These two magnitudes are interchangeable only if the distances traversed are proportional to the (uniform) velocities, and this is the case in statics. In the case of a lever in equilibrium, the distances from the point of application of the force to the fulcrum are proportional to the (virtual) speeds with which these points would move if the lever were tilted. And since the weights attached to the lever in equilibrium are inversely proportional to the distances from the fulcrum and to these velocities, the measures of force as either the product of weight and distance or as the product of weight and velocity are interchangeable. As long as such a generalized concept of "force" or "momentum" in statics and dynamics seemed unproblematical, the independence of the "momenta" from each other, which in fact underlies the different proofs for the law of the lever given since antiquity and which were never called in doubt, seemed unproblematical as well.[16]

Thus it is the concept of "momentum" as generalized from statics to mechanics that enabled Wallis to circumvent the question of whether simultaneously applied forces are independent of each other in producing their effects and may be added; Wallis adds weights on the one hand and velocities or distances on the other and "compounds" the momentum by multiplication.

Conclusion

Looking back on Fabri's criticism and on Wallis's rebuttal, the different positions all appear well grounded in the then existing shared conceptual system of mechanics. The arising disagreements and inconsistencies were not due to "mistakes" of one party, but were rooted in deficiencies of the conceptual system itself. Expressed in the language of classical mechanics the main problem involved was the introduction of one example ("force") of a new kind of magnitude, "vectors."

The introduction of force as a vector met with two difficulties: one common to all innovations, one specific to this development. Introducing a new concept effects a whole cluster of other concepts bound together both by semantic connections and by application to the same phenomena. The introduction of "force," for example, necessitated a reconstruction of the conceptual system involved to construct again a consistent whole, for instance, to reconcile the basic principles of conservation with the existing magnitudes (mv and mv^2). We can thus also determine how and when this controversy was resolved. It was first resolved by Leibniz, who further introduced the magnitude *vis-viva* (mv^2), thus integrating both concepts, the scalar and the vector magnitudes referring to "the force of a body in motion" in a consistent theory that was governed by basic conservation laws for both magnitudes.

And this touches on the specific problem on which this controversy rested: the necessity of recognizing that the very same phenomenon—the efficacy of a moving body—can be conceptualized in two very different ways, both referring to two distinct entities. On the background of the many discussions over the concept of "force" in which the episode involving Wallis and Fabri was embedded, it seems that it fulfills all requirements initially discussed to qualify as a controversy: the discussion was over a scientific issue, even an issue at the core of the system involved; it was not primarily over matters of fact and involved differences in meaning of the concepts ("impetus," "compounding") due to different "centers" of the conceptual systems (impetus physics in Fabri's case, statics in Wallis's); it also involved vital interests of the participants that could not be discussed in the present context: "Newtonians" (as they were to be called later) and "Cartesians," "British" and "Continental" loyalties were involved as well as different attitudes to so-called metaphysical issues (for example, conservation principles). Later stages of the controversy also involved issues of political commitments both of principle and of current politics.[17] Last but by all means not least, there indeed was no readily available solution to the controversy, and the stages of the resolution worked out in the framework of this and subsequent controversies in fact form a series of conceptual developments.

Notes

A previous version of this chapter has been criticized, without much avail, by Skúli Sigurdsson.

1. The wish to distinguish between a "scientific controversy" and a "controversy in science" is the motivation behind Mcmullin's criticism of E. Mendelson (see Mendelson 1989; Mcmullin 1989). Thus, Mcmullin declares that "[a] scientific controversy is concerned with matters of belief" (p. 51) and that "epistemic factors are more likely to be determinative" (p. 88). However, attempting to do justice to the "nonepistemic factors," he observes that a controversy is "a human act, a social episode," in contradistinction to explanation and justification, which are "expressible as propositional relationships" (p. 54). He therefore concludes that "the understanding of controversy belongs in the fist place to the historian, then, [and] not to the logician" (pp. 53–54) but later adds that the historian may choose to speak in the abstract of reasons and proofs rather than in terms of beliefs and decisions (see p. 58).

Thus all sinks anew into darkness: a scientific controversy is both a social and logical issue and the hostorian should be a social historian as well as a historian of ideas and a logician. Of course, in practice all these functions are usually performed by one person, but analytically they have to be clearly distinguished.

In Gil (1990) the term "controversy" is not discussed but rather used according to an intuitive understanding.

2. Thus an accepted correction of a mistake will not count as a "controversy."

3. For an example of this phenomenon, see Michael Ruse's chapter 13 in this volume. See also the discussion of "nonstandard epistemic considerations" in McMullin (1989, pp. 62ff).

4. For an example of this phenomenon, see Freudenthal (1986) and Shapin and Schaffer (1985).

5. See *Science in Context*, vol. 10, 1 (Spring 1997): *Models of Critique*.

6. I accept Davidson's (1973) conclusion (not necessarily his arguments) that a "conceptual scheme," in the sense that it is strictly intranslatably into another, does not make sense (p. 134) and that partial failure of translation between different conceptual schemes (forming part of a general common one) does make sense (p. 140). However, I differ from Davidson in two points. First, I see no reason to consider conceptual schemes in the established moderate sense as being "only words apart" (p. 134). The fascinating topic of scientific controversies is located exactly in the domain of partial failure of translatability and cannot be dispensed with as dealing merely with differences of words. Second, all of Davidson's examples of translatability, both the anthropological ones (into English by Benjamin Whorf) and the scientific ones (into "postrevolutionary idiom" by Thomas Kuhn) (p. 130) are cases of translations into "more advanced" languages. Hence, translatability may apply in one direction only, and in this case very strong claims both of intranslatability (from the advanced into the less advanced idiom) and of cognitive development could be attached to this observation. Davidson brushes off the issue with a short parenthetical remark: "(I shall neglect possible asymmetries)" (p. 131).

7. See Kitcher's chapter 1 in this volume, where he argues that the resolution of controversies consists in a more advanced consensual position.

8. Of course, emphasizing the contribution of controversies to scientific development does not mean that this is the goal of the participants. On the contrary, a controversy is not merely a form of disagreement, but also one form of *agonia*, of contest, of antagonism. The immediate aim of the parties is to win a victory over the adversary (concerning a cognitive disagreement and by arguments!), notwithstanding the existence of mediated aims such as the discovery of "truth" or the appointment to a position. The agonistic character of a controversy becomes clear when considering that if both sides in a discussion decide to explore the issue jointly this cooperative effort does not count as a controversy.

9. Evidently, I am very much in accord with the intertion to interpret controversies as due to differences in meaning (Papineau 1977). Papineau's adoption of a holistic theory of meaning lock, stock, and barrel *completely* levels the differences between empirical and theoretical as well as between the core and peripheral components of a theory (Papineau 1977, pp. 114–115). As a result thereof and of the implicit presupposition that a theory consists of a limited set of explicit clear and distinct propositions, "controversy" is equated with a "competition between two theoretical systems" (p. 116) to be decided according to idealized criteria of empirical adequacy, and so on, and is not considered any more as a specific phenomenon.

Thus tentative as it is, my present attempt to clarify the notion of "controversy" necessarily refers to and depends on positions concerning fundamental and controversial philosophical issues such as meaning. Needless to say, I cannot argue adequately for these positions here.

10. Denique & de vi Elastica agit, ostenditque ex ea oriri, guod corpora inter se concurrentia post concursum non simul procedunt, sed a se invicem dissiliunt. Neque enim in moto incepto vim nescio quam esse collocandam, nex in resiliendo motum novum absque nova cuasa incipere. Et quoniam ad ostendendam aequalitatem anguli incidentiae & reflexionis, assumerat motus obliqui compositionem ex perpendiculari & parallelo ad superficiem excipientem; respondet Honorato Fabro, qui in Tractatu de Motu lib. 6 & Dial. Phys. 2. hanc compositionem in dubium revocaverat. (Leibniz 1969, p. 255)

The mathematician William Neile was another scientist who had raised plausible doubts concerning the compounding of motions and who was engaged in a long controversy with Wallis, a controversy of which Leibniz may have known. On June 12, 1671, Oldenburg informed Leibniz that Wallis believed William Neile's views to be similar to those Leibniz expressed in his *Hypothesis physica nova* (see Leibniz 1849, I:19–23; Oldenburg 1965, VIII:99–104; Wallis letter to oldenburg, April 7, 1671: Oldenburg 1965, VII:559–564). Since Leibniz was in London from January 24, 1673, until the end of February and met at least twice with Oldenburg (shortly after his arrival and on February 13) and even attended two sessions of the Royal Society (February 1 and 18), it is likely that he asked for more details concerning Neile's views, which allegedly were similar to his own, especially since Oldenburg had told him that "Neile's reflections on the principles and nature of motion . . . may be found lodged in the Society's archives" (Oldenburg 1965, VIII:103). Space does not permit me to discuss this controversy here.

11. Proposition I: "Every uniformly heavy body in natural motion will go more siftly the more it shall depart from its beginning or the more it shall approach its end."

12. Proposition II: "A uniformly heavy body in violent motion will go more weakly and slowly the more it departs from its beginning or approaches its end."

13. Drake and Drabkin comment (1969): "This line of reasoning, which remained the principal obstacle to an understanding of actual motions from the time of Aristotle to that of Galileo is not so weak or implausible as it may seem to us. Under the assumption that deceleration is an essential (not an accidental) characteristic of violent motion, and acceleration an essential characteristic of natural motion, the argument is most compelling" (pp. 80–81, fnt. 20).

14. The similar position of Descartes is discussed in detail in Damerow et al. (1991, chs. 2 and 4).

15. That compounding forces (accelerations) is not simply addition still vexed Russell: "But this composition is not truly addition, for the components are not *parts* of the resultant" (Russell 1903, p. 451). See also the discussion in Nagel (1979, pp. 386–397). On conceiving of impetus as transferred from the mover onto the projectile, see Wolff (1978). "Thus the impetuses determined along two different lines in such a manner struggle in proportion, thus those conflict less whose lines approach closer to coincidence; but those conflict more whose lines approach closer to opposition. And the proportions of the diagonals follow the same rule, all of which follows from what has been said" (Fabri 1646, p. 67, theorem 142).

16. Significantly, this origin of the early concept of force was stressed by Leibniz when he explicitly criticized this transposition of the concept form statics to dynamics. The "Brevis Demonstratio" (1686) begins thus: "Seeing that velocity and mass compensate for each other in the five common machines, a number of mathematicians have estimated the force of motion by the quantity of motion or by the product of the body and its velocity" (Leibniz 1969, p. 297; 1849, VI:117). Leibniz's whole argument in this and subsequent papers is that in accelerated motion the distances traversed and the velocities are not proportional to each other and that there is therefore a difference whether one defined force as ms or mv.

17. A study of these aspects of the controversy may reveal that the fact that Fabri was a Jesuit was also of importance. For differences of world-views comprising metaphysics, ethics, and social theory in the Newtonian-Cartesian debates, see Freudenthal (1986). For political interests at that time, see Shapin (1981).

References

Cartwright, N. (1983), How the Laws of Physics Lie, Oxford, (Claredon Press).

Damerow, P., Freudenthal, G., McLaughlin, P., and Renn, J. (1991), Exploring the Limits of Preclassical Mechanics, Berlin, Springer.

Davidson, D. (1973), On the Very Idea of a Conceptual Scheme, reprinted in J. Rajchman and C. West (eds.), *Post-Analytic Philosophy,* New York, Columbia UP 1985, pp. 129–143.

Drake, S., and Drabkin, I. E. (eds.). (1969), Mechanics in Sixteenth-Century Italy: Selections from Tartaglia, Benedetti, Tuido Ubaldo & Galileo, Madison, Wisconsin UP.

Fabri, H. (1669), Dialogi physici, Lyon.

——— (1646), Tractatus physicus de motu locali, Lyon.

Freudenthal, G. (1986), Atom and Individual in the Age of Newton, Dordrecht, Reidel.

Gil, F. (ed.). (1990), Controvérsias Científicas e Filosóficas, Actas do colóquio organizado pelo Gabinete de Filosofia do Conhecimento na Universidade de Évora, Lisboa, Fragmentos.
Leibniz, G. W. (1696), Review of J. Wallis, Opera Mathematica, Acta Eriditorum, June 1696: 249–260.
——— (1969), Philosophical Papers and Letters (L. E. Loemker, ed), Dordrecht, Reidel.
——— (1849), Mathematische Schriften (C. I. Gerhardt, ed.), Hildesheim, Olms 1971.
McMullin, E. (1987), "Scientific Controversy and its Termination," Scientific Controversy: Case Studies in the Resolution and Closure of Disputes in Science and Technology (T. H. Engelhart Jr. and A. L. Caplan, eds.), Cambridge, Cambridge University Press, pp. 49–91.
Mendelsohn, E. (1987), "The Political Anatomy of Controversy in the Sciences," Scientific Controversy: Case Studies in the Resolution and Closure of Disputes in Science and Technology (T.H. Engelhart Jr. and A. L. Caplan, eds.), Cambridge, Cambridge University Press, pp. 93–124.
Nagel, E. (1979), The Structure of Science, Indianapolis, Hackett.
Oldenburg, H. (1965–1973) The Correspondence of Vols I-IX A.R. Hall and M. Boas Hall, Madison, Wisconsin UP.
Papineau, D. (1977), The *Vis Viva* Controversy: Do Meanings Matter? Studies in History and Philosophy of Science 8: 111–142.
Russell, B. (1900), A Critical Exposition of the Philosophy of Leibniz, Cambridge, Cambridge UP.
——— (1903), The Principles of Mathematics, Cambridge, Cambridge UP.
Shapin, S., and Schaffer, S. (1985), Leviathan and the Air-Pump, Princeton, N.J., Princeton UP.
Tartaglia, N. (1537), La Nova Scientia. . . . Venice. [See Drake and Drabkin (1969).]
Wallis, J. (1695), Mechanica, in Opera Mathematica, vol. 2, Oxford.
Wolff, M. (1978), Geschichte der Impetustheorie, Frankfurt, Suhrkamp.

8

The Structure of a Scientific Controversy

Hooke versus Newton about Colors

MAURIZIO MAMIANI

The controversy between Robert Hooke and Isaac Newton about colors may be regarded both from a factual point of view and from a theoretical one. Newton's *New Theory about Light and Colours* was read at the meeting of the Royal Society on 8 February 1672, and Hooke was asked to peruse it and bring a report to the society. A week after, at the next meeting, Hooke read his brief report. A copy of it was immediately sent to Newton.[1] There arose a quarrel, which went on until the beginning of 1676.[2]

Despite its short duration and the restricted nature of the scientific topic under discussion, this quarrel is very significant theoretically, because of its gradual development toward more and more general issues concerning the meanings of the scientific research. For this reason, I believe that the quarrel between Hooke and Newton about colors includes some items typical of scientific controversies, which may be isolated from the historical background. Consequently, I suggest that the analysis of historical cases of scientific controversies is also the key to detecting their theoretical structure. To this end, I shall try to answer the following questions: On what grounds and by which means do alternative scientific theories confront one another? What are the conditions under which the disputing parties would give up the controversy? I shall distinguish the historical phases of the quarrel between Hooke and Newton from the theoretical levels of the discussion; on the other hand, I shall try to establish some intrinsic, not casual, relations between these levels.

The first phase of the controversy was characterized by the absence of personal competition. Both the disputants were animated by a genuine quest for truth and were not divided by precedent altercations. Moreover, both Hooke and Newton testified in their letters that they wanted to come to a mutually satisfactory arrangement. Hooke's criticism focuses on Newton's doctrine of colors and its epistemological justification.

It may be convenient to subdivide the items under discussion in two classes, the first corresponding to Newton's theory of colors (T) and the second to his epistemological justification (J). The first class T contains the following assertions:

1. Colors are original and "connate" properties of the rays of light.
2. There are two sorts of colors: one original and simple, the other compounded of these.
3. There is no sort of ray that can exhibit whiteness, because it is always compounded of all the primary colors mixed in a due proportion.
4. Whiteness is the usual color of light.[3]

The second class J contains the following assertions:

1. The theory of colors, above summarized, is not a hypothesis, but the consequence of experiments concluding directly and without any suspicion of doubt.
2. The science of colors becomes mathematical, and it is as certain as mathematical demonstrations.[4]

I want to underline that the items under discussion in the first class are not matters of fact according to Hooke, but according to Newton they are. By consequence, the controversy arises because of Newton's pretension of deriving theoretical assertions directly from experiments, so that the theory may be regarded as certain, as matter of fact. Moreover, Newton claims that theoretical assertions may be also demonstrated mathematically.

In this phase the level of controversy is critical. Both disputants accept that experiments are the ground on which alternative theories confront one another, but disagree about the effective link between matters of fact and theory. This is a general consideration, which may be extended to every scientific controversy. Whenever matter of fact is invoked, the empirical data are not under discussion really; it is only their theoretical interpretation that is being questioned.

Thus, it is certain that the prisms of Hooke produced colors in the same way as those of Newton, and that both opponents achieved the same experimental evidence. In particular, Hooke performed all of Newton's experiments about colors and found them correct. But in making the experiments, Hooke never met with the propositions of class T or J. As a consequence, he could not see a necessary connection between the experimental phenomena and Newton's theory:

> I have perused the Excellent Discourse of Mr Newton about colours and Refractions, and I was not a little pleased with the niceness and curiosity of his observations. But though I wholy agree wth him as to the truth of those he hath alledged, as having by many hundreds of tryalls found them soe, yet as to his Hypothesis of salving the phaenomena of Colours thereby I confesse I cannot yet see any undeniable argument to convince me of the certainty thereof.[5]

Hooke confirmed also the so-called *experimentum crucis,* in which a colored beam of light persists unchanged under further refraction, but denied Newton's interpretation of this fact, for example, that the colored ray preexists in the light before refraction.[6]

Newton is confident that Hooke's confirmation of his own experiments will overcome the theoretical disagreement of his antagonist. In a draft of his answer to Hooke, Newton wrote:

> I am much pleased that so ingenious & experienced a person as Mr Hook hath taken my discourse about refractions & colours into consideration & I desire that my thanks may be returned to him for his observations & *more especially for his confirmation of the experiments.*[7]

The hope of Newton is that Hooke's exceptions derive from the use of "oblique & glancing expressions"[8] in Newton's discourse, and he is convinced that the dissent is due to a verbal misapprehension. Newton indeed is very angry; however, he declares himself to be ready "to give way to ye mitigation of whatsoever ye Heads of ye R. Society shall esteem personall."[9]

The controversy is now entering a phase in which personal feelings are involved. So Newton's reply to Hooke's considerations is very harsh: he charges Hooke with partiality, disloyalty, dullness, and guilty negligence.[10] However, Newton fails to see the point of Hooke's criticism, insisting on the difficulties of his fundamental supposition, grounded on the waves or vibrations of ether, because of the impossibility of propagating the rays of light in straight lines. Later Hooke reminded Newton of the inflexion phenomena.[11] On experimental ground the match was even.

From the beginning, Hooke had carried on the discussion at the level of the justification of the theoretical assertions. He declares that propositions of classes T and J are hypothetical, and the fact that they "salve all the phenomena" is not sufficient to establish their truth, since "the same phaenomenon will be salved by my hypothesis as well as by his without any manner of difficulty or straining: nay I will undertake to shew an other hypothesis differing from both his & myne, yt shall do the same thing."[12] Therefore, all the propositions of class T are uncertain, and the propositions of class J are simply meaningless. Hooke maintains that there is not a direct and unquestionable conclusion from the experiments, and a fortiori that no physical theory may be as certain as mathematical demonstrations.

Apparently influenced by Descartes, Hooke thinks that a scientific theory must be "easy to be conceived."[13] The simpler a theory is, the easier it is to conceive. Simplicity is the only criterion of choice among alternative theories. Thus, Hooke underlines that in absence of a necessary connection between matter of fact and theory, there is no such thing as an *experimentum crucis.*

In general terms, Hooke brought the controversy to the most radical opposition, in which there may be no disagreement about matters of fact. Thus, the root of the quarrel is about the creative power of thought itself, for example, the indefinite variety of ways in which we may conceive and represent the phenomena. So Hooke is also convinced that there is no such thing as a demonstration of a physical theory:

> I doe not therefore see any absolute necessity to believe his theory demonstrated, since I can assure Mr Newton, I cannot only salve all the Phaenomena of Light and colours by the Hypothesis, I have formerly printed and now explicate yt by, but by two or three other, very differing from it, and from this, which he hath described in his Ingenious Discourse. Nor would I be understood to have said all this against his theory as it is an hypothesis, for I doe most Readily agree wth him in every part thereof, and esteem it very subtill and ingenious, and capable of salving all the phaenomena of colours; but I cannot think it to be the only hypothesis; not soe certain as mathematicall Demonstrations.[14]

On the contrary, Newton thought that there was a direct link between phenomena and their definitions, and never accepted that theoretical propositions are not reducible to matter of fact. Hooke disagrees with Newton notwithstanding his own agreement both with the matter of fact and with the hypothetical assumptions of the theory itself. All that Hooke rejects is the absolute necessity of it. It could be otherwise.

Scientific controversies usually do not reach this paradoxical unity of both agreement and disagreement. On the contrary, I suspect that in the history of science not many cases of scientific controversy that rise fully to this level of divergence may be found. And yet I think that every scientific controversy is a particular instance of this kind of divergence. We may perhaps consider the case of Hooke's discussion as the most general level of dissent to which a scientific controversy can aspire.

Newton's reply to Hooke is not candid. At first he denies he has maintained that light is a body: "'Tis true that from my Theory I argue the corporeity of light, but I doe it without any absolute positivenesse, as the word *perhaps* intimates, & make it at most but a very plausible consequence of the Doctrine, & not a fundamental supposition."[15] However, Newton claims that de facto colors are original and immutable qualities of the rays that exhibit them,[16] and if the rays of light were not corporeal, they could not have primary qualities. Hooke is right when he writes to Lord Brouncker: "That I have mistaken Mr. Newtons hypothesis I am sorry. But yet I believe as many others as heard it were mistaken too."[17]

The defensive position of Newton is subtler:

> I knew that the Properties wch I declared of light were in some measure capable of being explicated not onely by that, but by many other Mechanicall Hypotheses. And therefore I chose to decline them all, & speake of light in generall termes, considering it abstractedly as something or other propagated every way in streight lines from luminous bodies, without determining what that thing is, whether a confused mixture of difform qualities, or modes of bodies, or of bodies themselves, or of any virtues powers or beings whatsoever.[18]

In this manner the objection of Hooke, for example, that no scientific hypothesis is absolutely necessary, is neutralized. However, in order to neutralize Hooke's objection, Newton is compelled to propound a new method of scientific research, where hypotheses have no place. On one hand, there is abstraction (in preference, when it is possible, mathematical abstraction), for example, general terms without physical meaning; on the other hand, there are experiments, for example, matters of fact. But where is the theory? The theory is nowhere. Hooke is furious. He cannot grasp his adversary on this new ground. Newton claims that his conclusions do not derive from hypotheses, and so there is no room to dispute.

Newton's reply is as radical as Hooke's objection. Newton may defend himself only rejecting the scientific legitimacy of the use of hypotheses. He is convinced (and he is not wrong) that Hooke's position is fundamentally skeptical. Emphasizing skepticism about hypotheses, Newton claims to decline them all.

In a contemporary letter to Gaston Ignace Pardies, Newton replies to Hooke indirectly, writing: "And if anyone make a guess at the truth of things from the mere possibility of the hypotheses, I don't see how to determine any certainty in any science; if indeed it is permissible to think up more and more hypotheses, which will be seen to raise new difficulties."[19] And to Henry Oldenburg: "In the meane while give me leave to insinuate that I cannot think it effectuall for determining truth to examin the severall ways by wch Phaenomena may be explained, unlesse where there can be a perfect enumeration of all those ways."[20] Hooke had maintained that many other hypotheses could explain the same phenomena, this or that hypothesis being not absolutely necessary. The enumeration of hypotheses is anything but perfect.

From a historical point of view, it is curious that both Newton and Hooke are greatly influenced by Boyle's *Experiments and Considerations Touching Colours* (1666), in which we may find the same arguments (lack of necessity in the use of hypotheses and hence abstraction from them) utilized by Hooke and by Newton.[21] Why did Boyle not quarrel with himself? Because Boyle was not concerned with a theory of colors. On the contrary, Hooke and Newton were primarily oriented toward the theoretical explanation of the phenomena. Both opponents are therefore leaving the first level of the controversy, where particular experiments are reported and in which the match was stalemated. So scientific theories do not confront one another only by means of experimental data. There is a second level in controversies, which pertains to the justification of the theory on the grounds of methodological rules (simplicity, enumeration, etc.). It is this second level that conveys conviction to the first. To Hooke we owe the first awareness of what is now called the "experimenter's regress,"[22] or circle. A good experiment is what produces the matter of fact that the experimenter tries to establish according to his theory. Other experiments are bad.

It is easy to recognize the historical phase that corresponds to this second level. The main argument used by Hooke against Newton's theory is a methodological rule: no multiplying entities without necessity. Newton's theory is useless, because if also (as Newton claims) we may conceive the white to be compounded of all the other colors, as any uniform motion may be compounded of thousands of motions in the same manner as Descartes had explained refraction, then there is no physical necessity for it.

Newton confutes this argument with the same rule: if light is supposed (as Hooke claims) to be caused by certain definite waves of subtle matter, and colors correspond to their diffusion and expansion, then these inequalities may suffice to differentiate them. Therefore, there is no reason to seek other causes of these effects, unless one multiplies entities without necessity. So also at this second level of the controversy, it is possible to support different theories by means of the same arguments.

When a scientific controversy reaches this level, it is difficult to find a resolution. The opponents share the same rules, but apply them in different ways. Newton declares that Hooke "knows well yt it is not for one man to prescribe Rules to ye studies of another."[23] So the controversy is impelled toward yet another new level, which requires some arbitration. And it was what just happened. Hooke appeals to the authority of Francis Bacon, as the supreme warranty of philosophy.[24] And Newton suggests one criterion ad hoc to judge the adequacy of a scientific theory.[25]

Before discussing this new level of the controversy, it may be useful to consider several points. The first two levels of the controversy concerned experiments and theoretical reasoning, respectively. We may feign that the two opponents could also quarrel with one another on a desert island. But at the third level of the controversy, the scientific community is involved intrinsically. Hooke appeals to the authority of

Francis Bacon because the Royal Society of London was willing to recognize it. Newton suggests his criterion in order to obtain the assent of the Royal Society. The scientific community is charged often with the choice of one of the two opponent theories. We could expect rhetorical devices to be used at this level. And it is so. Rhetorical devices are also present in every level of the controversy, for example, when Hooke uses the metaphor of the organ pipes[26] to ridicule Newton's theory. The achievement of conviction by means of rhetorical devices is always present in a dispute. However, at this level the object to be persuaded changes: in the first two levels, it is the opponent; in the third, the community.

How does the scientific community make the choice? What are the efficacious arguments? I think that only the history of science can reply to these answers exhaustively. However, there is also room for general considerations. The arguments of the third level of controversy admit no sort of evidence. They appeal to the consensus of the majority of the scientific community. It is this consensus that gives authority to an experimental report or to a methodological rule. For this reason, we may find rhetorical devices employed to flatter the community. We could say that in absence of absolute evidence, or in order to create it, there is a moral consensus. Scientific theories are chosen ultimately by the same process of growth that establishs ethical or political laws, religious beliefs, and aesthetic preferences.

Hooke's appeal to the authority of Bacon is counterbalanced by Newton's claim that his theory is both mathematically demonstrated and derived from experiments: the authority of Bacon versus the authority of mathematics and experience. In one way or another, authority requires a moral acknowledgment, which is the guide of consensus.

It is significant that at this level the controversy is inoculated against experimental evidence. When at last Hooke produces the "razor-edge" experiment which shows that there is a deflection of light from a straight line and so contradicts Newton's corpuscular theory directly, he is defending himself and his wave hypothesis in order to maintain its equivalence with the corpuscular one.[27] Hooke did not attack Newton on the experimental ground because he was convinced that if experiments do not suffice to prove the truth of a theory, neither do they to prove its untruth. Even though Newton's criterion was wholly empirical, Hooke was not willing to use it. Much later the Royal Society will not permit Rizzetti,[28] an Italian mathematician, to use it.

Newton's criterion is the exact opposite of Hooke's claim: because they prove a theory, the experiments must invalidate all objections taken from hypotheses. Hence, the opponent must produce other experiments that contradict the theory directly. Newton adds: "If any such may seem to occur." Now, the "razor-edge" experiment was such. What did Newton do? He first discredited the moral reliability of Hooke before the Royal Society, insinuating that he was a plagiarist: "I thought I had seen the Experiment before in some Italian Author. And the Author is Honoratus Faber in his Dialogue *de Lumine,* who had it from Grimaldo."[29] Then Newton encounters no difficulty in showing that the strange phenomenon is but a new kind of refraction. There is no necessity to think that light is propagated in curved lines as sound is. In so doing, Newton acknowledged practically that Hooke was right in claiming the equivalence of hypotheses!

At this level, I suppose, the controversy does not admit stronger arguments than moral ones. When Newton read to the Royal Society a new discourse entitled *An Hy-*

pothesis explaining the Properties of Light discoursed of in my severall Papers (1675), Hooke charged his adversary with piracy. After hearing the second part of the discourse, Hooke remarked that "the main of it was conteined in his *Micrographia,* which Mr Newton had only carried further in some particulars."[30]

The reaction of Newton was resolute:

> As for Mr Hook's insinuation yt ye summ of ye Hypothesis I sent you had been delivered by him in his Micrography, I need not be much concerned at the liberty he takes in yt kind. Yet because you think it may do well if I state ye difference I take to be between them, I shal do it as briefly as I can, & yt ye rather that I may avoyd ye savour of having done any thing unjustifiable or unhansome towards Mr Hooke. But for this end I must first (to see what is his) cast out what he has borrowed from Des Cartes or others.[31]

So irony succeeded where neither experiment nor reason had succeeded. Newton described Hooke as a superficial dilettante who had borrowed all from others, in particular from the "continental" Descartes. Then irony is coupled with sarcasm: "I desire Mr Hooke to shew me therefore, I say not only ye summ of ye Hypothesis I wrote, wch is his insinuation, but any part of it taken out of his *Micrographia:* but then I expect too that he instance in what's his own."[32]

Coming back to general considerations, I have shown that it is possible to hold quite different theories by means of the same experiments, and that it is also possible to maintain different theories by means of the same theoretical arguments. There are neither unquestionable crucial experiments nor crucial arguments. Lastly, the controversy appeals to the moral acknowledgment of the scientific community.

We may now ask ourselves under what conditions the disputing parties would give up the controversy. At the first level, the experimental one, Newton asks for an experiment that contradicts his theory directly. Hooke asks for this *experimentum crucis:* to make a white body by means of all the colored bodies in the world! Hooke's claim is absurd. But Newton's demand is absurd, too. When does an experiment contradict a theory directly? Newton will reply that "razor-edge" experiment contradicts his theory only indirectly, and so on.

At the second level, the theoretical one, the resort to methodological rules prevails. Here Newton maintains that he made no hypotheses about the substantiality of light. Hooke answers that, if so, he has nothing more to say. The controversy could terminate. However, Newton adds that the substantiality of light is not a hypothesis, but a matter of fact, and that hypotheses must be banished from science. The controversy starts again.

At the third level, the moral one, the appeal to the scientific community prevails. Hooke alleges the authority of Bacon to sustain the use of hypotheses. Newton states an empirical criterion that permits the use of crucial experiments in the first level. Both the opponents charge each other with plagiarism. Newton claims that Hooke is a follower of Descartes, not of Bacon and that the "knife-edge" experiment is of Grimaldi, not of Hooke. The acme of the controversy is reached. Table 8.1 summarizes the elements of the controversy we may consider typical items.

It seems to me that a scientific controversy need not pass through all these levels historically. For example, the brief dispute of Newton with Pardies stopped at level 1, when Newton sent him a scheme for his *experimentum crucis.* Pardies wrote to Oldenburg that he was satisfied.[33] The dispute with Huygens ended similarly at level 2,

Table 8.1 Typical Items of the Hooke-Newton Color Controversy

Levels	By which means do alternative scientific theories confront one another?	Under what conditions would the disputant parties give up the controversy?
1. Experiments, or matter of fact	Crucial experiments	Experiments that contradict the questioned theory directly
2. Theoretical arguments	Methodological rules and ways of reasoning	Acceptance of the same rules and of the manner of applying them
3. Consensus of the scientific community	Moral arguments	Authority received from the scientific community

when, in two remarkable letters, Newton criticized Huygens's method of experimental analysis. Huygens gave up the controversy after the first letter, but he was not satisfied. He never attended to colors again.

Hooke's controversy reached the third level explicitly, because of the antithetical grounds on which it arose. But charging Newton with plagiarism, Hooke gave him the palm of victory unwittingly. Newton easily showed his own innocence and alleged Hooke's charge as a moral proof of his consent.

However, the dispute stopped without Hooke's surrender. It is significant that both opponents were willing to embrace private correspondence, and so leave the community level of the controversy. Hooke suspects that Oldenburg conspires against him, and perhaps he was right. Oldenburg often openly takes sides with Newton. At the third level the controversy is more and more like a kind of war, and hence alliances are useful. Also useful is a strategic retreat; thus, Hooke claims:

> Your Designe and myne I suppose aim both at the same thing wch is the Discovery of truth and I suppose we can both endure to hear objections, so as they come not in a manner of open hostility, and have minds equally inclined to yield to the plainest deductions of reason from experiment.[34]

With a play of words, Hooke tries to take back the dispute to the first level:

> This way of contending I believe to be the more philosophicall of the two, for though I confess the collision of two hard-to-yield contenders may produce light yet if they be put together by the ears of other's hands and incentives, it will produce rather ill concomitant heat which serves for no other use but . . . kindle cole.[35]

Newton's answer is as much conciliatory as Hooke's:

> There is nothing wch I desire to avoyde in matters of Philosophy more then contentions, nor any kind of contention more then one in print: & therefore I gladly embrace your proposal of a private correspondence. What's done before many witnesses is seldome wthout some further concern then that for truth: but what passes between friends in private usually deserve ye name of consultation rather then contest, & so I hope it will prove between you & me.[36]

Then Newton adds the famous sentence: "If I have seen further it is by standing on ye sholders of Giants."[37]

Contest may become consultation by avoiding the witness of scientific community. The armistice had been asked by Hooke, and signed by Newton. But the war about light and colors only ended with the death of Hooke, the dominion of human passions being stronger than concern for truth.

Notes

1. Hooke was not informed that his report was sent to Newton. Later Hooke expressed to Lord Brouncker his regret for this circumstance: "The truth is I never intended it for Mr. Newtons perusall which If I had I might possibly have been more choice in my expression, and have taken more time in the penning thereof. for I doe assure your Lordship I had not above three or 4 hours times for the perusall of Mr. Newtons paper and the writing of my answer, though I confesse I heard it read once before in the Society." Hooke to Lord Brouncker, June 1672. *The Correspondence of Isaac Newton,* ed. H. W. Turnbull, J. F. Scott, A. R. Hall, L. Tilling (Cambridge: Cambridge University Press, 1959), I, 198.

2. I shall consider the following papers, whose chronological order corresponds to the different phases of the dispute: Newton's reply on 11 June 1672 to Hooke's report; the *Memorandum* by Hooke on 19 June 1672; Hooke's letter to Lord Brouncker on June 1672; Newton's letter to Oldenburg on 7 December 1675 (*An Hypothesis explaining the Properties of Light discoursed of in my severall Papers*); Newton's letters to Oldenburg on 21 December 1675 and 10 January 1676 against Hooke's charge of piracy; Hooke's letter to Newton on 20 January 1676; Newton's reply to Hooke on 5 February 1676.

3. Newton to Oldenburg, 6 February 1671/2, propositions 1, 5, 7, and 8. Newton, *Correspondence,* I, 97–98.

4. Ibid., 96–97.

5. Hooke to Oldenburg, 15 February 1671/2, ibid., 110.

6. "In order to examine, whether severall colours after the first Refraction would have severall refractions, I made severall experiments . . . ; wch seems at first sight much to confirm Mr Newtons Theory of colours & light; but yet I think it not an *Experimentum crucis.*" Memorandum by Hooke, 19 June 1672, ibid., 195.

7. University Library, Cambridge, MS Add. 3970, f. 445. Italics added.

8. Newton, *Correspondence,* I, 193.

9. Ibid.

10. "I must confess . . . I was a little troubled to find a person so much concerned for an *Hypothesis,* from whome in particular I most expected an unconcerned & indifferent examination of what I propounded. . . . But he knows well yt it is not for one man to prescripe Rules to ye studies of another, especially not without understanding the grounds on wch he proceeds. . . . [Mr Hooks Considerations] consist in ascribing an Hypothesis to me wch is not mine; in asserting an Hypothesis wch as to ye principall parts of it is not against me; in granting the greatest part of my discourse if explicated by that Hypothesis; & in denying some things the truth of wch would have appeared by an experimentall examination." Newton to Oldenburg, 11 June 1672, ibid., 171–173.

11. Hooke to Lord Brouncker, June 1672, ibid., 200–201.

12. Ibid., 111.

13. Ibid.

14. Ibid., 113.

15. Ibid., 173.

16. Ibid., 264.

17. Ibid., 199.

18. Ibid., 174.

19. Ibid., 169.

20. Ibid., 209.

21. Cf. M. Mamiani, *Il prisma di Newton* (Roma-Bari: Laterza, 1986), 17–47.

22. Cf. S. Schaffer, *Glass works: Newton's prisms and the uses of experiment. The Uses of Experiment,* ed. D. Gooding, T. Pinch, S. Schaffer (Cambridge: Cambridge University Press, 1990), 69–70.

23. Newton, *Correspondence,* I, 172.

24. "I see noe reason why Mr. N. should make soe confident a conclusion that he to whome he writ did see how much it was besides the busness in hand to Dispute about hypotheses. for I judge there is noething conduces soe much to the advancement of Philosophy as the examining of hypotheses by experiments & the inquiry into Experiments by hypotheses. and *I have the Authority of the Incomparable Verulam to warrant me.*" Hooke to Lord Brouncker, June 1672, ibid., 202. Italics added.

25. "And therefore I could wish all objections were suspended, taken from Hypotheses or any other Heads then these two; Of showing the insufficiency of experiments to determin these Queries or prove any other parts of my Theory, by assigning the flaws & defects in my Conclusions drawn from them; Or of producing other Experiments wch directly contradict me, if any such may seem to occur. For if the Experiments, wch I urge be defective it cannot be difficult to show the defects, but if valid, then by proving the Theory they must render all other Objections invalid." Newton to Oldenburg, 6 July 1672, ibid., 210.

26. "But why there is a necessity, that all these motions, or whatever els it be that makes colours, should be originally in the simple rays of light I doe not yet understand the necessity; noe more than that all those sounds must be in the air of the bellows which are afterwards heard to issue from the organ-pipes." Hooke to Oldenburg, 15 February 1671/2, ibid., 111.

27. "And if Mr. Newton try ye Experiment I am very confident he w[ill] ascribe that deflection of ye rays neither to Reflection nor Refraction soe that hereby I hope the seeming impossibility of ye Fundamentall supposition will be removed." Hooke to Lord Brouncker, June 1672, ibid., 201. The "Fundamentall supposition" is Hooke's theory. The hopes of Hooke were incautious: when Newton tried this experiment, he ascribed the deflection of the rays to a new kind of refraction! Discussed further below.

28. Cf. Schaffer, *Glass works,* 96–100.

29. Newton, *Correspondence,* I, 384.

30. Ibid., 407.

31. Ibid., 405.

32. Ibid., 408.

33. "Je suis tres satisfait de la derniere réponse que M. Newton a bien voulou faire à mes instance. le derniere scrupule qui me restoit touchant l'*experience de la croix,* a esté parfaitement levé. Et je conçois tres-bien par sa figure ce que je n'avois pas compris auparavant. L'experience ayant esté faite de cette façons je n'ay rien á dire." Pardies to Oldenburg, 30 June 1672, ibid., 205–206.

34. Ibid., 412–413.

35. Ibid., 413.

36. Ibid., 416.

37. Ibid.

9

Scientific Dialectics in Action

The Case of Joseph Priestley

PIERLUIGI BARROTTA

> As you would not, I am persuaded, have your reign to resemble that of *Robespierre,* few as we are who remain disaffected, we hope you had rather gain us by persuasion, than silence us by power.
>
> J. Priestley to C. L. Berthollet et al.

On a Historiographical Problem

Priestley's tenacity in defending the phlogiston chemistry has often puzzled historians. In his works, Priestley very often underlines that hypotheses are at best heuristic instruments that one must be ready to abandon in any moment. As he proudly writes in one of his works: "[W]henever I have drawn general conclusions too soon, I have been very ready to abandon them."[1] Indeed, his scientific activity was perfectly consistent with this view, and it is well known that he was prone to change his mind with puzzling frequency. Notwithstanding, he never abandoned phlogiston.

Priestley's attitude especially represents a puzzle for what we can call the "traditional historiography" of science. If the implausible idea that Priestley did not adhere to the new chemistry because he lacked competence is rejected, then the only available solution to the traditional historiography is to claim that he was blinded by his prejudices in the face of overwhelming empirical evidence. Facts were clear and he simply refused to see them. Partington claims, for example, that "in chemistry he was most orthodox, and clung tenaciously to the last to the theory of phlogiston which his own experiments, if they had been properly interpreted, would have been the first to overthrow."[2] This is what we can call the thesis of Priestley's *psychological dogmatism.*

The most recent historiography of science has been more sympathetic to Priestley's work. As in traditional historiography, here too we can easily find an underlying philosophical view. Priestley did not reject Lavoisier's system of chemistry because he was psychologically dogmatic. Priestley, like any creative scientist, had his own peculiar world-view, which was strictly related to his scientific theories. He did not refuse to look at facts; rather, observations were interpreted in accordance with a conceptual scheme embodied in his world-view. Undoubtedly he was dogmatic, but one should realize that dogma plays an important epistemological role in scientific research. As is well known, this view is associated with Thomas Kuhn and partially with Stephen Toulmin, both of whom devoted some attention to Priestley.[3] Generally speaking, Kuhn's idea is that "[w]hen paradigms enter, as they must, into a debate about paradigm choice, their role is necessarily circular. Each group uses its own paradigm to argue in that paradigm's defence."[4]

The idea that empirical argumentations between Priestley and Lavoisier were circular has also been upheld by some historians who specialize in eighteenth-century chemistry. Robert E. Schofield, for instance, notes that it is nonsensical to blame Priestley for not accepting the oxygen theory. In fact, the vast majority of chemists who linked their names to phlogiston—such as Henry Cavendish, Jeremias Richter, and Carl Scheele—did not adhere to the new chemistry. Also, those who accepted it—such as Richard Kirwan—gave up chemical studies altogether. "One finds," Schofield writes, "in the examination of such instances as these, reaffirmation of the old paraphrase, 'Old theories never die, their supporters just fade away.'"[5] Therefore, against the thesis of Priestley's psychological dogmatism, the new historiography opposes the thesis of Priestley's *epistemological dogmatism.*

Of course, the division of the studies on Priestley into an "old" and a "new" historiography is an oversimplification. Many distinctions should be drawn within each.[6] Yet, such a simplification allows us to identify a philosophical presupposition that is shared by both historiographical approaches, that Priestley represents an example of dogmatism because both approaches focus only on the relation between scientific theory and the facts pertinent to it. Both implicitly ask the question, Did Priestley face a binding and compelling piece of evidence against phlogiston theory?" In the past years historians took for granted a positive answer to this question, whereas there is more recently a general tendency to respond negatively. In each case the final answer to the question has always been that Priestley represents a case of dogmatism, although for deeply different reasons.

In this chapter I try to show that Priestley's conduct cannot be considered a case of dogmatism. I maintain that the above-mentioned question is historically misleading. I shall replace that question with one that I hope to show is much more productive in understanding Priestley's work: Was the dialogue between Priestley and his opponents fruitful for scientific progress? There is much historical evidence that suggests the answer must be positive. Although Priestley always refused to adhere to the new chemistry, he put forward interesting and penetrating criticisms that required an answer from his opponents. His objections proved to be fruitful for scientific progress in chemistry, since they *led to empirical discoveries and more precise conceptual specifications.* In this sense, Priestley cannot be considered an instance of dogmatism.

Of course, my historical reconstruction, too, refers to a specific philosophical view. Through Priestley's case I want to show that science is a *dialectical* enterprise, based on the dialogue among interlocutors and not simply on the logical relationships between facts and theories.[7] Thus, this chapter has both a historiographical and a philosophical aim. In fact, the study of a historical episode that seems to contrast dramatically with the idea that science is no more than a case of Socratic dialogue, yields strong support for the dialectical view of science.

Theology, Philosophy of Matter, and Chemistry

According to Priestley, science has both a theoretical-cognitive value and a theological-practical value. The idea that nature shows the benevolent divine design is particularly apparent in Priestley's interest in those phenomena that corrupt and restore the salubrity of atmospheric air. This issue is strictly relevant to understanding his chemical research, since all vitiation processes were considered to be phlogistication processes and, symmetrically, those of restoration to be dephlogistication processes.

A good example of this attitude is given by the discovery that vegetation purifies the air corrupted by animals' breathing. In 1772, from a series of experiments Priestley drew the conclusion that animals corrupted atmospheric air. Consistently with phlogistic interpretation, he rejected the hypothesis that the death of the animals put in sealed boxes was caused by "the want of any *pabulum vitae,* which has been supposed to be contained in the air."[8] On the contrary, he maintained that the air's noxiousness was caused by something—identified with phlogiston—that irritated animals' lungs. Such an identification is indeed quite natural, since there is a strict similarity (recognized by modern theory, too) between the phenomena of combustion and respiration, and that of calcination or oxidation, in all of which there is a contraction of the air closed in a vessel.[9]

In the same year, through a long series of observations, he regarded as "exceedingly probable that the putridum effluvium is in some measure extracted from the air by means of the leaves of plants,"[10] which in this way made the air wholesome again. These observations suggested to him the following moral remark: "[I]t can hardly be thought, but that it may be a sufficient counterbalance to it, and that the remedy is adequate to the evil."[11]

This remark is interesting because it shows how Priestley was ready to see the "harmonies" of nature. In fact, at that time the evidence supporting the role of vegetation in the purification of air was far from compelling, as Priestley himself honestly admitted. Only in 1778 was he able to claim that his initial hypothesis was correct, and he was certainly delighted by this conclusion. As we read in one of his theological works:

> It is a considerable evidence of the goodness of God, that the inanimate parts of nature, as the surface of the earth, the air, water, salts, minerals, etc. . . , are adapted to answer the purposes of vegetable and animal life, . . . and the former of these is evidently subservient to the latter; all the vegetables that we are acquainted with either directly contributing to the support of animal life, or being, in some other way, useful to it.[12]

In theology Priestley was a convinced upholder of a rationalistic reading of the Holy Scriptures.[13] To him, if the Scriptures were correctly interpreted, no contradiction could arise between them and scientific knowledge. It is not by chance that his *Disquisitions Relating to Matter and Spirit* (1777) begins with the exposition of the Newtonian rules of correct philosophizing.

The interpretation of a Christianity "in accordance with reason" leads Priestley to reject the dualism between matter and spirit. Priestley devotes much effort to showing that dualism is alien to the Scriptures (for instance, to the doctrine of the Resurrection) and leads to insoluble problems, related to the difficulty of maintaining at the same time the interaction between matter and spirit, and their absolute extraneity.

The "modern" theology (Priestley uses this term to mean the progressive departure from the Scriptures) sees matter as endowed—together with extension—with solidity and impenetrability, and devoid of any powers including those of attraction and repulsion. This is not tantamount to saying that attraction and repulsion are alien to matter, but rather that they are inessential to its subsistence.

This philosophy of matter is untenable for Priestley, who believes that any solid body "must necessarily have some particular *form* or *shape*." Priestley argues that such a solid body cannot subsist

> unless the parts of which it consists have a *mutual attraction,* so as either to keep contiguous to, or preserve a certain distance from each other. This power of attraction, therefore, must be essential to the *actual existence* of all matter, since no substance can retain any *form* without it.[14]

Without such a power of attraction, "the solidity of the atom entirely disappears."[15]

To the power of attraction, Priestley adds the power of repulsion. He argues, in fact, that all phenomena concerning the impenetrability and resistance of bodies can be explained through that power. Consequently, by applying the rule of not accepting more causes that those necessary for the explanation of phenomena, it follows that a correct way of philosophizing "obliges us to suppose, that the cause of *all* resistance is repulsive *power,* and in no case whatever the thing that we have hitherto improperly termed *solid,* or *impenetrable matter.*"[16]

A philosophy of matter with these characteristics has direct consequences for chemical analysis. If the fundamental properties of matter are the powers of attraction and repulsion, then volumetric measurements acquire considerable epistemic importance. In fact, the volumes of masses are directly determined by the attraction and repulsion between the constitutive parts of matter, and in particular, the internal power of repulsion can be modified by the usual laboratory instruments. As Priestley notes: "[T]hat the component particles of the hardest bodies do not actually touch one another, is demonstrable from their being brought nearer together by cold, and by their being removed farther from each other by heat."[17]

To Schofield this "metaphysics of powers" arises from a peculiar form of atomism, whereas more recently, J. G. McEvoy has upheld that this view is related to Priestley's phenomenalistic epistemology. In any case, both of them maintain that such a view represents the metaphysical foundation of Priestley's phlogiston chemistry.

I would like to add here not only that volumetric measurements acquire much importance, but also that the very concept of mass is not precisely defined in that philosophy of matter. I could not find throughout all the *Disquisitions* a clear concept of the mass of a body. This problem is made even more puzzling because Priestley also rejects, although with some hesitation, the concept of inertia. In a passage that significantly appears only in the first edition (and therefore it cannot be found in the 1972 collection of Priestley's theological works edited by John Rutt), Priestley rejects the idea that "matter is possessed of a certain *vis inertia,* and is wholly indifferent to a state of rest or motion, but as it is acted upon it by a foreign power,"[18] although this exposes him to criticism that his philosophy of matter is incompatible with Newtonian mechanics.

In the discussion between Priestley and Richard Price—published by Priestley in 1778 with the title *A Free Discussion of the Doctrines of Materialism and Philosophical Necessity*—we can find some more precise pieces of information about Priestley's ideas. With some justification, Price objects that Priestley's concept of mass is not compatible with Newton's laws of motion. Furthermore, Price notes that the powers of attraction and repulsion must be powers of something. Hence, what is this something? It cannot be mere empty space, since it is not easy to imagine how a void can have powers; neither can it have a specific shape, because in this case it should have other smaller parts that, following Priestley's kind of reasoning, could only subsist thanks to the power of attraction, and the same would be true for the parts of these parts and so on, ad infinitum. Priestley's answer is particularly interesting: "The *laws of motion* are only general rules, to which the facts relating to the approach of bodies to each other, and their receding from each other, are reducible, and are consistent with any *cause* of such approaching or reciding."[19] In modern terms, Priestley tries to answer the criticism by adopting an instrumentalistic view of scientific theories, which makes Newton's theory compatible by definition with any philosophy of matter.

Following Priestley's reasoning, the role, if any, played by the law of conservation of matter is far from clear: during a chemical reaction is the sum of the mass of the substances employed unchanged? Although the law of the conservation of matter is not, strictly speaking, incompatible with Priestley's philosophy of matter, neither can it be considered a natural consequence of that philosophy.

Lavoisier himself did not claim to have established the law by means of experimental *facts* gathered through laboratory analysis. In *La philosophie de la matière chez Lavoisier* (1935), Hélène Metzger persuasively shows that Lavoisier used the invariance of masses as an "indispensable criterion of [empirical] verification"[20] and never thought to ascribe to himself its "discovery."

Historians typically agree that for Lavoisier the law of the conservation of matter is a part of a wide epistemological program. I cannot develop this point here, but it is worth emphasizing that, even though it is stated only once and incidentally in his *Traité* (1970),[21] the law plays an important role in every work of the French chemist. The basic point is that the conservation of matter yields the justification for the reformulation of chemistry in terms of a rigorously physical and quantitative science. As Lavoisier polemically writes: "This science, left apart for a long time from all the others, believed that it could do without the weights and measurements to which physics owes its most beautiful discoveries."[22] It is not certainly by chance that Lavoiser's first supporters were not chemists but physicists.[23] Doubtless, Toulmin certainly showed a remarkable historical perspicacity when he emphasized—most likely without knowing Priestley's theological works—the importance of volumetric measurements in Priestley's opposition to Lavoisier's chemistry.

Should we conclude that chemical revolution was a fight between two different and incompatible metaphysics? Undoubtedly, it was this, too. However, before concluding that during the revolution empirical argumentations were "circular," it is convenient to look at the specific chemical research in greater depth. Although I do not want to emphasize this point, it is worth mentioning, after all, that Priestley's *Disquisitions* appeared in 1777. As Priestley himself informs us in the preface of his work, he had earlier held a much more traditional philosophy of matter, and some of his most important discoveries were made precisely in those years.

Ontological Continuities and Discontinuities: The Rise of Pneumatic Chemistry

In a specific sense there is some foundation to the claim that empirical arguments between oxygen and phlogiston theories are circular. In fact, the two theories seem to determine the language in which chemical phenomena are described. Hence, every empirical description seems to presuppose as being correct the very interpretation challenged by the opposing party.

The arguments supporting this idea are well known and I only illustrate them briefly here. I have already noted that Priestley thought of processes concerning the vitiation and volumetric reduction of air as phlogistic processes. Thanks to this interpretation, he could consider, for instance, the phenomena of calcination (for example, oxidation), combustion, and respiration as being similar, and conversely, those of metal revification and plant photosynthesis. All these phenomena are classified as being similar within the oxygen theory, as well. However, oxygen language is incompatible with that of phlogiston. Whenever in the latter there is liberation of phlogistic substances, in the former there is absorption of oxygen (and vice versa). As a consequence, they presuppose two alternative and incompatible ontologies. For instance, calces and water are simple substances in phlogiston theory, whereas they are compounds in oxygen theory.

Of course, no one would deny that facts have some relevance, but in no way can they contradict a language in the strict sense. To give a famous example, consider the well-known phenomenon that during metal oxidation there is an increase in the weight of metals. It could be argued that that phenomenon contradicts phlogiston theory, but such a conclusion would be naive. In fact, a host of hypotheses can be put forward to explain that phenomenon that have all the characteristics required to continue using phlogiston language.[24] Furthermore, in the case of Priestley we should remember that for him volumetric measurements are epistemically more important than ponderal analysis. Hence, another reason why ponderal arguments proved to be circular was that they refer to epistemic values that are not shared by the other party.

This is the basic point argued by Kuhn and Toulmin, and Schofield seems to make the same point when he claims that "the phlogiston interpretation of respiration is, in many ways, the logical and 'mathematical' equivalent of Lavoisier's oxidation interpretation."[25] To strengthen their argument, it could be noted that even some chemists at that time argued that the characteristics of phlogiston and oxygen theories did not allow any *experimentum crucis* between the two theories. For instance, after his discovery that water is produced when a mixture of inflammable and dephlogisticated airs have been made to explode, Cavendish argues:

> [A]s adding dephlogisticated air to a body comes to the same thing as depriving it of its phlogiston and adding water to it, and as there are, perhaps, no bodies entirely destitute of water, and as I know no way by which phlogiston can be transferred from one body to another, without leaving it uncertain whether water is not at the same time transferred, *it will be very difficult to determine by experiment which of these opinions is the truest;* but as the commonly received principle of phlogiston explains all phenomena, at least as well as Mr. Lavoisier's, I adhere to that.[26]

I agree that this approach captures important aspects of the scientific revolution in chemistry. The idea that empirical arguments were circular however, is, unfounded. In fact, together with the evident ontological discontinuity, there is also a clear ontological continuity that must be emphasized at least as much.

The new historiography of science correctly underlines that in the final analysis the controversy between Priestley and Lavoisier arises from a different way of understanding chemical research. However, it neglects to emphasize with equal force that both Priestley and Lavoisier belonged to a common research tradition dating from the beginning of that century known as *pneumatic chemistry,* meaning the study of airs or gases.

Early chemical research had overlooked the function of air as a chemical agent. Air was typically considered to be a mere instrument, only capable of facilitating the liberation of phlogistic substances.[27] Only after Stephen Hales's work did chemists become aware of the fundamental role of air in explaining chemical processes. In *Vegetable Staticks,* published in 1727, Hales demonstrated that air could be fixed into, and liberated from, bodies. By analyzing a host of organic and inorganic substances, he discovered that phosphorus and sulfur increase in weight when burned, since they absorb air, thus anticipating Lavoisier's results. As a consequence, Hales polemically asked why contemporary chemists did not accept "this now fixt, now volatile, Proteus among the chymical principles."[28]

Pneumatic chemistry was born, although not yet in its full form. In fact, Hales did not give much importance to the qualitative differences—such as color, smell, and solubility—of the several isolated airs. Instead he focused on their quantitative differences, such as density or elasticity. This peculiar approach of chemistry arises from a Newtonian conception of matter, but—and this is much more important for our issue—it was also coherent with the idea, at that time very common, that atmospheric air was a simple substance, not made up of chemically different airs. Hales agreed with this ancient view, and consequently when clear qualitative differences might have led him into thinking that several gases existed, he wrote that atmospheric air was—to use his terminology—"infected" by "noxious vapours" or "sulphureous spirits." To him "atmospheric air is a *Chaos* consisting not only of elastick, but also of unelastick particles" that float in large numbers in this fluid.[29]

The real pneumatic chemistry was born only when this ancient view was abandoned and chemists recognized the existence of an indeterminate number—to be specified through empirical investigation—of chemically different gases. This step did not take long. In 1741, William Brownrigg read the an essay at the Royal Society of London that was subsequently revised and published in 1765, together with an abstract of the original paper. Here, Brownrigg reported the isolation of a "mineral elastic spirit" that he recognized as "a particular kind of air, or permanently elastic fluid."[30] He ascribed to this volatile principle the wholesome qualities of mineral waters and he stressed that "it is highly probable that the elastic particles, which are emitted from various kinds of dense bodies, do thus vary one from another; and that they oft-times compose elastic fluids, which differ as much from each other, as those bodies differ from which they are produced."[31]

In 1756, Joseph Black published his famous essay on *magnesia alba.* He noticed that this substance (carbonate of magnesium) released "fixed air" (carbon dioxide) and recognized that it was different from common air and was "dispersed thro' the atmosphere, either in the shape of an exceedingly subtile powder, or more probably in that of an elastic fluid."[32] In 1766 Cavendish discovered another gas: "inflammable air" or hydrogen. In order to release "factitious airs" (namely, any kind of air contained in other bodies in an unelastic state), Cavendish caused some metallic substances (zinc, iron, and tin) to react with diluted acids (vitriolic [sulfuric] acid and spirit of salt [hydrochloric acid]; in modern terms, the acids reacted with metals and released hydrogen, which burned in the presence of oxygen). The discovery of inflammable air was of considerable importance during the controversy over phlogiston. In fact, Cavendish, in line with phlogiston theory, argued that inflammable air came from metals and hence identified it with pure phlogiston.[33] Until Cavendish synthesized water from inflammable and dephlogisticated airs, this experiment was a serious puzzle for Lavoisier.

Thanks to these discoveries, the best chemists of that time realized that a new and promising field of inquiry had been born. Hence, it is not by chance that both Priestley and Lavoisier recognized that their investigations took root in Hales's work. In the first volume of his *Experiments and Observations* (1774–1777), Priestley reminds the reader that

> Dr. Hales, without seeming to imagine that there was an material difference between these kinds of air and common air, observed that certain substances and operations *generate* air, and others *absorb* it; imagining that the diminution of air was simply a taking away from common mass, without any alteration in the properties of what remained.[34]

Later on, he made it clear that his whole research was a part of pneumatic chemistry:

> There are, I believe, very few maxims in philosophy that have laid firmer hold upon the mind, than that air, meaning atmospherical air . . . is *a simple elementary substance,* indestructible, and unalterable, at least as much so as water is supposed to be. In the course of my inquires, I was, however, soon satisfied that atmospherical air is not an unalterable thing; for that the phlogiston with which it becomes loaded from bodies burning in it, and animals breathing it, and various other chemical processes, so far alters and depraves it, as to render it altogether unfit for inflammation, respiration, and other purposes to which it is subservient.[35]

Granting the ontology determined by pneumatic chemistry—namely, the existence of chemically different gases—Priestley explains their diversity by means of phlogiston theory.

Lavoisier's research, too, depends on Hales's work. In his *Pli cacheté* (1772),[36] Lavoisier announces the discovery that through burning, sulfur and phosphorus do not decrease, but on the contrary increase in weight, and this is so because of the great quantity of air that is fixed during the combustion. Later on, "with Hales's apparatus" he repeats the experiment using litharge (a lead oxide) and notices a similar phenomenon. Hales's influence on Lavoisier is even more apparent in Lavoisier's *Memorandum* (1773) and *Opuscules physique et chimique* (1774). In the *Memorandum*, for instance, we read:

> [I]t is certain that under a large number of circumstances an elastic fluid is liberated from bodies. . . . Some scholars, such as Mr. Hales and his followers, thought that this was the very atmospherical air. . . . Hales maintained that this fluid was different from what we breath only because it was more loaded with noxious or wholesome substances according to the nature of bodies from which is obtained. Other physicists, followers of Hales, noted such large differences between the air from bodies and the air we breath that they maintained it was another substance, which they called fixed air.[37]

In the *Opuscules,* Lavoisier describes Hales's work as an "almost inexhaustible source of meditation."[38]

It is clear from these quotations that both Priestley and Lavoisier settled their research in the scientific tradition determined by pneumatic chemistry, although they interpreted this tradition very differently. Although Priestley assumed in his investigations the existence of qualitatively different airs, he continued to believe that such differences were due to a *unique* principle, that of phlogiston: "[T]here is a regular gradation from *dephlogisticated air,* through common air, and *phlogisticated air,* down to nitrous air; the last species of air containing the most, and the first-mentioned the least *phlogiston* possible."[39] Lavoisier's divergence from tradition was more radical. Not only did he not resort to phlogiston to explain chemical phenomena, but he also gave a new definition of simple or elementary chemical substances. In a very well-known passage, we read:

> All that can be said upon the number and nature of elements is, in my opinion, confined to discussions entirely of a metaphysical nature. . . . [B]ut if we apply the term *elements* or *principles of bodies,* to express our idea of the last point which the analysis is capable of reaching, we must admit, as elements, all the substances into which we are capable, by any means, to reduce bodies by decomposition."[40]

Historians of science have particularly focused on the "operationalistic" definition of chemical elements propounded by Lavoisier. They typically underline such a definition as an important part of Lavoisier's conceptual revolution. Undoubtedly, it is true that Lavoisier suggested the idea that each gas was composed of a specific principle and caloric (so, for example, oxygen was made up of a caloric and oxygen base), but caloric plays a radically different role with respect to phlogiston. Caloric cannot explain chemical differences, but rather thermal phenomena together with the different states of a body. Hence, caloric is a principle that belongs to physics, not, like phlogiston, to chemistry.

Lavoisier's definition of chemical elements is fundamental for the creation of chemistry understood as the combining of chemical elements, which in turn is related to the idea of a chemistry based on physical-quantitative methods. In an essay on the action of acids on metals, Lavoisier also tried to formalize the language of chemistry, although he recognized that "we are still very far from being able to introduce mathematical precision in chemistry."[41] For Lavoisier the definition of element and the law of conservation of matter were one and the same thing in his view of chemistry as a physical and quantitative science.

I shall return to the differences between Priestley and Lavoisier later. At the moment it is more important to stress that albeit with many differences, both focused their investigations on the properties of gases, and with such investigations they linked their names. Notwithstanding their dramatically different approaches, they were basically two pneumatic chemists.

Ontological Levels and Explanatory Theories

Above I discussed how oxygen and phlogiston theories determine two different and incompatible ontologies, yet both the ontologies refer to a shared ontology, brought about by the rise of pneumatic chemistry. Hence it is opportune to distinguish between two different ontological levels.

The first ontological level, as defined by pneumatic chemistry, tells us that *there exist different gases or airs showing different chemical properties.* Both Lavoisier and Priestley—and in general all pneumatic chemists—devoted much empirical investigation to determine the largest possible number of new gases together with their chemical properties. The locution "first level" does not mean an intrinsically more elementary level. What I mean is that the ontological structure determined by pneumatic chemistry was never questioned during the controversy; nor could it be questioned, since, as I have shown above, both Priestley and Lavoisier based their research on such an ontology. In this sense we can call it the "fundamental ontology."

Priestley's phlogiston theory and Lavoisier's oxygen theory determined a different ontological level, which presupposes the first. For instance, in the former we have "dephlogisticated air" (for example, air deprived of most of its phlogiston) and in the latter we have "oxygen" (which is chemically a simple element); in the former we find "calces" (which are simple elements) and in the latter we find "oxides" (which are compounds); and so on. As languages, they cannot be refuted by means of experimental facts, since facts are described by using either of the two languages. As languages, and only in this sense, as I shall show, we can call phlogiston and oxygen theories two *natural interpretations,* since they determine what does and does not exist in nature.

The distinction between the two ontological levels sheds some light on a well-known problem among both historians and philosophers of science: Who was the first to discover the compound nature of water? And similarly: Who was the first to discover oxygen? Kuhn has argued that is impossible to give a precise answer to these questions, since the distinction between facts and their interpretations is epistemologically unfounded. For instance—I take this to be his argument—if we adopt a phenomenological viewpoint, we should claim that oxygen was discovered by the person who isolated it first. However, it is very likely that oxygen had already been isolated by Hales and by who knows how many other people, who did not recognize it as a new entity. Hence, the phenomenological approach leads to the nonsensical consequence that oxygen was discovered when no one even suspected its existence. On the other hand, the idea that oxygen was discovered by whoever first correctly *interpreted* the experimental result also gives rise to conceptual problems. In fact, as already noted,

Lavoisier considered oxygen to be compounded by oxygen base and caloric. This interpretation is not correct, since caloric has subsequently been abandoned by physics. Hence, the claim that oxygen was only discovered after the abandonment of caloric leads to the counterintuitive consequence that it would have been discovered when all the world knew its existence. Similar considerations apply regarding the discovery of the compounded nature of water. Kuhn's argument has embarrassed even those who are not sympathetic to his epistemology. For instance, to the question, Who did first discover oxygen? A. Musgrave cannot find better answer than "I do not know and I do not care."[42]

On the contrary, I maintain that the question is perfectly sensible and legitimate *as soon as the ontological level that is appropriate to the given situation is singled out.* To discover means to find out something that exists in the world, and we have appreciated that questions concerning the "existence" partially depend on our conceptual schemes. If one believes that the appropriate ontology is given by pneumatic chemistry, then there is no doubt that oxygen was discovered by Priestley and the composition of water by Cavendish. In fact, Priestley first recognized that dephlogisticated air was a different chemical substance from atmospheric air, and Cavendish first discovered that water was the result of the explosion of two gases and not the product of the preexisting humidity in the atmospheric air (questions concerning the dates have been fairly well clarified by historical studies).[43] However, if we accept the ontology determined by Lavoisier's natural interpretation, which subsequently prevailed, then it is likewise apparent that those discoveries must be ascribed to Lavoisier, since he first clarified that oxygen is a chemically simple substance, and that water is actually compounded by two different substances.

If we project ourselves into the "situational logic" in which protagonists moved, it seems to me reasonable to claim that the appropriate ontology is the one determined by pneumatic chemistry. Their contemporaries noticed that ascribing the above-mentioned discoveries to Lavoisier implied giving an illegitimate meaning to the term "discovery." At most it could be claimed that Lavoisier suggested an original *interpretation* of Priestley's and Cavendish's discoveries. In fact, they rejected precisely the ontology propounded by Lavoisier. Once more, only the ontology of pneumatic chemistry was beyond question at that time.[44]

Yet, what seems clear, if we project ourselves into the situational logic, becomes much more confused if we adopt a later viewpoint. In fact, later on it was the ontology propounded by Lavoisier that became common ground for chemists. Indeed, such confusions started quite soon. In 1840, Jöns Jacob Berzelius writes:

> It has been said that the differences in view between Watt or Cavendish and Lavoisier were only a matter of language, being sufficient to translate the former in the terms of the latter to obtain the same idea. However this is not the case. Watt and Cavendish considered oxygen, hydrogen and water as different states of the very same ponderable substance, whereas Lavoisier proved that water was compounded by two specific ponderable bodies; and this is precisely his discovery.[45]

Berzelius correctly points out the ontological implications of languages. However, he adopts the viewpoint of the wrong ontology, that is, the ontology accepted in *his* age. Leaving aside easy nationalisms, I maintain that the failure to specify the appropriate ontological level plays an important role in the pointless controversies about the priority of discoveries among historians.

Together with the fundamental ontology and the two natural interpretations, during the controversy we find two different systems of *explanatory theories.* Clearly enough, each system depended on one of the two natural interpretations. Lavoisier was perfectly aware that his theories involved the very language in which phenomena were described (although, of course, he did not state this point in the radical way presented here, which is the result of a very different cultural context). In the *Discours préliminaire* to his *Elements of Chemistry* (1790) he writes: "[W]hile I thought myself employed only in forming a nomenclature, and while I proposed to myself nothing more than to improve the chemical language, my work transformed itself by degrees, without my being able to prevent it, into a treatise upon the elements of chemistry."[46]

On the contrary, phlogiston theorists typically held a different view. For instance, Cavendish argued against Lavoisier's nomenclature that "the only way to avoid false opinions is to suppress as reasoning as much as possible . . . and can anything tend more to rivet a theory in the minds of learners than to form all the names which they are to use upon that theory."[47] In a similar vein, James Keir complains that the language of anti-phlogistonists "is formed on the supposition of their system being certain."[48] In all his works, Priestley remained consistent with his phenomenalistic epistemology, and consequently carefully distinguished facts from theoretical hypotheses. For the same epistemological reason, he never cared to specify a precise system of explanatory hypotheses. At their very best, they were heuristic devices, useful in the discovery of new facts.[49] Thus, it is not surprising that Priestley was prone to abandon them very quickly.

It is worth remembering that Priestley—here, too, in line with his epistemology—held that phlogiston was only a hypothetical substance. He maintained that phlogiston was "an unknown cause of certain well known effects."[50] Furthermore, in several works, he claimed that he was willing to abandon phlogiston had experiments shown evidence against its existence.[51] By referring to these passages, McEvoy—who more than any other scholar has emphasized the importance of Priestley's epistemology for his work as a chemist—writes that "contrary to prevalent historiography, Priestley was conscious of the hypothetical nature of phlogiston theory as he was of the oxygen theory. Though satisfied with phlogiston theory he was always ready to reject it should it be shown to be false."[52]

However, this leads us to a question that we have met already: Why was Priestley so tenacious in his defense of phlogiston, although he proved himself very ready to abandon any other kind of hypothesis? It is impossible to answer this question if we consider phlogiston a hypothesis on a par with the many other hypotheses that he put forward throughout his scientific career. Actually, phlogiston was a very peculiar "hypothesis," since it provided the "natural interpretation" through which he gave a meaning to phenomena and structured chemical research. As I have shown, phlogiston was the chemical counterpart of his theological interest for the phenomena concerning both the vitiation of air and its volumetric properties. Although, in accordance with his phenomenalistic epistemology, phlogiston was a hypothetical substance, it could not be abandoned as any other hypothesis. More precisely, although Priestley never gave a privileged status to his explanatory theories, they were based without exception on the fundamental idea that in all processes showing the contraction in the volume of a given quantity of air (like in the calcination or breathing phenomena) there was a phlogistication of that air, such that the air was no longer suitable to support the process.

The "traditional" historians of science, such as Partington, charged Priestley with incoherence, since he was tenacious in his defense of phlogiston yet at the same time too willing to abandon his explanatory hypotheses. However, this charge overlooks the fact that phlogiston and explanatory hypotheses had different functions: phlogiston provided the natural interpretation of chemical phenomena, whereas explanatory hypotheses *presupposed* and *were built upon* such a natural interpretation. In fact, hypotheses such as "vegetation absorbs phlogiston from atmospheric air," "dephlogisticated air is produced from all kinds of earth mixed with spirit of nitre," and "inflammable air is pure phlogiston" presuppose in an obvious way the conceptual scheme of phlogiston. Priestley never hesitated in changing this kind of hypothesis.

In order to understand the controversy over phlogiston, it is important to focus our attention on the interaction among these three conceptual levels: the fundamental ontology determined by the rise of pneumatic chemistry, the two different and incompatible natural interpretations, and the two systems of explanatory theories. More specifically, what I examine next is the development of the explanatory theories that the progression of pneumatic chemistry prompted. As I shall show, the debate was fascinating, the criticisms proved to be penetrating, and the dialogue never became dogmatic.

Priestley's Opposition

In 1791, with Kirwan's conversion to the new theory, the scientific revolution was practically concluded. The reasons underlying the change of consensus have been extensively studied, and I shall not dwell much on this issue. However, I would like to emphasize that the progress of pneumatic chemistry led to the discovery of an impressive number of new gases, the different chemical properties of which were not easily explicable through the use of a single universal principle. Whoever reads Priestley's *Observations Relating to Theory* (1790), where he tries to "ascertain the number of *elements* that are necessary to constitute all the substances with which we are acquainted, and especially the different kinds of air,"[53] can readily appreciate the difficulties he had to face. Indeed, historians typically agree that this work is very modest scientifically, although I believe that it is historically significant.

Lavoisier's natural interpretation was undoubtedly more appropriate in explaining the new empirical discoveries. Furthermore, as is well known, the synthesis of water solved many problems for Lavoisier. In a 1783 *Mémoires* on the composition of water[54] Lavoisier was able at the same time to offer a new explanation to Cavendish's 1776 experiments on metals and acids, and Priestley's (1783) experiments on *minium,* which up to that time had provided strong support for phlogiston. From this, should we conclude that the opposition to scientific revolution was irrational or, at the very least, unreasonable? As I shall show, even from the most "liberal" methodologies we cannot derive such a conclusion.

Let us consider Musgrave's reconstruction of the controversy. He maintains that with Cavendish's synthesis of water the two theories were to be considered empirically equivalent. However, he argues, the equivalence was reached only through ad hoc moves, since phlogistonists adjusted phlogiston theory to the discoveries derived by the oxygen research program. As he claims, the chemists of the late eighteenth century "saw that 1784 phlogiston theory merely accommodated known facts, many of which had been discovered by testing predictions made within the oxygen program."[55] To Musgrave, the methodology of Imre Lakatos's concept of research programs perfectly explains the success of oxygen over phlogiston.

Historians of science quite often blame philosophers for misrepresenting the real history of science. Reconstructions such as Musgrave's seem to show that they are somewhat justified in this charge. Many chemists of that time rejected—adducing very good reasons—precisely what seems apparent to Musgrave. For instance, Charles Blagden argued that the hypothesis of the compound nature of water was ad hoc:

> From Mr. Lavoisier's own account of his experiment, it sufficiently appears, that at that period he had not yet formed the opinion that water was composed of dephlogisticated and inflammable airs; for he expected that a sort of acid would be produced by their union. . . . *Mr. Lavoisier's present theory . . . perfectly agrees with that of Mr. Cavendish; only that Mr. Lavoisier accommodates it to his old theory, which banishes phlogiston.*[56]

Blagden has good evidence supporting his claim. It is well known that the term "oxygen" originally stood for the principle of acidity. In an essay read in 1779, Lavoisier stated his theory of acidity in the most precise way.[57] On the basis of his experiments on several acids, and in particular the phosphoric, vitriolic, and nitrous (actually, nitric) acids, he claims that "the purest air or eminently breathable air is the constitutive principle of acidity,"[58] which he calls *principe oxygine*. Hence, it is not surprising that he imagined that from the explosion of a mixture of inflammable air and oxygen the result would have been an acid. In fact, together with P. Gengembre, he repeatedly tried to identify it.

From a philosophical viewpoint, Musgrave's mistake lies in his conviction that it is always possible to specify which of two or more theories entails a given empirical result. If this were so, it would be easy to determine which theory is empirically progressive and which simply "accommodates" the discoveries derived from an alternative theory. Yet, in this case we do not have such an algorithm, since neither of the two theories strictly entailed the synthesis of water from inflammable air and oxygen.

However, it would be overly hasty to think of the controversy as a clash between two dogmas, where empirical arguments proved to be circular in the sense explained above. On the contrary, the existence of a common ontology—provided by pneumatic chemistry—allowed an open and fruitful discussion. The impossibility of binding or compelling empirical arguments does not entail the inexistence of empirically *persuasive* arguments. Indeed, Priestley's criticisms required answers from his interlocutors, and in this way led to empirical discoveries and better conceptual specifications.

As I shall show, Priestley did not follow Cavendish and Blagden in criticizing Lavoisier's theory of acidity. However, he was perfectly aware that the synthesis of water not only did not yield a conclusive proof of oxygen theory, but also gave rise to serious problems for Lavoisier. To Lavoisier the decomposition of water could explain why in several experiments inflammable air or hydrogen had been isolated. This idea was indeed crucial in the system of explanatory theories upheld by Lavoisier. Therefore, Priestley quite rationally referred to this idea as being *the* anti-phlogistic theory.

Priestley put forward several arguments against anti-phlogistic theory. Here I focus on two of these that seem to me particularly significant. The first argument directly concerns the inflammable air or hydrogen. Priestley obtained from charcoal and finery cinder (an oxide of iron) a new kind of inflammable air and held that this constituted a decisive argument against Lavoisier's theory. In fact, there being no water in either substance, where does the inflammable air come from? As he claimed:

> This experiment seems to be decisive against the hypothesis of Mr. Lavosier, and others, who say that the inflammable air procured by means of iron and charcoal, comes from water. . . . Had this been the case, there was nothing in either of the materials made use of in this experiment from which the inflammable air could possibly come, there being no *water* contained in either of them.[59]

For a long time, Priestley's criticism puzzled his opponents, who found a satisfactory answer only in 1801, thanks to William Cruickshank, who explicitly referred to Priestley's criticism. For reasons that I have examined already, Priestley did not care that the inflammable air from metals and the inflammable air from charcoal had different weights. Continuing Priestley's research, Cruickshank found that the inflammable air isolated was "but a very little lighter than common air,"[60] while hydrogen was much lighter. Studying the new kind of inflammable air, he recognized a new kind of carbon oxide, namely, carbon monoxide.

Thus, Priestley's criticism led to the discovery of a new kind of gas and, perhaps more surprisingly, required a modification of anti-phlogistic theory. As Schofield himself notes,[61] the latter theory claimed that the production of inflammable air arose from the decomposition of water, and this idea had to be abandoned. Furthermore, in this theory oxides were the final product of combustion, and consequently Priestley legitimately seized the opportunity to ask how the final product of one combustion could burn again. In fact, responding to Cruickshank he notes, in one of his latest works, that "this air from finery cinder and charcoal, though called an *oxyd* must be essentially different from all other oxyd, none of which are combustible, being substances already saturated with oxygen."[62]

Priestley was confident that Cruickshank's experiments would have led to the overthrow of oxygen theory. He writes that Cruickshank's statements "directly [militate] against the system that he endeavours to support."[63] In connection with the early experiments with Voltaic piles he claims that they are incompatible with the idea that water is a compound and, probably referring to Cruickshank's experiments, notes that "there are other symptoms of [the] decline [of the new system of chemistry] in point of credit, and I doubt not that in a few years it will be universally exploded. *Sic transit gloria mundi.*"[64]

Priestley was partly wrong and partly right. Lavoisier's natural interpretation—according to which chemistry is the science of the combination of chemical elements, founded on the same model as physics—would prove to be an irreversible change. Yet, the anti-phlogistic theory put forward by Lavoisier would be refuted before long.

Priestley was not content with Lavoisier's explanation of Cavendish's 1766 experiments on metals and acids. As I have shown, after the discovery of the composition of water, Lavoisier claimed that the inflammable air or hydrogen came from water, not from metals, as Cavendish had suggested. However, precisely the ponderal measurements—so important for Lavoisier and his school—suggested to Priestley that this issue needed further clarification. Priestley carefully tackled this problem in his last book, *The Doctrine of Phlogiston Established* (1800):

> [T]he question to be solved is from which of the substances present comes the *inflammable air* that is procured in the process. The phlogistians say it comes from *iron,* and the anti-phlogistians from the *water.* But to this I object that, since, according to their own hypothesis, water consists of about six times as much oxygen as it does of hydrogen, there must be a large deposit of oxygen in the vessel, and I cannot find it there."[65]

Strangely enough, Priestley did not realize that the very same argument seriously weakened Lavoisier's theory of acidity.[66] Consequently, he missed the opportunity fully to develop his criticism. Nonetheless, he raised a crucial problem.

In a letter to S. Mitchill he emphasized that the difficulty in Lavoisier's interpretation of Cavendish's 1766 experiments was founded on the very "doctrine" of the composition of water,[67] and he was certainly right, since Lavoisier agreed that oxygen constituted most of the weight of water.[68] A correct explanation of the problem raised by Priestley was only to be found after Humphry Davy's experiments.

Priestley and Davy knew and respected one another, as Davy's frequent quotations from Priestley's work and their correspondence demonstrate.[69] But, more interesting than this, Davy clearly took phlogiston theory and its upholders' criticisms seriously. Both his published works and his *Memoirs* provide evidence that he thought seriously of resurrecting the phlogiston theory. Although it is beyond the scope of this chapter to examine the metaphysical roots of his dissatisfaction with French chemistry,[70] it is relevant to understanding the kind of phlogiston theory to which he referred. Let us examine the following passage, taken from his notes:

> If the electrical energies of bodies be examined, oxygen, and all bodies that contain a considerable portion of oxygen, appear to be *negative;* hydrogen, the metals, and all combustible bodies, *positive.* . . . A new phlogistic theory might be established, which would explain all the phenomena, as well as the antiphlogistic. . . . If we suppose water *simple;* then, oxygen will be water–, hydrogen water+, the metals, . . . unknown bases, water+; charcoal, sulphur, phosphorus, nitrogen, unknown bases, water+."[71]

This is no more than a specific version of Cavendish's 1784 phlogiston theory, to which Priestley also adhered till the end of his days.

Note that Davy does not reject the view of chemistry as a physical science, and that phlogiston, being now related to the concept of electric charge,[72] acquires new properties in his theory. This attitude is not surprising. Explanatory theories, although generated and justified within a given natural interpretation, can be reinterpreted, with some modifications, in different research contexts. Furthermore, natural interpretations themselves can be modified—also in some important characteristics—without losing their identity.[73]

In the last years of Priestley's life, chemical investigations took a new turn thanks to the use of Voltaic piles. The rise of electrochemistry allowed new discoveries to be made that brought forth new conceptual problems. Specifically, Davy succeeded in isolating the alkaline metals and decomposing alkaline earths. The peculiar characteristics of alkaline metals pushed Davy, although with great caution, to suggest the hypothesis that "a phlogistic chemical theory might certainly be defended, on the idea that metals are compounds of certain unknown bases with the same matter as that existing in hydrogen,"[74] and his experiments on ammonia seemed to strengthen the conjecture that "metals and inflammable solids, usually called simple, [are] compounds of the same matter as that existing in hydrogen, with peculiar unknown bases."[75]

These phlogistic conjectures did not succeed. However, they are historically important, since they show that in his research Davy bore both phlogiston and oxygen theories in mind and used one to criticize the other. Furthermore, independent of his view, the discovery that strong alkalis, such as potash and soda, contained oxygen objectively weakened Lavoisier's theory that oxygen was the principle of acidity.[76]

The attention which Davy paid to phlogistonists' criticism led him to take into serious consideration an anomaly concerning Lavoisier's theory of acidity. And in this case the phlogistonists' suggestions proved to be correct. Lavoisier's theory of acidity implied that muriatic acid (hydrochloric acid) contained oxygen, but this acid was recalcitrant in liberating its oxygen. Furthermore, Lavoisier's theory led to problems concerning the interpretation of the experiments.

Scheele isolated chlorine and, conforming with phlogistic theory, called it "dephlogisticated acid of salt" (namely, HCl from which hydrogen—for example, phlogiston—had been subtracted). Needless to say, anti-phlogistonists maintained that, on the contrary, this was an oxygenation process. Lavoisier noted that "oxygenated muriatic acid" (actually, chlorine) lost its acid properties,[77] but this did not induce him abandon his theory of acidity. Indeed, this anomaly did not appear to be particularly important. Not only were all known acids proven to contain oxygen, but chlorine preserves combustion, a property that at that time was related to the presence of oxygen.[78]

Phlogistonists had a different attitude, and also thanks to them the problem was duly emphasized. As Cavendish writes:

> Mr. Lavoisier endeavours to prove . . . that dephlogisticated air is the acidifying principle. From what has been explained it appears, that this is no more than saying, that acids lose their acidity by uniting to phlogiston. . . . But as to the marine acid and acid of tartar, it does not appear that they are capable of losing their acidity by any union with phlogiston.[79]

Priestley never thought of attacking Lavoisier's theory of acidity.[80] Nevertheless, he did not fail to raise the question: "I concluded from many experiments on substances containing marine acid, that it differs essentially from both the vitriolic and nitrous in this, that it cannot, by any combination whatever, be made to yield dephlogisticated air, at least with the degree of heat that I was able to apply."[81] These remarks did not lead to the rebirth of phlogistic natural interpretation, but they involved the overthrow of anti-phlogistic theory. In fact, Davy showed that Lavoisier's interpretation of Cavendish's experiments on metals and acids was wrong and that Priestley was right in stressing its weakness.

Davy begins his essay on "oxymuriatic acid," read in 1810, by recalling phlogistic theories: "The illustrious discoverer of the oxymuriatic acid considered it as muriatic acid freed from hydrogen; and the common muriatic acid as a compound of hydrogen and oxymuriatic acid; and on this theory he denominated oxymuriatic acid dephlogisticated muriatic acid."[82] Throughout his investigations he proved that "oxymuriatic acid" (actually, as I have already said, the element chlorine) is converted into muriatic acid through its combination with hydrogen. Consequently, Davy claims that it was the phlogistic theory that was correct, not Lavoisier's:

> Scheele's view (though obscured by terms derived from a vague and unfounded general theory) of the nature of the oxymuriatic and muriatic acids, may be considered as an expression of facts; whilst the view adopted by the French school of chemistry, and which, till it is minutely examined, appears so beautiful and satisfactory, rests in the present state of our knowledge upon hypothetical ground.[83]

Despite much resistance,[84] both Lavosier's view of acidity and the anti-phlogistic theory had to be abandoned. Thanks to Davy, it was shown that Priestley's objections concerning the experiments on acids and metals were correct: the hydrogen liberated could *not* come from water. In fact, Davy established that it came from acid: "When muriatic acid gas is acted upon by mercury, or any other metal, the oxymuriatic acid is attracted from hydrogen by the stronger affinity of the metal."[85] Although Priestley's theory was wrong (hydrogen was not from metals), his criticisms were right also from the viewpoint of contemporary science.

Conclusion

Priestley has often been seen as a dogmatic scientist, deaf to the demand for changes. The new historiography has changed only in claiming that dogmas are inherent to the progress of scientific research itself. Actually, even the case of Priestley shows that scientific research is more like a Socratic dialogue than a dialogue between deaf people. Traditional historiography of science has been criticized on the basis that facts are subject to various and mutually incompatible interpretations. Yet the new historiography of science, too, should be revised.

The new historiography pointed to Priestley's and Lavoisier's different conceptions of chemical research in claiming the existence of a radical ontological discontinuity between the chemistry of phlogiston and that of oxygen. According to this view, during the revolution empirical arguments unavoidably became circular, since they presupposed the conceptual schemes that were the object of the controversy. This thesis, however, overlooks important aspects of the scientific debate. In fact, the two natural interpretations were founded on a common ontological framework, determined by the rise of pneumatic chemistry. All the chemists—independently of their acceptance of phlogiston or oxygen—were involved in the growth of pneumatic knowledge, namely, in both the number and properties of the gases experimentally isolated. Thus, it is not surprising that the debate was fruitful and never a mere clash between two different dogmas.

The interest in Priestley's position does not consist in some of his speculations—such as the connection between chemical phenomena and electricity—which he was not able to develop. What I want to emphasize is rather that his criticisms proved to be useful for scientific progress. He found some people willing to listen to him, and the dialogue proved to be productive.

The existence of heterodox positions in science is a philosophical and historiographical problem only if we claim that scientific rationality is based exclusively on the cogency of facts (asserted by the traditional and denied by the new historiography). But in so doing, we neglect that progress requires—to use Priestley's words—a "free discussion,"[86] where "heretical" positions may have an important role in stimulating the dialogue, thus provoking conceptual specifications and even new empirical discoveries.

Notes

1. Priestley (1779–1786), vol. 1, p. xi.
2. Partington (1962–1964), vol. 3, p. 246.
3. In Kuhn (1970) there are several remarks on Priestley, and Toulmin (1957) devoted a seminal essay to Priestley's defense of phlogiston.
4. Kuhn (1970), p. 94.
5. Schofield (1964), p. 287. Schofield (1957) has also persuasively argued that Priestley's scientific knowledge was not amateurish.
6. For a detailed overview of the studies of Priestley, see Schaffer (1984). Note also that I use the term "new historiography" in a very broad sense. Actually the "new" historigraphy is about thirty years old! Among the most recent scholars, it is worth mentioning J. G. McEvoy, who, unlike Schofield, has emphasized the role played by Priestley's *epistemological* convictions. McEvoy appears to be less committed to what I have called the "new historiography of science." Nonetheless, he clearly agrees that Priestley's attitude can only be explained within the context of his philosophical background. See, for instance, McEvoy (1978–1979). In Barrotta (1994) I provide a more detailed historical account supporting my interpretation of the controversy between Priestley and Lavoisier.
7. For a defense of the role of dialectics in science see, for instance, Pera (1994).
8. Priestley (1774–1777), vol. 1, part 1, sec. 4, p. 71.
9. On this issue see also Priestley (1776), where he writes, "That respiration, however, is, in reality, a true phlogistic process, cannot, I think, admit of a doubt, after its being found, that the air which has served for this purpose is left in precisely the same state as that which has been exposed to any other phlogistic process" (p. 238).
10. Priestley (1774–1777), vol. 1, part 1, sec. 4, p. 88.
11. Ibid., pp. 93–94.
12. Priestley, *Institutes of Natural and Revealed Religion,* 1772–1773, 2nd ed. 1782, in Priestley (1972), vol. 2, p. 8.
13. On this issue see, for instance, McEvoy and McGuire (1975).
14. Priestley, *Disquisitions Relating to Matter and Spirit,* 1777, 2nd ed. 1782, in Priestley (1972), vol. 3, p. 223.
15. Ibid., p. 224.
16. Ibid., p. 227.
17. Ibid.
18. Priestley (1777), p. 3.
19. Priestley, *A Free Discussion of the Doctrines of Materialism and Philosophical Necessity,* 1778, in Priestley (1972), vol. 4, p. 19.

20. Metzger (1935), p. 21. English translation is mine.

21. Lavoisier (1790), p. 41, where he writes: "We may lay it down as an incontestible axiom that, in all the operations of the art and nature, nothing is created; an equal quantity of matter exists both before and after the experiment: the quality and quantity of the elements remain precisely the same and nothing takes place beyond changes and modifications in the combination of these elements."

22. Lavoisier, *Développement des dernières expériences sur la décomposition et la recomposition de l'eau,* 1786, in Lavoisier (1862–1893), vol. 5, p. 333. English translation is mine.

23. This point has been noted by Lavoisier's contemporaries as well. Fourcroy describes the periodic meeting in Lavoisier's house as follows: "[A]mong the important advantages of these meeting, what most struck me, and influenced very soon the *Académie* and later on all the work in physics and chemistry published in France in the following twenty years, was the agreement between geometers' and physicists' way of reasoning. The precision and austerity of the language of the former passed step by step into the mind of the latter." Fourcroy (1796), pp. 33–34. English translation is mine.

24. The most famous, and very often ridiculed, hypothesis is that of the absolute levity of phlogiston (on the history of this hypothesis, see Partington and McKie (1937–1939). Priestley never accepted such a hypothesis. See, for instance, Priestley (1774–1777), vol. 2, where he writes: "I never had any faith at all in the doctrine of the principle of levity. On the contary, . . . I consider the difference of weight between a metal and a calx, which has given occasion to that doctine, as wholly owing to the fixed air and water imbibed by the latter in the act of calcination" (p. 312). Also, after Lavoisier's experiments on the *mercurius calcinatus* (see Lavoisier, *Mémoire sur la nature du principe qui se combine avec les métaux pendant leur calcination, & qui en augmente le poids,* 1775, in Lavoisier 1862–1893, vol. 2, pp. 122–128) that refuted Priestley's explanation, phlogistonists could resort to another possible explanation. For instance, in 1782, Kirwan claimed: "It is true, that the mercurial calx, and also the calces of lead, and many others, yield dephlogisticated air; but then the mercury is always revived, so that is evident, it retakes the phlogiston from the fixed air, of which nothing then remains but the dephlogisticated part, which accordingly appears in the form of dephlogisticated air." Kirwan (1782), p. 228. Priestley adopted Kirwan's explanation. See Priestley (1790), vol. 3, pp. 541–542.

25. Schofield (1966), pp. 189–190.

26. Cavendish, *Experiments on Air,* 1784–1785, in Cavendish (1921), vol. 2, p. 180–181. Italics are mine.

27. For an extensive overview on this issue see, for instance, Abbri (1984), chap. 1.

28. Hales (1727), p. 315.

29. Ibid.

30. Brownrigg (1765), p. 219.

31. Ibid., p. 238.

32. Black (1777), pp. 69–70.

33. As he writes: "It seems likely . . . that, when either of the above mentioned metallic substances are dissolved in spirit of salt, or the diluted vitriolic acid, their phlogiston flies off, without having its nature changed by the acid, and forms the inflammable air." Cavendish, *Experiments on Factitious Air,* 1766, in Cavendish (1921), vol. 2, p. 79.

34. Priestley (1774–1777), vol. 1, p. 4.

35. Ibid., pp. 30–31.

36. Lavosier, *Pli cacheté,* 1772, in Lovoisier (1955–1964), vol. 2, pp. 389–390.

37. Lavoisier, *Memorandum,* 1773; in Berthelot (1890), pp. 46–47. English translation is mine.

38. Lavosier, (1862–1893), vol. 1, p. 456. English translation is mine.

39. Priestley (1775), p. 392.

40. Lavosier (1790), p. 3.

41. Lavoisier, *Considérations générales sur la dissolution des métaux dans les acides,* 1782, in Lavoisier (1862–1893), vol. 2, p. 515. English translation is mine.

42. Musgrave (1976), p. 195

43. Lavoisier's original essay (*Mémoire sur la nature du principe qui se combine avec les métaux pendant leur calcination, & qui en augmente le poids,* 1775) generically referred to atmospherical air. Only after the publication of Priestley's work did he speak about the "purest part of air" (for example, oxygen). As far as the composition of water is concerned, Cavendish in his essay *Experiments on Air* (1784–1785) informs us that his experiments on the synthesis of water started in 1781 (see Cavendish 1921, vol. 2, p. 171). He delayed the publication of his work because he was puzzled by the presence of nitrous acid, which he correctly suspected to be inessential to the composition of water. (See also, for instance, Partington [1928], p. 17 ff).

44. Through the notion of "situational logic" I maintain I can provide a satisfactory answer to the query of one of the anonymous referees, who writes, "[A]lthough Barrotta has an interesting approach to understanding scientific discovery and fixing priorities, he ultimately begs the question by relegating priority disputes to a choice of appropriate ontological level that itself is left arbitrary." The choice is not arbitrary because the ontology given by pneumatic chemistry was the only ontology that was not questioned by the whole scientific community at that time.

45. Berzelius (1840), p. 28. English translation is mine.

46. Lavoisier (1790), p. 1.

47. Cavendish, *Unpublished works,* in Cavendish (1921), vol. 2, p. 325.

48. See Schofield (1963), p. 292.

49. On this issue, see McEvoy (1978–1979), part 1, p. 32ff.

50. Priestley (1774–1777), vol. 1, p. 328.

51. This claim is repeated in one of his latest works (see Priestley 1800, p. 3). Priestley tell us that after Lavoisier's experiments on the *mercurius calcinatus* and, later on, after the synthesis of water from inflammable and dephlogisticated airs, he thought seriously of abandoning phlogiston. See Priestley (1783, p. 400; 1785 p. 291).

52. McEvoy (1978–1979), part 1, p. 52.

53. Priestley (1790), vol. 3, p. 533.

54. Lavosier, *Mémoire dans lequel on a pour objet de prouver que l'eau n'est pas une substance simple, un élément proprement dit, mai qu'elle est susceptible de décomposition et de recomposition,* 1783, in Lavoisier (1862–1893), vol. 2, pp. 334–359.

55. Musgrave (1976), p. 205.

56. This passage is from a letter addressed to Loronz Crell and published in *Chemische Annales* in 1786. The letter has been reprinted in Cavendish (1921), vol. 2, pp. 41–42. Italics are mine.

57. Lavoisier, *Considérations générales sur la nature des acides et sur les principes dont ils sont composés,* 1779, in Lavoisier (1862–1893), vol. 2, pp. 248–260.

58. Ibid., p. 249. English translation is mine.

59. Priestley (1779–1786), vol. 3, p. 111n.

60. Cruickshank (1801), p. 3.

61. Schofield (1966), p. 326.

62. Priestley (1803a), p. 25. See also Priestley (1802, 1803b).

63. Priestley to W. Morgan, 1802, in Schofield (1966), p. 317.

64. Priestley to S. L. Mitchill, 1801, in Schofield (1966), p. 309.

65. Priestley (1800), p. 5.

66. As Crosland (1973) has argued, Lavoisier's theory of acidity is based on the *quantity* of oxygen present in the given compound: only after a certain degree of oxidation could the acid properties emerge. The fact that water—the neutral substance *par excellence*—contains that large quantity of oxygen cannot easily be rendered compatible with Lavoisier's theory.

67. See Schofield (1966), p. 293.

68. See Lavoisier, *Dévelopment des derniéres expérience sur la décomposition e la recomposition de l'eau,* 1786, in Lavoisier (1862–1893), vol. 5, p. 327, where he writes "to constitute 100 pounds of water requires no less than 81 and no more than 87 pounds of that constituent part that has been fixed in iron, increasing by this its weight." English translation is mine.

69. In an 1801 letter from Priestley to Davy we read: "I have read with admiration your excellent publications, and . . . I thank you for the favourable mention you so frequently make of my experiments." Schofiled (1966), pp. 313–314. See also Davy (1839–1840), vol. 3.

70. See, for instance, Siegfried (1964).

71. Davy, *Memoirs of the life of Sir H. Davy,* in Davy (1839–1840), vol. 1, pp. 155–157.

72. Now hydrogen would be made up of water plus positive electricity, and oxygen of water plus negative electricity. During the process the two substances neutralize and water is deposited. It is worth mentioning that Priestley himself had suggested some "speculations arising from the consideration of the similarity of *electric matter* and *phlogiston*" (Priestley 1774–1777, vol. 1, p. 274). Priestley, however, was never able to develop such "speculations."

73. On this issue see, for instance, Laudan (1977), ch. 3.

74. Davy, *The Bakerian Lecture, on some new Phenomena of chemical Changes produced by Electricity, particularly the Decomposition of the fixed Alkalies, and the Exhibition of the new Substances which constitute their Bases; and on the general Nature of alkaline Bodies,* 1808, in Davy (1839–1840), vol. 5, pp. 89–90n.

75. Davy, *Electro-Chemical Researches on the Decomposition of the Earth; with some Observations on the Metals obtained from the Alkaline Earth, and on the Amalgam Procured from Ammonia,* 1808, ibid., p. 132. Ammonia is actually peculiar in its chemical properties, since by reacting with water it forms ammonium hydroxide, by analogy with metals.

76. On this issue, see Crosland (1973).

77. Lavoisier, *Rapport sur l'acid marin déphlogistiqué,* in Lavoisier (1862–1893), vol. 4, p. 421, where he writes that that acid "loses almost all its acid properties." In his *Traité,* Lavoisier returns to that anomaly and writes that the peculiar characteristics of oxygenated muriatic acid justify an exception in chemical nomenclature. See Lavoisier (1790), p. 27.

78. See Abbri (1984), p. 350. Moreover, Lavoisier had a good theoretical reason to explain the difficulty in liberating oxygen from muriatic acid: "It is known to chemists that the more bodies . . . are simple . . . the more the methods of composition and recomposition are difficult. Thus it is clear that the decomposition and recomposition of acids must involve greater difficulties than the analysis of salts, which enter into the composition of acids." *Considérations générales sur la nature des acides et sur les principes dont ils sont composés,* 1779, in Lovoisier (1862–1893), vol. 2, p. 250. English translation is mine.

79. Cavendish, *Experiments on Airs,* 1784, in Cavendish (1921), vol. 2, p. 181.

80. In several passeges, Priestley calls oxygen the "principle of acidity." See, for instance, Priestley (1790), vol. 3, p. 535.

81. Ibid., vol. 2, p. 152.

82. Davy, *Researches on the Oxymuriatic Acid,* 1810, in Davy (1839–1840), vol. 5, p. 284.

83. Ibid., p. 291.

84. See Partington (1962–1964), vol. 4, pp. 55–57.

85. Davy, *Researches on the Oxymuriatic Acid,* 1810, in Davy (1839–1840), vol. 5, p. 291.

86. Priestley (1800), p. 2.

References

Abbri F. (1984), *Le terre l'acqua le arie. La rivoluzione chimica del Settecento,* Il Mulino, Bologna.

Barrotta P. (1994), *Dogmatismo ed eresia nella scienza: Joseph Priestley,* Franco Angeli, Milan.

Berthelot M. (1890), *La révolution chimique,* Alcan, Paris.

Berzelius J. (1840), *Rapport annuel sur les progrès des sciences physiques et chimiques,* 1840, trans. M. Plantamour, Paris 1841.

Black J. (1777), "Experiments upon Magnesia Alba, Quicklime, and some other Alcaline Substances," in *Philosophical Society of Edinburgh,* 1756, 2nd ed., W. Creech, Edinburgh 1777.

Brownrigg W. (1765), "An Experimental Enquiry into the Mineral Elastic Spirit, or Air, contained in Spa water; as well as into the Mephitic Qualities of this Spririt," in *Philosophical Transactions of the Royal Society of London,* 55, pp. 218–235.

Cavendish H. (1921), *The Scientific Papers,* ed. E. Thorpe, 2 vols., Cambridge.
Crosland M. (1973), "Lavoisier's Theory of Acidity," in *Isis,* 64, pp. 306–325.
Cruickshank W. (1801), "Some Observations on different Hydrocarbonates and Combinations of Carbone with Oxygen, in reply to some of Dr. Priestley's late Objections to the New System of Chemistry," in *Journal of Natural Philosophy, Chemistry, and the Arts,* 5, 1801.
Davy H. (1839–1840), *Collected Works,* 9 vols., London.
Fourcroy A.F. de. (1796), *Notice sur la vie et les travaux de Lavoisier,* Paris.
Hales S. (1727), *Vegetable Staticks,* London.
Kirwan R. (1782), "Continuation of the Experiments and Observations on the Specific Gravities and Attractive Powers of various Salines Substances," in *Philosophical Transactions of the Royal Society of London,* 72, pp. 179–236.
Kuhn T. (1970), *The Structure of Scientific Revolutions,* 2nd ed., University of Chicago Press, Chicago.
Laudan L. (1977), *Progress and Its Problems,* University of California Press, Berkeley.
Lavoisier A.L. (1790), *Elements of Chemistry,* trans. R. Kerr, University of Chicago, Chicago 1952.
——— (1862–1893), *Oeuvres,* 6 vols. Paris.
——— (1955–1964), *Oeuvres de Lavoisier. Correspondence,* ______, Paris.
McEvoy J.G. (1978–1979), "Joseph Priestley, 'Aerial Philosopher': Metaphysics and Methodology in Priestley's Chemical Thought," in *Ambix,* 25, 1978, part 1, pp. 1–55; 25, 1978, part 2, pp. 93–116; 25, 1978, part 3, pp. 153–175; 26, 1979, part 4, pp. 16–38.
McEvoy J.G. and McGuire J.E. (1975), "God and Nature: Priestley's Way of Rational Dissent," in *Historical Studies in the Physical Sciences,* ed. Russell McCormmach, Princeton University Press, Princeton, N.J., pp. 325–404.
Metzger H. (1935), *La philosophie de la matière chez Lavoisier,* Hermann, Paris.
Musgrave A. (1976), "Why Did Oxygen Supplant Phlogiston?" in *Method and Appraisal in Physical Sciences,* ed. by C. Howson, Cambridge University Press, Cambridge.
Partington J.R. (1928), *The Composition of Water,* London.
——— (1962–1964), *A History of Chemistry,* vols. 3 and 4, Macmillan, London.
Partington J.R., and McKie D. (1937–1939), "Historical Studies on the Phlogistic Theory," in *Annals of Science,* 2, 1937, pp. 361–404; 3, 1938, pp. 1–58; 4, 1938, pp. 337–371; 5, 1939, pp. 113–149.
Pera M. (1994), *The Discourses of Science,* University of Chicago Press, Chicago.
Priestley J. (1774–1777), *Experiments and Observations on different Kinds of Air,* London, vol. 1, 1774; vol. 2, 1775; vol. 3, 1777.
——— (1775), "An Account of further Discoveries on Air," in *Philosophical Transactions of the Royal Society of London,* 65, pp. 384–394.
——— (1776), "Observations on Respiration and the Use of the Blood," in *Philosophical Transactions of the Royal Society of London,* 66, pp. 226–248.
——— (1777), *Disquisitions relating to Matter and Spirit,* London.
——— (1779–1786), *Experiments and Observations relating to various Branches of Natural Philosophy; with a Continuation of the Observations on Air,* vol. 1, London 1779; vol. 2, Birmingham 1781; vol. 3, Birmingham 1786.
——— (1783), "Experiments relating to Phlogiston," in *Philosophical Transactions of the Royal Society of London,* 73, pp. 398–434.
——— (1785), "Experiments and Observations relating to Air and Water," in *Philosophical Transactions of the Royal Society of London* 75, pp. 279–309.
——— (1790), *Experiments and Observations on different Kinds of Air and other Branches of Natural Philosophy, connected with the Subject, in three volumes, being the former Six Volumes abridged and methodized, with many additions,* Birmingham.
——— (1800), *The Doctrine of Phlogiston Established and that of the Composition of Water Refuted,* Northumberland.
——— (1802), "A Reply to Mr. Cruickshank's Observations in Defence of the new system of Chemistry", in *Medical Repository,* 5, pp. 390–392.

——— (1803a), "Remarks on Mr. Cruickshank's Experiments upon Finery cinder and Charcoal," in *Medical Repository,* 6, pp. 24–26.

——— (1803b), "Additional Remarks on Mr. Cruickshank's Experiments on Finery cinder and Charcoal," in *Medical Repository,* 6, pp. 271–273.

——— (1972), *The Theological and Miscellaneous Work's,* ed. John Towill Rutt, 25 vols., Kraus Reprint, New York.

Schaffer S. (1984), "Priestley's Questions: An Historiographic Survey," in *History of Science,* 22, pp. 151–183.

Schofield R.E. (1957), "The Scientific Background of Joseph Priestley," in *Annals of Science,* 13, pp. 148–163.

——— (1963), *The Lunar Society of Birmingham,* Clarendon Press, Oxford.

——— (1964), "Joseph Priestley, the Theory of Oxidation and the Nature of Matter," in *Journal of History of Ideas,* 25, pp. 285–294.

——— (ed.) (1966), *A Scientific Autobiography of Joseph Priestley, 1733–1804: Selected Scientific Correspondence, with Commentary,* MIT Press, Cambridge, Mass.

Siegfried R. (1964), "The Phlogistic Conjectures of Joseph Priestley," in *Chymia,* 9, pp. 117–124.

Toulmin S. (1957), "Crucial Experiments: Priestley and Lavoisier," in *Journal of the History of Ideas,* 18, pp. 205–220.

10

Controversies and the Becoming of Physical Chemistry

KOSTAS GAVROGLU

That controversies have had a rather pronounced role in disciplinary politics is something implicit to the character of the controversies themselves. But controversies were quite decisive in defining (sub)disciplinary boundaries as well. In this chapter I assess the role of controversies in defining such boundaries through their role in forming the discourse of a particular (sub)discipline. Assessing the dynamics of a controversy between British chemists and physicists at the end of the nineteenth century concerning claims over the atom and understanding the controversy surrounding the discovery of argon in 1894–1895 will clarify issues related to the processes of legitimizing the language of physical chemistry and the praxis of its practitioners. I also comment on quantum chemistry.

The controversies I shall be discussing have revolved around the rather involved processes of *appropriation*, *misappropriation*, and *reappropriation* of concepts, techniques, and entities. Undoubtedly, a discipline's assertiveness and strength greatly depend on the successful appropriation of techniques and concepts from another discipline. But this is not enough. The process of appropriation, on the one hand, has to be supplemented by an equally significant process of a reappropriation of concepts, of reestablishing, that is, a lost jurisdiction over certain concepts. And, on the other hand, a subdiscipline's becoming depends significantly on the misappropriation of concepts that appear to be undermining the autonomy of the discipline. All these processes were fraught with disputes, and it was their resolution or, rather, their public negotiation that gave rise to the constitutive consensus among the practitioners of the newly emerging physical chemistry during the closing decade of the nineteenth century.

Before discussing the specific controversies related to the emergence of physical chemistry, let me make some points about chemistry itself. In examining the ways chemists attempted to deal theoretically with the classic problems of chemistry, the historian is invariably confronted with the chemists' particular attitude on how to construct a theory in chemistry, how much one can "borrow" from physics, and what is the methodological status of empirical observations for theory building. The choices made by the chemists and the schemata they proposed brought into being new research traditions, articulated new strategies of experimental manipulation, implied a different role for mathematics in each tradition, and gave rise to different styles of research within these traditions. It was, among other things, the confluence of all these processes that eventually became decisive in forming the chemists' culture.

Such historical considerations, however, involve a number of intriguing philosophical and theoretical dimensions closely related to the implications of reductionism and realism for chemistry. The use of rules, for example, which is a constitutive aspect of the chemists' culture, has led to the formation of a framework where it is possible to have more than one theory or theoretical schema that the practitioners of chemistry can use in a complementary manner. Furthermore, and notwithstanding the more or less violent invasions of mathematics into chemistry during the last two centuries and the remarkable successes of such invasions, the chemists' culture has been the culture of the laboratory, their theoretical constructs were always very sensitive to the exigencies of the laboratory, and theory building has always strongly dependend on using as inputs experimentally measured values of various parameters. In other words, the use of semi-empirical methods in constructing theoretical schemata has always had a far stronger legitimacy in chemistry than in physics.

In the discussions about scientific realism, there is an implicitly shared set of values that undermine the possible contributions chemistry can provide to these philosophical issues. It is believed, not unjustifiably, that the special role of mathematics in physics delineates the problems of scientific realism clearly. It is claimed that physics deals with the fundamental entities of the world and that there are no intrinsic limitations as to how deep it could probe. Whether it studies the planets, billiard balls, atoms, nuclei, electrons, quarks, or superstrings, it is still physics, and the change of scale does not oblige the change, as it were, of the discipline itself—as would be the case in biology and chemistry. Although, on the whole, I agree with this view, I have two reservations. First, it is not too clear to me that when we are discussing quarks or superstrings or the big bang we are doing theoretical rather than mathematical physics. Mathematical physics is often confused with theoretical physics, and I do not think that mathematical physics is the mathematically more exact treatments of problems in physics. During various periods, and especially among the British physicists of the dynamic tradition during the second half of the nineteenth century, mathematical physics has been practiced without necessarily a reference to the underlying ontology. And even in those cases where it was possible to construct a model, this was taken to be a feature that further justified the mechanical explanation, rather than the model itself being the depiction of the underlying physical structure.

My second reservation is that the view which confines the study of realism predominantly to the problems of physics is more a matter of convenience and it does not have a serious justification. Furthermore, such an attitude neglects the theoretical particularity of chemistry. And what is much more important, it supposes an absolute reductionism of chemistry to physics. If nothing else, these undeclared assumptions deprive philosophy of science of a vast area where issues about the ontological status of theoretical entities and the criteria for empirical adequacy for the acceptance of a theory have been discussed quite systematically.

The history of chemistry is a history of chemists' attempts to establish its relative autonomy with respect to physics. Hence, unlike the physicists, the chemists are obliged to proceed to ontological commitments that are *unambiguous and clearly articulated*, and they have little or no tolerance for an attitude that stipulates these to be temporary commitments. Otherwise, the chemist would be at a loss about the underlying ontology, and would never be sure whether chemistry should be doing the describing and physics the explaining. The chemists have passionately debated these issues, and the myth of the reflective physicist and the more pragmatic chemist is, if anything, historically untenable.

At the center of all these theoretical issues are the many faces of the emergent discourse of chemistry and the changes incurring as a result of the becoming of its many and new subdisciplines. Some of the issues related to the emergent discourse of chemistry are the ontological status of theoretical entities, the collective decision of where to stop when searching for building blocks, the role of empirical data in theory building, the coexistence of more than one theory or theoretical schema, the question of their corroboration and falsification, the explanatory strength of the theoretical schemata in chemistry, the legitimacy of the ensuing discourse, and its reliance less on establishing "objective" criteria to which theories should conform and more on the consensus of the community and the metamorphosis of the praxis of its practitioners.

It is often the case that 1885 is considered the "foundation date" of physical chemistry. It is the year when Wilhelm Ostwald, together with Jacobus van't Hoff and Svante Arrhenius, started the publication of *Zeitschriftfür Physikalishe Chemie* Instead of emphasizing foundation dates, I think it is much more useful to consider the rather long period of gestation and becoming of physical chemistry. This period started in the early 1880s and continued, in effect, until the early 1930s. The beginning of this period was marked by van't Hoff's attempt to formulate a theory of osmosis without reducing it to a physical theory, but by justifying it theoretically because of its *analogy* with a physical theory. The close of the period was marked by the successful application of the Schrödinger equation to chemical problems. This fifty-year period witnessed a multifarious public negotiation concerning the degree of autonomy of physical chemistry with respect to both physics and chemistry. It was a public negotiation about the character of theory, about the theoretical status of empirical inputs, about the ontological status of theoretical entities. Above all, this debate determined the legitimizing procedures and concensual activities to be collectively adopted by the members of the emergent community of physical chemists.

The public debate was between different groups of chemists, and it was suggestive of the way each chose to map the undefined and undelineated middle ground that some called physical chemistry and some called molecular physics. In a manner analogous to the situation in spectroscopy, the question was whose domain physical chemistry was. Was it an activity for physicists or chemists? How would the boundaries of this newly emerging area be drawn? What would be the character and extent of the practitioners' allegiances to physics and chemistry? These issues, which bore an immediate relation to the whole question of the *status* of physical chemistry, would be discussed and disputed well into the interwar years, even after the successes of quantum mechanics in chemical problems.

It is often the case that a necessary prerequisite in understanding the formation of disciplinary boundaries is the study of the particularities of local schools, of their global influences and of the discourse developed by each school. Examining, for example, the formation of physical chemistry necessarily obliges one to study the ways the discipline was viewed by the practitioners in the various local schools. The differences both methodological and ontological between, say, Ostwald's group in Leipzig and Alfred Noyes's group at MIT, have been as decisive in defining the boundaries of physical chemistry as was their common commitment in using physics for the analysis of chemical phenomena and, more specifically, in exhausting the possibilities provided by thermodynamics for chemistry.

Another example is quantum chemistry. The discourse developed by all those following the approach of Walter Heitler and Fritz London to chemical valence, and the discourse developed by all those following the more pragmatic approach of Linus Pauling and Robert Mulliken, implied a different set of constitutive criteria for quantum chemistry.

Understanding the role of controversies in forming subdisciplinary boundaries will also help in understanding differentiations in the practitioners' praxis at a level intermediate to the two extremes. Such praxis neither is transcultural and ecumenical nor is marked exclusively by the particularities of strictly local schools, some of which were formed around certain individuals. Though I am not averse to the suggestion that such a level may, in fact, be the "national level" where particular philosophical, cultural, and aesthetic trends have been "condensed" into the practice of the community, I would prefer to talk in terms of different types of scientific discourse.

It is within such a context that I attempt to show the role of the two controversies I mentioned in delineating the autonomous status of physical chemistry. These controversies are by no means the only ones and, at the end, I shall briefly refer to the case of quantum chemistry, which has been analytically discussed elsewhere.[1]

Whose Is the Atom?

From about the mid-1880s the British chemists were expressing an increasing interest in the possibilities provided by the dynamic approach of the British physicists to understand more and more the behavior of atoms. This interest gradually turned into a demand to reappropriate the atom, which had been "snatched" from them by their fellow physicists. The views put forth by the energeticists and the developments in physics, where the understanding of the behavior of the atoms was at the expense of the "billiard ball" ontology so appealing to the chemists, in a way obliged the chemists to reacquire what they felt was theirs in the first place. The chemists were starting to reassert their presence in the discussions from which they had been left out. But British chemists at the time were a lonely lot, sandwiched between two ontologies, both betraying the vision of *their* great Dalton. The British chemists found themselves between a world of ethereal vortices articulated by their fellow physicists and a world of anti-atomic energetics propagandized by some German physical chemists. The discussions between themselves and their fellow physicists was at times difficult and often soul-searching. The chemists' attempt to convince the physicists was becoming progressively more urgent as the energeticists' viewpoint was gaining, if not momentum, at least respectability. The latter could only be regarded as undermining the best of British traditions, by substituting the dynamical principles and atomism, by considerations related to energy conservation and the phase rule. These successes of the German physical chemists coincided with the British physicists' further entrapment in the mechanisms of ether, which made them progressively less and less receptive to the demands of the chemists. But by 1906 the energeticists had lost their momentum, and in a couple of years their guru, Ostwald, would make a public declaration that he had been wrong all along and that he now believed in atoms. Having endured the attacks of the energeticists, with less than enthusiastic support of their case by the physicists, the British chemists abandoned the talk about consensus and now demanded the implementation of their own terms.

To underline the various issues of this public dispute, I mainly use the proceedings of the British Association for the Advancement of Science (BAAS) as well as the various addresses delivered at professional organizations. I choose the BAAS because of the two neglected aspects of these remarkable gatherings. Apart from their explicitly political role, the BAAS meetings were a quite unique forum where the content and level of the presidential addresses at the various sections, at least, were deliberately kept at a simpler level than an article in a professional journal would warrant, yet they were by no means popular lectures. These were lectures primarily aimed at the relevant community and toward those scientists closest to the particular community. Mathematicians had an eye toward the physicists, astronomers toward mathematicians, chemists toward physicists, physicists toward everyone else, and so on. And such was the situation because of the second aspect of the BAAS meetings: they comprised a public forum for debate among the disciplines. Complaints were aired, allies were sought, campaigns were launched, the merits of methods were debated, the character of laws was articulated, and new research directions were announced. And what remained invariant throughout the years was that on the whole, the rhetoric involved in these proceedings was the rhetoric of disciplinary politics.

By the late 1870s the atom had been thoroughly appropriated by the physicists. This process of appropriation was quite complex, but since I am interested in British mathematical physics, let me just mention the way the grand perpetrator, James Clerk Maxwell, felt about the problem. In fact, from early on, Maxwell did not consider that there was any qualitative difference between physics and chemistry. In an unpublished lecture at Kings College, London, in 1860, he considered the difference between physics and chemistry to lie more in the method of study rather than in the subject matter itself. If it would be possible to conceive the chemical elements as different arrangements of particles of one primitive kind of matter, then chemistry, asserted Maxwell, would be reducible to physics.

Maxwell's presidential address to the Mathematical and Physical Section of the BAAS meeting in 1873 was on molecules. He did acknowledge at the very beginning that the ideas embodied in molecules are those of chemistry, but felt that molecules should not remain the chemists' prerogative, since, he said, there was a universal interest in molecules. The law of definite proportions was considered to "have a high degree of cogency." But, as Maxwell quickly added, it was the result of a "purely chemical reasoning [and] not [of] dynamical reasoning. It is founded on chemical experience, not on the laws of motion."[2]

In his article on the atom for the ninth edition of the *Encyclopaedia Britannica*, Maxwell clearly differentiated between the metaphysical doctrine of atomism and the discontinuity of the constitution of matter. To elucidate the latter, he gave a long survey of Kelvin's vortex atoms in the ether and the possibilities provided by dynamical theory. Maxwell was the editor for the physical sciences for this edition, and it must not have been coincidental that not only was the article on the atom written by him, but also the article on the molecule was written by two of his associates, H. W. Watson and S. H. Burbury. The article is an exposition of the kinetic theory of gases, and only the last quarter of it was titled "the chemical molecule" and written by a chemist, Crum Brown. The terms of the appropriation of the chemists' molecules by the physicists were set by Maxwell, and the double ontology was there from the very beginning.

It would have been quite ungrateful, in fact, for the chemists to react to the appropriation of the building blocks of their science, when everything was done for such noble purposes. In fact, the chemists may have been so flattered that they must have failed to notice what, I think, was not a careless omission by Maxwell. Dalton, the British chemists' hero, was nowhere mentioned by name either in Maxwell's address at the BAAS on molecules or in the Britannica article on atoms. Everyone else related with the various developments of the atomic hypothesis was mentioned, but not Dalton.

Chemists under the Spell of Dynamical Theory

In the early 1880s it appeared that dynamical theory may even provide an explanation of why atoms are held together to form a molecule. In 1882 J. J. Thomson had shown that a complex mechanism of vortex rings could account for the mechanism of valence. In fact, Thomson, in this work which won him the Adams Prize for 1882, proposed that the pairing of two vortex rings is "what takes place when two elements of which these vortex rings are atoms, combine chemically."[3] It was this work that introduced the notion of pairing to account for chemical valence—a notion that gained an amazing heuristic value in the work of the American chemist G. N. Lewis and whose theoretical significance was eventually understood in 1927 with the classic paper of Heitler and London.

The need for cooperation with the physicists and the feeling of welcome in the joint attempts to understand chemical phenomena were the running themes in most of the public lectures by the chemists. The dominant figures of the chemical establishment would become the most effective propagandists of such an alliance. Henry Roscoe in 1884 stressed that one "of the noteworthy features of chemical progress is the interest taken by physicists in fundamental questions of our science."[4] In 1885 Henry Armstrong discussed one of the cardinal issues of chemistry, the nature of chemical change, a subject that, he emphasized, "requires the immediate earnest attention of chemists and physicists."[5] In 1888 Thomson published his lecture notes at the Cavendish under the title *Applications of Dynamics to Physics and Chemistry*. Though chemistry was cruelly reduced to physics, there did emerge a dynamical framework that could accommodate the physical *and* chemical phenomena together. In 1893 Emerson Reynolds placed great emphasis on the kinds of problems he called the "physico-chemical" problems.[6] In 1895 Ralph Meldola argued that the "one great desideratum of modern chemistry is unquestionably a physical or mechanical interpretation of the combining capacities of atoms."[7]

And, then, all such talk stopped. It was 1895 and Ostwald at the Lubeck meeting of the Society of German Scientists and Physicians had launched his campaign for energetics whereby chemical phenomena could be accounted by macroscopic considerations like energy conservation rather than by a recourse to the atomic hypothesis.

Tense Relations

In 1902 Edward Divers, vice president of the Chemical Society of London, delivered the presidential address to the chemical section of the BAAS. The title of his talk was "The atomic theory without hypothesis." He was responding, as he said, to the "need to treat atomic theory as being a true theory instead of as an hypothesis." Though it still had the defect of resting on a metaphysical basis, he was convinced that the Daltonian view was a theory rather than a hypothesis. He remarked that some people held the view that a chemical theory should be developed without reference to the atomic hypothesis, and this had been the belief of many eminent chemists. He formulated the Daltonian theory in such a way that everything was defined with respect to the central notion of chemical interaction, and thus Avogadro's law "if divested of all reference to the mechanical structure of gases" became another expression of the law of proportions. His point was to press for a consideration and discussion of the doctrines of chemistry and of the atomic theory itself as something concerned exclusively with experimental chemical facts.

And, then, he turned against the physicists who had snatched the chemists' atom and went about meddling in its quite straightforward ontology:

> Physicists have never been satisfied with the hard, indivisible ball of specific substance and definite mass which has served chemistry so well. They have given it bells, have made a vortex ring of it, and have indeed done much that few chemists can understand, to make it meet the exacting requirements of their science. But to us it has always remained the same; what we have done to it has been external . . . we have not meddled with its interior. . . .[8]

Divers argued that chemistry dealt with tangible truths and did not need the help of mechanical models to make its various laws legitimate. As I described above, only a few years earlier another figure of the chemical establishment, Ralph Meldola, had talked about the "one great desideratum of modern chemistry being unquestionably a physical or mechanical interpretation of the combining capacities of atoms." The difference between the two was not a difference of opinion about the reality of the atoms. It reflected the changing sentiments of some chemists in the way they would pursue their disciplinary politics after the turn of the century.

There were two aspects of Divers's talk that are particularly relevant to my argument. The first was his attempt to have the community agree on an unambiguous ontology if they were to counteract successfully the attacks of the energeticists. The second was the abandonment of the atom and the embrace of the molecule as the basic unit of chemistry, as a way of escaping the disconcerting mess that he felt the atom was facing with the electrons, X-rays, and radioactivity. And since the physicists did not appear to be particularly sensitive to the chemists' calls to find a common ground against the energeticists, then let them have the atom. The chemists would stick to the molecule. The long romance between the chemists and the physicists, and the not so subtle overtures for a permanent cohabitation, turned into improper insinuations. And as the chemists were aggressively reacquiring what had been appropriated by the physicists in the 1870s, relations between the two reached their lowest point.

But then, something utterly peculiar happened.

Ostwald in London

The chemists invited the wolf to their midst. In 1903, it was decided by the Chemical Society of London to invite the famous organic chemist Emil Fischer from Germany, who had just received the Nobel Prize, to deliver the Faraday Lecture of 1904. He was to have spoken about synthetical chemistry and its relation to biology. But in the beginning of 1904, Fischer informed them that he was too ill to travel to England and that he could not deliver the lecture. The invitation, upon Fischer's recommendation, was extended to Ostwald, the *enfant terrible* of physical chemistry, who was willing to prepare such an important address to be delivered in less than four months after Fischer informed the society that he could not come.

Ostwald came to London and gave a stunning performance. It was not a talk aimed at convincing his audience. It was a talk aimed at crushing his audience. Ostensibly, his purpose was to show that all the laws of chemistry that could be deduced from the atomic hypothesis could equally well be deduced from the theory of chemical dynamics, which he told them was the most significant achievement of modern chemistry. He claimed that chemical dynamics had made the atomic hypothesis unnecessary and had put the various chemical laws "on more secure grounds than that furnished by the atomic hypothesis."[9] He implored them to answer for themselves the question about the nature of elements and their compounds. He told them that every "generation of chemists must form their own views regarding this fundamental problem of chemistry." And then he asked them "from what store of ideas will a modern chemist derive the new materials for a new answer to this old question?" A physicist, he told them, will have a ready answer: he will construct the elements in a mechanical way or "if he is the modern type, he will use electricity as timber." The chemist, he reminded them, should look at these structures with due respect, but with some reserve: "Long experience has convinced the chemists (or at least, some of them) that every hypothesis taken from another science ultimately proves insufficient."

He reasoned that chemical culture necessitated the adoption of his approach. He felt that he was "stepping on somewhat volcanic ground." He knew that most people among his audience were quite satisfied with the atoms and that they did not in the least want to change them for any other conception. But he insisted that careful consideration should be given to his views:

> For I also feel assured that you will offer me the severest criticism which I shall be able to find anywhere. If my ideas should prove worthless, they will be put on the shelf here more quickly than anywhere else, before they can do harm. If, on the contrary, they should contain anything sound, they will be freed here in the most efficacious way from their inexact and inconsistent components, so as to take the shape fittest for lasting use in science.[10]

And then, as if he felt that his victory was not complete, he challenged his audience. He acknowledged that he was speaking in the very country that was the birthplace of the atomic hypothesis in its modern form, and only a short time ago the centenary celebration of the atomic hypothesis had reminded everyone of the enormous advance that science had made in this field during the last hundred years. Interestingly, he, like Maxwell, never mentioned Dalton's name. One could almost feel the audience fuming, prevented by etiquette from shouting against the sacrilege so unashamedly perpetrated by Ostwald at the very heart of London, at the Lecture Theatre, in fact, of the Royal Institution, where so much had been said about the real atoms all these past years.

Being Unpleasant to Former Allies

But little did they all suspect that this was Ostwald's swan song. Less than two years later, in 1906, Ostwald had indicated that he was having second thoughts about his denial of the existence of atoms. There was no time to be wasted. Without even waiting for Ostwald's official declaration, Arthur Smithells, the forceful spokesman of British chemistry, at the next meeting of the BAAS in 1907 took it upon himself to deliver the panegyric of the victorious.

The first half of his address as the president of the chemistry section of the BAAS is somber and technical, except for one rather unnerving detail. The subject he chose to discuss was the "chemistry of flame," on which, of course, he had been working for a long time. Might it be the case that British chemists were prepared to go *that* far? Reading the second half of the talk leaves one wondering.

He told his audience that he was very excited about the state of chemistry, and when compared to 20 years ago, everything appeared to be more promising. Though the discovery of radioactivity did mark a new epoch in the history of chemistry, there was not enough evidence to warrant any unsettlement of "the scientific articles of the chemists' faith." Radium was in a way an embarrassment, since it was elementary and it also broke into other elementary substances. Chemists were perplexed, though, not so much by the new ideas as by the invasion "of chemistry by mathematics. The exuberance of mathematical speculation of the most bewildering kind concerning the nature, or perhaps I should say the want of nature, of matter, is calculated to perturb a stolid and earthly philosopher.[11]

Smithells was, as he declared, the representative of the chemists, and he wanted to make some points, because in recent times, even before the advent of radium, events had caused chemists to ask whether chemistry was not beginning to drift away from them. He remarked that in the past years the most important developments had been on the physical side, and one great chemist remarked to him that he was feeling "submerged and perishing in the great tide of physical chemistry which was rolling up into our laboratories." It was precisely such men who should be preserved for chemistry. He noted that there seemed to be a solicitude in some quarters to make a chemist more than a chemist, a solicitude that, if granted, will make him something less than one. "Chemistry," he commanded, "should not be invaded by mathematical theorists."

Smithell's conservative backlash was complemented in two years by Armstrong's aggressive stand. Concerning his views about the energeticists, Armstrong boasted that, even though his attitude was one of "complete antagonism towards the speculations of the Ostwald school," he was nevertheless the first English chemist to publicly remark that Ostwald's investigations were of the highest importance. But now Ostwald had changed his mind, and Armstrong warned his fellow chemists in a most dogmatic manner about the dangers of dogmatism. He reminded his audience how Ostwald had

> Charged his test tubes with ink instead of chemical agents and by means of a too facile pen he has enticed chemists the world over into becoming adherents of the cult [of his school]—a cult the advance of which may well be ranked with that of Christian science, so implicit has been the faith of its adherents in the doctrines laid down for them, so extreme and narrow the views of its advocates. . . . The lesson we shall have learnt will be of no slight import . . . if it serve to bring home to us the danger of uncontrolled literary propagandism in science, if it but cause us always to be on our guard against the intrusion of authority and of dogmatism in our speculations.[12]

But that was not the end. There was one more account to be settled. Armstrong appealed to the physicists to make themselves more acquainted with the methods of the chemists and to stop speculating unnecessarily:

> Now that physical inquiry is largely chemical, now that physicists are regular excursionists into our territory, it is essential that our methods and our criteria be understood by them. I make this remark advisedly, as it appears to me that, of late years, while affecting almost to dictate a policy to us, physicists have taken less and less pain to make themselves acquainted with the subject matter of chemistry, especially with our methods of arriving at the root conceptions of structure and the properties as conditioned by structure. It is a serious matter that chemistry should be so neglected by physicists.[13]

Those same chemists who a few years back were speaking so enthusiastically about the prospects of a collaboration with the physicists were now rallying around their leaders who loudly and clearly were declaring the break of their relations with the physicists.

The Response by the Physicists

My final point is that a consensual framework was already emerging with the work of Joseph Larmor. Starting with his *Aether and Matter* in 1900 and culminating in his Wilde Lecture of 1908, Larmor's work expressed the confluence of the two divergent traditions of the British physicists and the British chemists. And, significantly enough, such a consensual framework was defined by partly breaking away from the tradition of *mathematical* physics and by introducing *theoretical* physics as the new mode for doing mathematical physics. It was Larmor's *Aether and Matter* that, more than anything else, can be considered a concrete response to the need for a concensual theoretical framework for both the physicists and the chemists. I think that Larmor was pursuing an agenda where it was clear to him that the *ontological issues* that had to be settled between the physicists and the chemists were closely related with the *methodological issues* involved in setting up a concensual theoretical framework. Decisions about the former necessitated decisions about the latter. The new consensus was a matter not only of devising a new theory as such, but also of articulating the novelty of the theoretical approach.

During the first years of the twentieth century, the conditions were becoming favorable for the slow abandonment of mathematical physics by the British physicists and for their timid approach to theoretical physics. The chemists were ceasing to view physical chemistry merely as the application of physical techniques to chemistry and were becoming more receptive to *its* autonomous language and to *its* mathematics. Armstrong, despite his aggressive stance against the physicists, kept on referring to Larmor's Wilde Lecture of 1908 titled "The Physical Aspect of Atomic Theory," devoted almost exclusively to chemistry.

But it was a very short-lived period. To paraphrase Ostwald's patronizing declaration, it was becoming progressively more and more urgent that every generation of physicists and chemists form their own views regarding the fundamental questions of their respective fields. In 1909 the presidential address to the mathematical and physical section of the BAAS was delivered by Ernest Rutherford. It was ostensibly about electrons and radioactivity. But it was also the first time in many decades that the word "ether" was not mentioned even once, and the first time Einstein and Planck are mentioned together. It was a message no one could afford to ignore.

Can a Chemical Element Not Combine with Other Chemical Elements?

Lord Rayleigh's measurements for the exact determination of the densities of gases had started in 1882 while he was the Cavendish Professor of Experimental Physics at the University of Cambridge. He continued them in 1888, having left Cambridge and having been appointed as the Professor of Natural Philosophy at the Royal Institution in London. It was a program aimed to test Prout's hypothesis by finding the atomic weights of gases and observing the extent to which they were multiples of the atomic weight of hydrogen. By 1892 Rayleigh found a curious discrepancy. In a letter to *Nature*[14] he noted that the density of nitrogen depended on the method used to isolate the gas. The nitrogen he derived by the two different methods he called "physical" nitrogen and "chemical" nitrogen. The former was isolated by removing the oxygen, moisture, and carbon dioxide from samples of atmospheric air. Chemical nitrogen was prepared from ammonia. It was found that physical nitrogen was heavier than chemical nitrogen by about 1/1000. Rayleigh's next step was to find ways to exaggerate this difference: "One's instinct at first is to try to get rid of a discrepancy, but I believe that experience shows such an endeavor to be a mistake. What one ought to do is magnify a small discrepancy with a view of finding out an explanation."[15]

Further improvements showed that chemical nitrogen was about 0.5 percent lighter than physical nitrogen. The first alternative Rayleigh entertained was that atmospheric nitrogen was too heavy because of the imperfect removal of oxygen from the atmospheric air, or chemical nitrogen was too light because when it was removed from ammonia it was contaminated with gases that were lighter than nitrogen. Further experiments by Rayleigh excluded both possibilities. It was also possible that the discrepancy was due to the dissociation of the nitrogen molecules and their subsequent formation into N_3, much like producing ozone from oxygen by silent discharge. Rayleigh ruled out this possibility, too, by showing that electrification and sparking had no appreciable effect in altering the densities of the two kinds of nitrogen. By the beginning of 1894 Rayleigh was convinced that the atmosphere contained a hitherto unknown constituent.

The methods used by Rayleigh in his experiments to isolate the new constituent were, in effect, very similar to the experiments performed by Henry Cavendish in 1785. Rayleigh tried first to remove the oxygen from the atmospheric air, then the nitrogen, and then the carbon dioxide and other similar gases. The difficulty, of course, was in the removal of nitrogen, since it chemically combines only with certain elements and only under specific conditions. Rayleigh's apparatus consisted of a Ruhmkorff coil connected to a battery and five elements of a Grove cell. The gases were in a test tube placed upside down in a container with a large amount of light alkalines. The current went through the wires, which passed inside two U-shaped glass tubes. The platinum ends were secured by being "glued" onto the test tubes. A short spark of about 5 milliseconds was found to be more efficient than a longer one. When the proportions of the gases were right, the absorption was about 30 cc/hour—thirty times better than Cavendish's apparatus.

A characteristic run is found in the very first page of Rayleigh's Notebooks.[16] He started with 50 cc of air and continuously added oxygen, and with the help of the sparks he could have a union of oxygen with nitrogen. The addition of oxygen continued until there was no noticeable contraction of the volume of the gas inside the test tube after sparking for one hour. What remained was transferred to another tube and found to be 1 cc. This was passed over alkaline pyrogallate, and the final product was 0.32 cc. What remained could not have been nitrogen since it did not decrease after continuous sparking, nor could it have come from somewhere else since repeated measurements had shown that it was proportional to the mass of the original intake of atmospheric air. Rayleigh called it the "residue."

In the meantime, William Ramsay, Professor of Chemistry at University College London, had proposed to Rayleigh that there may be a more efficient way to study the problem. His method consisted of a series of connected tubes that contained magnesium, copper oxide (which could unite with India rubber), preheated soda, lime soda (which will not contain water vapor), and phosphoric anhydride. The heated magnesium absorbed the nitrogen, and by repeating the process by recirculating the gas collected at the end of the previous run, Ramsay, starting with 1094 cc of nitrogen, was left with a residue of 50 cc, which nevertheless was still not very pure.

Up until the beginning of August 1894 Ramsay and Rayleigh were working independently, and at that time they decided to join forces and plan toward a joint publication. They were both convinced that atmospheric air contained either a new element or a new compound. The results were first presented during the BAAS meeting at Oxford in August 13, 1894. In a brief announcement, read by Rayleigh, reported that atmospheric nitrogen when purified from all the other known constituents of air was found to contain another gas, to the extent of about 1%, that was even more inert than nitrogen. The density of this gas was found to be between 18.9 and 20, and preliminary observations of its spectrum had found a characteristic line.

Right after the BAAS meeting, James Dewar wrote two letters to the London *Times* claiming that what was found by Rayleigh and Ramsay was the triatomic form of nitrogen. At the time Dewar was the Jacksonian Professor of Experimental Philosophy at the University of Cambridge and the Fullerian Professor of Chemistry at the Royal Institution. He held both posts until his death, and he was also the Director of the Davy-Faraday Research Laboratory of the Royal Institution. Dewar suggested that the allotropic form of nitrogen could produce spectra distinct from nitrogen, and in the case of Rayleigh and Ramsay, "the new substance is being manufactured by the respective experiments, and not separated from ordinary air."[17]

Dewar's laboratory notebooks[18] show that he had performed a series of experiments right before the BAAS meeting, and he drew his inferences from experiments involving liquefaction of air and the white deposits he always found in the otherwise transparent liquid. The conclusions he reached were through characteristicly chemical thinking. He suggested that the theoretical density of the new nitrogen compared to hydrogen should be 21, while the experimental numbers are between 19 and 20. He surmised that for such a body, "chemists would infer" that it ought to be characterized by great inertness, because phosphorus, the element most nearly allied to nitrogen, easily passes into an allotropic form known as red phosphorus, which, relative to the yellow phosphorus, was an inert body. If, therefore, such an active body as phosphorus could become, in condensed form, far less active chemically, then "by analogy, nitrogen, so inert to start with, must in the new form, become exceedingly active [WRONG]!"[19]

On December 6, 1894, Dewar presented to the Chemical Society of London his experiments concerning the liquefaction of nitrogen. The meeting had taken place less than a week after Lord Kelvin in his presidential address at the Royal Society of London had referred to the discovery of the new constituent as the greatest scientific event of the year. In his talk Dewar claimed that chemical and physical nitrogen liquefied at the same temperature and boiled off at the same rate. From this he inferred that the assumed new substance present in the atmospheric nitrogen does not condense at those temperatures when all other gases condense or that it behaves in exactly the same manner as nitrogen.[20] In an unsigned piece the next day reporting the meeting at the *Times*, it was remarked that "[c]hemists will appreciate the extreme singularity of a substance with the assigned density which fulfills either condition." The summary of the discussion of Dewar's paper was most probably written by Dewar's most fanatic ally, Armstrong.

Both Rayleigh and Ramsay did not attend the meeting. Dewar's announcement gave the opportunity to Armstrong to underline the case of his fellow chemists. He ventured to say that Rayleigh and Ramsay now could not hope to keep so remarkable a discovery to themselves much longer. He was adamant that "chemists could not be expected to remain . . . under the imputation that they had been eyeless during a whole century." And he concluded by talking about the "unquestionable rights of the chemists" to exercise entire freedom of judgment, and to critically examine the statements that had been made.[21] Apart from wishing to be absolutely certain before fully committing themselves to the suggested discovery, the other reason that Rayleigh and Ramsay were quite secretive about the details of their experiments was that they were planning to claim the Smithsonian Hodgkins Prize awarded to discoveries related to the atmosphere. This they received in 1895 after the final announcement of their discovery.

The final announcement was made at a meeting of the Royal Society at the Theatre of University College London on January 31, 1895. The paper was presented by Ramsay. He described all the different methods used to isolate atmospheric nitrogen and chemical nitrogen, and the difference of less that 1% in the measured densities of the two kinds of nitrogen. Then he presented the methods for removing the nitrogen and the different methods to induce chemical combinations with nitrogen. There was always a remaining residue that could not be gotten rid of. Ramsay, then, discussed a number of ways to isolate the new gas and to obtain it in relatively large quantities. Having achieved that, William Crookes and Arthur Schuster examined its spectrum and found that it did contain certain lines that were not contained in the nitrogen spectrum. This was one piece of convincing evidence that what was found was not N_3, but a new gas—argon. The other was the extreme inertness of argon, whereas most of the chemical evidence implied that it would be almost explosive. Ramsay continued describing the solubility of argon in water and its liquefaction, and a more detailed account was presented at the same meeting by Karol Olszewski. By measuring the velocity of sound in argon, Rayleigh and Ramsay managed to find the ratio of specific heats, 1.66. This implied that argon was monatomic and, hence, quite impossible to accommodate in the periodic table as that table was structured at the time.

Lord Kelvin chaired the meeting to which the councils of both the chemical and the physical Societies were invited. In front of 800 people Ramsay read the paper. Michael Foster, the other secretary of the Royal Society besides Rayleigh, was there, as were A. S. Balfour, Lyon Playfair, Henry Roscoe, George Stokes, James Paget, B. W. Richardson, Henry Gilbert, Philip Magnus, Henry Armstrong, Carey Foster, Arthur Rucker, Henry Odling, William Perkin, William Frankland, William Crookes, William Tilden, Sylvanus Thompson, Sydney Young, and Ralph Meldola. Dewar, however, was absent. After the end of the presentation, Armstrong and Rucker spoke. Though not as vitriolic as in his remarks after the Chemical Society meeting, Armstrong made a long speech questioning in effect the conclusiveness of the evidence brought by Rayleigh and Ramsay as to the inertness and the monatomicity of argon. Rayleigh said a few words at the end: "I am not without experience of experimental difficulties, but certainly I have never encountered them in anything like so severe and aggravating a form as in this investigation."[22]

After the formal announcement of the discovery of argon, *Nature* carried a detailed report of the meeting with various comments most probably written by Rucker, professor of physics at the Royal College in London. The report remarked:

> All that is known of argon was told to all. . . . As has been well said, the result is "the triumph of the last place decimals," that is, of work done so well that the worker knew he could not be wrong. . . . [and concerning the disagreements about the monatomicity of argon it was added that] [t]he courts of science are always open and every litigant has an unrestricted right of moving for a writ of error.[23]

Accommodating and Legitimating the New Element

In his presidential address at the Royal Society at the end of 1894, Lord Kelvin praised the new discovery of "the hitherto unknown and still anonymous fifth constituent of our atmosphere" as "the greatest scientific event of the past year." And then he reminded the audience of the comments he had made 23 years earlier:

> Accurate and minute measurement seems to the non-scientific imagination a less lofty and dignified work than looking for something new. But nearly all the greatest discoveries of science have been but the rewards of accurate measurement and patient long-continue labor in the minute sifting of numerical results.[24]

It is not uncommon, especially among scientists, to condider the argon story the paradigmatic case of such a culture of exact measurements. I think this is a greatly misplaced assessment. By considering the discovery of argon in such a context, one of its crucial elements is lost. The argon discovery is hardly one of the next-decimal-place view of physics. As Rayleigh was the first to point out, Cavendish, nearly a century earlier in his *Experiments on Air* (1784–85) while attempting to remove all the nitrogen from a jar, had noticed that there was a residue of less than 1/100th that he could not remove. Hence, it is rather unjustifiable to connect the discovery of argon to the culture associated with the specific type of measurements at the end of the nineteenth century. Furthermore, Rayleigh and Ramsay in the beginning of their paper put a quote from Augustus De Morgan's *A Budget of Paradoxes* (1869) as if to counteract any such attitude about "next-decimal-place":

> Modern discoveries have not been made by large collections of facts, with subsequent discussion, separation and resulting deduction of a truth thus rendered perceptible. A few facts have suggested an *hypothesis*, which means a *supposition*, proper to explain them. The necessary results of this supposition are worked out, and then, and not until then, other facts are examined to see if their ulterior results are found in nature.[25]

The discovery of argon happened during a time when physical chemistry was articulating its own autonomous language with respect to both physics and chemistry, when it was charting its own theoretical agenda and formulating its own theoretical framework, and has, to my mind, very little to do with next-decimal-place measurements.

I would also like to argue that the deep significance of the argon story is undermined unless it is considered a story involving a bitter public dispute over the legitimacy of a new chemical element whose most important characteristic was that by being chemically inert, it was negating the very notion of a chemical element. Argon forced chemists to reappraise some of the constitutive notions of their discipline. Similarly, physicists were obliged to rethink the boundaries between physics and chemistry and start coming to terms with the notion that chemistry, after all, might not all be reducible to physics.

It is not coincidental that Ramsay had discussed many similar issues related to physical chemistry with Ostwald and George FitzGerald. Ostwald had written to Ramsay that he would gladly publish his paper in the *Zeitschrift für Physikalishe Chemie*: "The fact is that I do not care very much for the new elements. But one so unexpected and almost impossible as that which you have found is something totally different from the trivial discoveries amongst the rare earths."[26] FitzGerald proposed that Ramsay make a determination of the specific heat at constant volume and a calculation of it from the value of the ratio of specific heats and the Pressure/volume/temperature relation, and thus decide whether it obeys the Dulong Petit law.[27] Ramsay was seeking FitzGerald's opinion about the peculiarities of the ratio of specific heats he had found for argon. The latter was convinced that such a calculation would lead to the atomic weight of 40. "This is certainly very mysterious." FitzGerald suggested that this may imply that the two atoms may have little or no independent motion and so the molecule behaved like a single atom. "I make this in the interests of chemistry because physically there can be no objection to an atomic weight of 40."[28] Ramsay had suggested to FitzGerald the possibility of a system of elements with zero atomicity and the latter, though very enthusiastic about the suggestion, warned Ramsay that the "[c]hemists will never believe in an element with no chemical affinity."[29] And Ramsay felt no scruples in telling Smithells that the implications of argon were such that "the whole fabric of chemistry is going to receive a shake."[30]

Different Laboratory Practices

The discovery of argon was far from being a joyous affair for chemists. It deeply insulted many distinguished British chemists. The two most vocal critics among the chemists were Armstrong, president of the Chemical Society, and Dewar. The extreme dislike both entertained against Ramsay is often given as the reason for this conflict. From all the evidence that has been coming to surface, it appears that Dewar had a rather strong predilection for disliking people generally, whereas Armstrong's tendency was an incomprehensible hero worship of Dewar. Though personal factors cannot be denied in trying to understand the reactions of the chemists to the discovery of argon, they are neither sufficient nor a substitute for understanding such conflicts and public disputes in the context of the dramatic developments taking place during the end of the nineteenth century in both chemistry and physics. But I think it is of interest to pursue Dewar's reactions further.

The conflict between Dewar and Ramsay was reminiscent of some aspects of Dewar's behavior when he was trying to liquefy helium after having successfully liquefied hydrogen in 1898. I think the helium problem was one, perhaps not even the most dramatic instance, of Dewar's failures. The way he went about the whole problem of argon is equally supportive of my argument that his *laboratory practice* in problems related to this new hazy area of physical chemistry led to a deadlock. As is evident from the entries in Dewar's laboratory notebooks, Dewar, in a frenzy of experimental activity at the end of November and all of December 1894, was studying the low-temperature behavior of chemical nitrogen and atmospheric nitrogen and desperately trying to establish that Rayleigh and Ramsay had mistaken the new gas with N_3, since he had found that both kinds of nitrogen liquefied at the same temperature.[31] Thus, the argon story raises a host of interesting historical questions, and one of them is the difference in outlook about the intermediate region of physical chemistry—not between a chemist and a physicist as, for example, was the case between Dewar and Kamerlingh Onnes, but between the two chemists Ramsay and Dewar. The contrast in the narrative and the problems being discussed by Ramsay in his address at the International Congress of Arts and Sciences at St. Louis in 1904 or in the introduction of the series he edited on physical chemistry and Dewar's 1902 presidential address, is also quite striking in this respect.

Dewar never transcended the notion of physical chemistry as a way of adopting physical techniques for chemistry. This, at the beginning, appeared a convenient and promising approach. But such a view became a deadlock. Dewar's *experimental practice* lacked a theoretical component. I want to stress the fundamental difference between the experimental practice of Dewar associated with liquefying gases by focusing solely on the possibilities provided by the Joule-Thomson effect and that of Kamerlingh Onnes, whose focus was the determination of isotherms. Kamerlingh Onnes's liquefaction of helium was not simply a triumph of a superior technique and improved instrumentation; it also displayed Kamerlingh Onnes's ability to assimilate such techniques and instrumentation in the newly articulated theoretical framework. Kamerlingh Onnes had adapted his laboratory practice to the exigencies of the newly formed subdiscipline of physical chemistry. It is not simply a question of who among the protagonists was theoretically more sophisticated. The overall dispute involved different groups of chemists, and it expressed a clash between different laboratory practices. The disagreements were suggestive of the way each laboratory tradition chose to articulate its own agenda within the newly emerging subdiscipline of physical chemistry. The disagreements so aggressively expressed by many chemists around and about the new element had mainly to do with the threatening emergence of physical chemistry as a distinct new subdiscipline rather than personal enmities. And it was exactly against this new framework that strong phobias were expressed by many chemists. In the end, after the dust had settled, it appeared that argon "belonged" to those physicists who for a moment felt like chemists and to those chemists who started realizing that physical chemistry was not simply a way of enriching chemistry with techniques borrowed from physics.

Quantum Mechanics for Chemists?

Right after the formulation of quantum mechanics by Werner Heisenberg, Max Born, Pascual Jordan, and Erwin Schrödinger, most chemists became aware of the amazing explanatory power of the new quantum mechanics, yet it was difficult to see how this newly developing explanatory framework would be assimilated into the chemists' culture. Many chemists were apprehensive that such an assimilation may bring lasting, and not altogether welcome, changes to their culture. But for some, it was a risk worth taking. Neville Sidgwick in his influential book *The Electronic Theory of Valency* (1927), published just after the dramatic developments of 1926, would have no inhibitions about letting the new quantum mechanics invade the realm of chemistry.

Indeed, he expressed an unreserved enthusiasm for it. Faced with the full development of the new mechanics by Heisenberg and Schrödinger, but not yet with an application of the theory to a chemical problem, Sidgwick in the very first lines of the preface to his book attempted to clarify the methodological stumbling block that he sensed will be in the way of his fellow chemists. He urged them to adopt concepts of atomic physics, but he also warned chemists that they "must accept the physical conclusions in full, and must not assign to these entities properties which the physicists have found them not to possess."[32]

In 1966 at a conference commemorating half a century of valence theory, Charles Coulson gave the closing talk. He was a mathematician by training and a writer of what became the standard textbook on valence. At this conference he suggested that "[f]ifty years of valence theory really means fifty years of changing ideas about a chemical bond. . . . The first third of the period . . . was necessarily concerned with . . . escaping from the thought forms of the physicist."[33] "Escaping from the thought forms of the physicists" had really been the dominant trend, not only in the first years of quantum chemistry, but also in much of the history of physical and structural chemistry in the latter part of the nineteenth century as well.

The beginnings and the establishment of quantum chemistry involved a series of issues that transcended the question of the application of quantum mechanics to chemical problems. Quantum chemistry developed an autonomous language with respect to physics, and the controversies during the first years of its development concerned the collective decision of the chemical community about methodological priorities and ontological commitments. The outstanding issue to be settled in the community turned out to be the character of theory for chemistry and, therefore, a reappraisal of the praxis of the chemists. As a rule, disputes and disagreements were as much about getting the correct solution to a problem as they were part of a rhetoric about how to go about solving similar kinds of problems. In other words, during the 1930s the discussions and disputes among chemists were to a large extent about the new legitimizing procedures and concensual activities to be incorporated to the chemists' culture.

In 1927 Heitler and London arrived in Zurich hoping to work with Schrödinger. Collaborating with others was not Schrödinger's style. They did not get too disappointed, and decided to tackle the outstanding problem of chemistry. They tried, in other words, to see whether the new quantum mechanics could explain the homopolar bond, that is, the bond between two neutral atoms to form a molecule. Hydrogen was the simplest case, and they proved that the mechanism could only be understood quantum mechanically and that it was exclusively due to the relative spin orientations of the two electrons. Right after its publication, it became quite obvious that the Heitler-London paper was opening a new era in the study of chemical problems. Such a "distinction is characteristically chemical," and its clarification marks the "genesis of the science of sub-atomic theoretical chemistry," remarked Pauling.[34] A similar view, with a slightly different emphasis, was put forward by John van Vleck, who talked about the "beginnings of a science of 'mathematical chemistry.'"[35]

Soon afterward, in 1931 Pauling started formulating his theory of the chemical bond. Concerning the chemical bond, Pauling proposed a set of rules. Not all of these rules were derived from first principles. Pauling exploited maximally the quantum mechanical phenomenon of resonance and was eventually in a position to formulate a comprehensive theory of chemical bonding. The success of the theory of resonance in structural chemistry consisted in finding the actual structures of various molecules as a result of resonance among other "more basic" structures.

In 1944 George Willard Wheland, who was a student of Pauling's and one of the strongest propagandists of the theory of resonance, published *The Theory of Resonance and Its Application to Organic Chemistry*. Appropriately, the book was dedicated to Pauling. Wheland's view was that "resonance is a man made concept in a more fundamental sense than most other physical theories. It does not correspond to any intrinsic property of the molecule itself, but instead it is only a mathematical device, deliberately invented by the physicist or chemist for his own convenience."[36]

At the time, Pauling did not seem to disagree with such an assessment. But when a later edition appeared in 1955, a lively correspondence ensued between the two about the actual character of resonance theory. Wheland thought that resonance was not an intrinsic property of a molecule. He wrote to Pauling that he believed that resonance was something deliberately added by the chemist or physicist who is talking about the molecule: "In anthropomorphic terms, I might say that the molecule does not know about the resonance in the same sense in which it knows about its weight, energy, shape and other properties that have what I would call real physical significance."[37] Pauling disagreed: "I feel that in your book you have done an injustice to resonance theory by overemphasizing its man-made character."[38]

Their correspondence continued and neither appeared to be convinced by the other. What Pauling greatly emphasized was not the arbitrariness of the concept of resonance, but its immense usefulness and convenience, which "make the disadvantage of the element of arbitrariness of little significance."[39] For Pauling this was his constitutive criterion for theory building in chemistry. It was the way, as he had noted, to particularize Percy Bridgman's operationalism in chemistry. In his analysis of resonance, Pauling expressed in the most explicit manner his views about theory building in chemistry. He asserted that the theory of resonance was a chemical *theory*, and in this respect, it had very little in common with the valence-bond *method* of making approximate quantum mechanical calculations of molecular wave functions and properties.

Though the method of molecular orbitals was first introduced by Friedrich Hund, it was Mulliken who provided both the most thorough treatment of the different kinds of molecules and the theoretical and methodological justifications for legitimizing the molecular orbital approach. Holding the view that the concept of valence itself is one that should not be held too sacred, Mulliken proceeded to formulate his "molecular" point of view where the emphasis was on the existence of the molecule as a distinct individual and not as a union of atoms held together by valence bonds. Using data from his exhaustive studies of molecular bond spectra, Mulliken was able to devise an *aufbau* principle for molecules in a manner analogous to Niels Bohr's assignment of quantum numbers for electrons in atoms. Mulliken claimed that one could talk about electrons in chemistry only when the electrons could be placed in molecular orbitals. And the way he devised the relevant quantum numbers was through purely empirical means from the data of molecular spectra.

Heitler and London were not really following all these developments very closely. The fateful events in Germany in 1933 had forced both to leave the country. Heitler went to Bristol and was working on quantum electrodynamics, and London went to

Oxford and had formulated, with his brother Heinz, the first successful theory of superconductivity. In 1935, as if somehow to make up for lost time, they started frantically writing to each other. The correspondence between Heitler and London is quite revealing. It shows the attitude of each about the possible development of the approach laid down in their common paper, the tension between them, and the search for means to consolidate their theory at a time when the American chemists appeared to be taking over the field of quantum chemistry.

They discussed the possibility of writing an article for *Nature* to present their old results and to include some new aspects that had not been emphasized properly in their earlier papers. Heitler felt that the importance of the theories of Pauling and Mulliken had been "monstrously overrated in America."[40] London adopted a more aggressive stand: "The chemist is made out of hard wood and he needs to have rules even if they are incomprehensible."[41]

The American quantum chemists were referring to the work of Heitler and London less and less, preferring the approaches by Pauling and Mulliken. Heitler and London were progressively realizing that part of the problem was their isolation, and this realization bred even more frustration. The fact that they had not even been attacked was not an indication of the acceptance of their theory. Their feeling was that their theory may have even been forgotten or that it "can be combated much more effectively by the conscious failure to appreciate and avoid mentioning it."[42] Heitler thought that some of the reasons why "some people in America" did not do the calculations properly was because they were "silly and lazy" and did not think that most of their colleagues were rascals. "And we should accept that our theory was quite complicated."[43]

Heitler visited London at Oxford in early December 1935. They were now fully aware that the Americans had articulated a schema that was antagonistic to their own. As soon as Heitler was back in Bristol, he read a paper by Wheland. Heitler was vitriolic in his response to London, telling him that "[i]f you cannot restrain me, I think, I will write a very clear letter to this Pauling (he should give a better upbringing to his students). It would be really good to write something which will mostly have those things that they are stealing in America."[44]

All about Theory

The discussions, debates, and disputes among chemists and physicists about quantum chemistry were very similar to the analogous discussions between chemists and physicists about physical chemistry during the last quarter of the nineteenth century. The settling of these issues affected deeply the laboratory practices and research agendas and were never snubbed by the chemists as a whole. The issues were suggestive of the various ways available to map the undefined and undelineated middle ground that some called physical chemistry, some molecular physics. Some considered it an application of quantum mechanics to chemical problems, and some considered it quantum chemistry. And most important, the question as to the character of theory in physical chemistry or quantum chemistry dominated the minds of many chemists. The net outcome of these controversies, debates, and discussions was that both physical chemists and quantum chemists could articulate their own autonomous language with respect to both physics and chemistry, chart their own research agenda, and formulate their own theoretical framework. The beginnings and the establishment of physical chemistry involved a series of issues that transcended the question of the application of physics to chemistry. Analogous undertones had the debates concerning the application of quantum mechanics to chemical problems. In both cases, the outstanding issue to be settled in the community turned out to be the character of theory for chemistry.

Some Concluding Remarks

In this chapter my preoccupation has been neither with the way controversies arise nor in the way they are resolved. Controversies, more than any other aspect of the scientific discourse, cannot be analyzed in terms of general principles. The significance of the controversies about ontological claims at the end of the nineteenth century, and about the character of the constituent elements of chemistry, lies in their varied cognitive content. I have attempted to understand this content to the extent that these controversies influenced the drawing of boundaries of the newly emerging physical chemistry and contributed to forming the discourse of physical chemistry. And the many controversies that came into being during the long developmental period of physical chemistry are but indications of the cultural pluralism that determined such a development.

Notes

Discussions with T. Arabatzis, A. Baltas, G. Freudenthal, I. Hacking, E. Hiebert, P. Kitcher, P. Machamer, A. Simoes and S. Schaffer have helped me clarify many of the issues I deal with in this chapter. I thank them all.

1. See K. Gavroglu (1995) Fritz London: A Scientific Biography (Cambridge: Cambridge University Press); K. Gavroglu, A. Simoes (1994) "The Americans, the Germans and the beginnings of quantum chemistry: the confluence of diverging traditions," Historical Studies in the Physical Sciences, 25: 47–110; K. Sopka (1988) Quantum Physics in America, (New York: American Institute of Physics); S. Schweber (1990) "The young Slater and the development of quantum chemistry," Historical Studies in the Physical Sciences, 20: 339–406.
2. J. C. Maxwell (1873) Presidential Address to Section A—The Mathematical and Physical Sciences, Proceedings of the British Association for the Advancement of Science, 211–230, p. 213.
3. J. J. Thomson (1914) "The forces between atoms and chemical affinity," Philosophical Magazine, 27: 1655–1675, p. 1673.
4. H. Roscoe (1884) Presidential Address to Section B—Chemistry, Proceedings of the British Association for the Advancement of Science, 659–669, p. 666.
5. H. Armstrong (1885) Presidential Address to Section B—Chemistry, Proceedings of the British Association for the Advancement of Science, 945–964, p. 957.
6. E. Reynolds (1893) Presidential Address to Section B—Chemistry, Proceedings of the British Association for the Advancement of Science, 708–715.
7. R. Meldola (1895) Presidential Address to Section B—Chemistry, Proceedings of the British Association for the Advancement of Science, 639–655, p. 643.
8. E. Divers (1902) Presidential Address to Section B—Chemistry, Proceedings of the British Association for the Advancement of Science, 557–575, p. 568.

9. W. Ostwald, "The Chemical Elements" (1904) The Journal of the Chemical Society, 85, 506–514, pp. 508–509.
10. Ibid., p. 509.
11. A. Smithells (1907) Presidential Address to Section B—Chemistry, Proceedings of the British Association for the Advancement of Science, 469–479, p. 477.
12. H. Armstrong (1909) Presidential Address to Section B—Chemistry, Proceedings of the British Association for the Advancement of Science, 420–454, p. 423.
13. Ibid.
14. Lord Rayleigh (1892) "Density of Nitrogen," Nature, 46:512–513.
15. Lord Rayleigh (1895) "Argon," Proceedings of the Royal Institution, 14:524–538, p. 525.
16. Rayleigh's notebooks are among his papers kept at the Hanscomb Air Base, Bedford, Mass.
17. J. Dewar, *The Times*, London, August 18, 1894.
18. James Dewar's laboratory notebooks are in the Dewar Archives at the Royal Institution, London.
19. See entries in Dewar's laboratory notebooks throughout November 1984.
20. J. Dewar (1984) "The relative behaviour of chemically prepared and of atmospheric nitrogen in the liquid state," Chemical News, December 6, 222–225.
21. H. Armstrong (1984) Chemical News, December 6, as reported to the Chemical Society, London.
22. From the report in *Nature*, February 7, 1895, 337.
23. Ibid.
24. Lord Kelvin, (1894) Chemical News, December 14, pp. 291–292.
25. Lord Rayleigh, W. Ramsay (1895) "Argon, a New Constituent of the Atmosphere," Philosophical Transactions of the Royal Society of London, 186A:187–241.
26. W. Ostwald to W. Ramsay, December 24, 1894. William Ramsay Papers, University College Library, London.
27. G. F. FitzGerald to W. Ramsay, December 14 and December 20, 1894, ibid.
28. G. F. FitzGerald to W. Ramsay, December 28, 1894, ibid.
29. G. F. FitzGerald to W. Ramsay, January 8, 1895, ibid.
30. W. Ramsay to A. Smithells, March 11, 1895, ibid.
31. Dewar notebooks, entries for August 9, August 14, November 21, November 27, November 29, December 3, December 14, December 20, 1894.
32. N. V. Sidgwick (1927) The Electronic Theory of Valency (Oxford: Clarendon University Press), p. iii.
33. C. Coulson (1970) "Recent developments in valence theory—symposium, fifty years of valence," Pure and Applied Chemistry, 24:257–287, p. 259.
34. L. Pauling (1928) "The application of the quantum mechanics to the structure of the hydrogen molecule," Chemical Reviews, 5:173–213, p. 174.
35. J. van Vleck (1928) "The new quantum mechanics," Chemical Reviews, 5:467–507, p. 506.
36. G. Wheland The Theory of Resonance and Its Applications to Organic Chemistry (New York: Wiley, 1940), p. 31.
37. G. W. Wheland to L. Pauling, January 20, 1956, G. Wheland Papers, University of Chicago Library.
38. L. Pauling to G. Wheland, January 26, 1956. G. Wheland Paters.
39. L. Pauling to G. Wheland, February 4, 1956. G. Wheland Papers.
40. W. Heitler to F. London, November 12, 1935, Fritz London Archives, Duke University, Durham, N.C.
41. F. London to W. Heitler, October or November 1935, ibid.
42. Ibid.
43. W. Heitler to F. London, November 22, 1935, ibid.
44. W. Heitler to F. London, beginning of December 1935, ibid.

11

Anthropology

Art or Science? A Controversy about the Evidence for Cannibalism

---

MERRILEE H. SALMON

Since anthropology's inception as an academic discipline, participant-observation has stood as *the* method for obtaining knowledge of cultural, social, linguistic, and physical features of other peoples. Franz Boas and Bronislaw Malinowski, who are primarily responsible for this state of affairs, were both trained as physical scientists before they turned to anthropology. Not surprisingly, they and their followers regard fieldwork as the anthropological equivalent of a scientific laboratory.

Needless to say, the analogy is imperfect. Fieldwork involves talking with subjects, often with the aid of an interpreter, and trying to make sense of what they say as well as classifying their nonverbal behavior in terms of its meaning. Thus, field observations have a greater interpretive component than judgments made about instrument readings in scientific laboratories. In addition, most—though not quite all—anthropological fieldwork is confined to what John Stuart Mill called "natural experiments." Replication is problematic because so many anthropological experiments are natural rather than artificial.

The analogy between fieldwork and laboratory work breaks down when considering attitudes toward replication. In laboratory investigations the ability to replicate results is essential. In part for practical reasons, however, cultural and social anthropologists assign a low priority to replication. Traditional societies, whose ways of life are the focus of many anthropological studies, are rapidly vanishing as industrial societies expand. Most anthropologists prefer to use their limited resources to gather data on endangered cultures instead of trying to replicate previous studies. In these circumstances, a tradition has developed in which anthropologists, working individually or in teams, establish themselves as authorities on a given society. Because their interpretations are regarded as statements of fact, replicating their work seems irrelevant.

Recent books by William Arens (1979) and Derek Freeman (1983) challenge this cozy state of affairs. Freeman criticized Margaret Mead's *Coming of Age in Samoa* (1928), a study that introduced generations of college students to anthropology. Mead tried to show in this work that cultural influences during adolescence override biological forces. Freeman's criticisms are Popperian. He says that Mead did not seek readily available information that could falsify her thesis. Moreover, he claims that Mead's Boasian training biased her in favor of "cultural determinism" and led her to accept false reports from her youthful informants about their sexual behavior.

When Freeman's book appeared, the popular press as well as the academic community took up various defenses of Mead's work. Three lines of defense were as follows. (1) Mead, as a young woman, concealing her married status to win the confidence of her informants, did not have access to the male power structure from which Freeman received his information. (2) When Mead herself—long in advance of Freeman's work—became aware of the defects of her early study, she decided to let it stand as an authentic field document of its time rather than to change it with the wisdom of hindsight. (3) Seeking data that would falsify a theory is more appropriate to physical than to social sciences, where trust between informant and investigator is a sine qua non.

Although all these defenses can be countered, and Freeman's factual claims about Samoan society at the time of Mead's fieldwork are now generally accepted, Mead's stature as an anthropologist remains undiminished and her early work is not repudiated. Perhaps because Mead herself is something of a cultural icon, a productive intellectual debate about the nature of anthropological evidence failed to develop in this context.

Arens's book, *The Man-Eating Myth: Anthropology and Anthropophagy* (1979), which received far less attention than Freeman's from the popular press, offers a stronger challenge to standards of evidence in anthropology. Arens argues that despite anthropologists' claims that cannibalism occurred in many human societies, only hearsay and biased evidence support the view that cannibalism was culturally sanctioned in any society, past or present.

The controversy about cannibalism seems to meet Gideon Freudenthal's criteria for a scientific controversy: the existence of cannibalism as a culturally accepted practice is a substantive issue affecting the content of anthropology; it resists resolution because of differences in interpreting the concept of evidence for anthropological claims, and it differentially affects the interests of contending parties. In contrast, while Freeman's claim that Mead could have obtained a better picture of Samoan sexual mores if she had adopted a Popperian methodology remains in dispute, few problems surround the interpretation or acceptance of the evidence for trouble in paradise that Freeman uncovered.

Arens notes that accusations of cannibalism have always carried political implications. Herodotus attributed cannibalism to desert dwellers at the edge of the then-known world. His image of the Androphagi's exotic behavior and strange language accents by way of contrast the cultural unity of the Greeks, one of his favorite themes. Arens tells us that when Columbus reached the New World, the Arawak speakers told him that their Carib (or Canib) enemies ate human flesh. The Aztecs of Mexico accused their enemies of eating captives. Europeans, however, used an illustration in Sahagún's *Florentine Codex* of an Aztec warrior in a cooking pot to show that the Aztecs ate *their* enemies (Vaillant 1965). Montaigne's famous essay on cannibalism (originally published in 1580) attributed the practice to indigenous Brazilians.

European explorations of the African and South American continents in the nineteenth and early twentieth century brought additional reports of cannibalism. Closer to our own time, D. C. Gajdusek, winner of a Nobel prize for his work on kuru, claimed that cannibalistic mortuary practices in new Guenia caused the spread of this rare progressive viral infection of the nervous system (similar to "mad cow disease"). In 1972, newspapers reported that survivors of a plane crash in the high Andes had resorted to cannibalism before their rescue.

Archaeological remains have been used to show that our hominid ancestors were cannibals. Broken and burned bones of Pekin man (*Homo sinanthropos*) found in a cave at Choukoutien were described as the remains of a prehistoric cannibal feast (Coon 1963). A Neanderthal cranium, pierced at its base and surrounded by a ring of stones, found in the Grotta Guattari near Naples, was interpreted as evidence for ritual cannibalism (Stiner 1991). Cannibalism has been diagnosed at Fontbrégoua (Villa et al. 1986) and at Krapina (Trinkaus 1985), both Neolithic sites. Archaeologists have also found evidence for cannibalism in fourteenth-century pueblo sites in the American southwest (White 1992).

While most anthropologists uncritically accepted the "fact" of cannibalism, they differed about how to interpret specific instances of the practice. Some emphasized the symbolic aspects of the behavior, while others took a more ecological and materialistic approach. Harner (1977) and Harris (1977) both say that Aztec cannibalism, long assumed to be purely ritualistic, was a functional response to shortages in protein-rich foods. They argue that where food was scarce, eating captured warriors made more sense than feeding them. Other anthropologists challenged this functional explanation. They used archaeological and historical data to show that the Aztec diet, while short on meat, was not protein deficient. The appearance of Arens's book, however, shifted discussion from how to interpret alleged episodes of cannibalism to questions about how to define the practice and how to support claims that a group engaged in cannibalism. In other words, the standards of anthropological evidence as well as the existence of cannibalism itself entered the controversy.

Definition and Classifications of Cannibalism

If "cannibalism" means ingestion by humans of any part of a human body, different purposes determine at least four types of cannibalism: (1) to satisfy hunger or provide a supplement to the regular diet (gastronomic cannibalism); (2) to prevent or cure disease (medicinal cannibalism); (3) to maintain continuity with one's dead relatives (mortuary cannibalism); and (4) to propitiate gods, enact revenge, or gain the strength of an enemy (sacrificial cannibalism). Further refinements are possible. Cannibalism to satisfy hunger in conditions of extreme starvation (survival cannibalism) is reasonably distinguished from other forms of gastronomic cannibalism.

A different classification refers to the relationship between eater and eaten: exocannibalism occurs when those who are eaten are outsiders, especially enemies; endocannibalism is the eating of relatives. Sagan (1974), a psychological anthropologist, divides cannibalism into aggressive and affectionate forms that correspond roughly to exocannibalism and endocannibalism. Other classifications depend on the degree of ritual attending the practice. For example, according to Poole (1983), the Bimin-Kuskusmin of New Guinea admit to a highly ritualized form of selective cannibalism while scorning their neighbors who, they say, treat human flesh as ordinary food. Most interesting to anthropologists, and the focus of Arens's challenge, is any *institutionalized,* or culturally approved, form of cannibalism. Arens disregards survival cannibalism, since he believes that is an aberration similar to pathological acts of cannibalism by disturbed individuals.

Lines of Evidence

Both Arens and Freeman raise serious questions about the nature of evidential support of anthropological claims. Freeman not only urges anthropologists to exercise critical judgment in dealing with informants, but also warns them about the dangers of theoretical bias and failure to seek counterevidence for their theories. Few anthropologists would disagree with this methodological advice. Arens argues that anthropologists did not adhere to their own standards of evidence in evaluating claims about cannibalism. Nevertheless, his work provoked anthropologists to defend what they saw as attacks on traditional ways of reasoning in anthropology. Below I examine several lines of evidence in support of the existence of cannibalism.

Eyewitness Accounts

Eyewitness accounts of cannibalism are rare. Arens claims that no culture acknowledges cannibalism as an approved practice among its current members. At the same time, almost every group has been accused of cannibalism by some other group. For example, British travelers and missionaries feared being cooked and eaten by the Azande, while Africans saw their belief in the British thirst for human blood confirmed by blood drives organized during World War II. Sahagún, describing Aztec cannibalism during the 1500s, says:

> And then, the offering made, the merchant had the body carried back to his house, and a meal of maize and human flesh prepared for his kin, sharing out the flesh of his "captive" just as the real warriors did. (Clendinnin 1991, p. 138)

Highland New Guinea allegedly remains an area where cannibalism is practiced, and where at least one anthropologist has claimed to witness acts of cannibalism. Yet accounts of human sacrifice or mortuary rites in New Guinea rarely contain eyewitness reports. Many such accounts conclude with the remark that the body was taken away, to be eaten by the women. (These descriptions by men thus manage to attribute cannibalism that occurs in their own culture to the Other.) Despite gruesome depictions of cannibalistic feasts that illustrate early works of anthropology as well as many fairy tales (Warner 1994, 1997), the act is almost always hidden from the eyes of the outsider who records the practice.

Responses to Arens's complaint about the lack of eyewitness accounts reveal the crucial evidentiary role of the eyewitness in anthropology. Some anthropologists say that eyewitness reports constitute the best evidence, but offer excuses for their absence in the case of cannibalism. Springer (1980), in his review of Arens, says that the only eyewitnesses to cannibalism were indigenous informants of European chroniclers, such as the Huron allies of the French Jesuits in America. The Huron's illiteracy, however, precludes a first-hand written account. Springer insists upon the reliability of the Jesuit accounts of torture, human sacrifice, and cannibalism among the Iroquois, although the Jesuits were eyewitnesses only to torture and sacrifice. He justifies his position by saying that in the human sciences some trust of sources is required and that the standards of proof demanded in the physical sciences do not apply in anthropology. He does not specify any standards, but the context suggests that he believes that direct observation of phenomena constitutes the strict standard that would correspond to an eyewitness account in human sciences.

Sanday's (1986) review of Arens also defends the reliability of the Jesuit reports of cannibalism. Instead of appealing to trust in sources as a stand-in for eyewitness accounts, however, she appeals to the detailed nature of the chronicles and the conviction of their authors. She quotes an especially moving account of the torture and death of a Jesuit's two priestly companions. The Jesuit affirms his trust in his Huron informant, but never claims to have observed the reported consumption of roasted hearts.

French and Spanish accounts of cannibalism in the New World, often written by clerics, tend to regard cannibalism as an inevitable accompaniment of human sacrifice. The association between the two practices may reveal more about Europeans than the people they study, however. The symbolic sacrifice of Christ's body and blood and its consumption in the Mass both suggest and reinforce the association.

Another tie links human sacrifice with cannibalism. Stocking (1993) points out that cannibalism, along with human sacrifice and headhunting, has long served as a standard mark of the Other. Characterizing a group as the Other, as Herodotus recognized, has obvious political significance, and all too often precedes attempts to justify their subjection or slaughter.

Neither Springer nor Sanday disputes eyewitness reports as the ultimate authority for the existence of a custom. They simply excuse their absence and suggest other forms of evidence as suitable substitutes. Yet seeing an occurrence of some action is clearly not the same as witnessing a custom. To call an action an instance of a culturally approved practice is to embed it in a framework not open to direct observation.

Although Tuzin (1983) does not explicitly address the issue of culturally approved practice, his account of conflicting interpretations of cannibalism on the part of Japanese soldiers exposes the problem. The Arapesh of New Guinea, with whom the soldiers were quartered during World War II, claimed that the soldiers had eaten human flesh during the last months of the war. Their stories are confirmed by archival evidence that when some Japanese soldiers were captured in early 1945, their packs contained human body parts. Japanese and American analysts interpreted the situation as an example of survival cannibalism in a period of extreme food shortage. The Arapesh, in contrast, regarded the soldiers' behavior as a psychotic aberration, brought on by their despair at losing the war. The images most frightening to the Arapesh, who often face severe food shortages, are those of parents eating children or children eating parents. Tuzin argues that the Arapesh simply could not believe that the soldiers who had lived with them as family members could commit the ultimate antisocial act. This suggests that the context of the observation is as important as what is seen. Moreover, the context of observation includes the background knowledge of the observer.

Judgments by a Trained Observer

Tuzin and Brown (1983), co-editors of a set of papers from a 1980 American Anthropological Association symposium on cannibalism, agree that most ethnographic evidence for cannibalism is woefully inadequate. Instead of blaming the absence of literate eyewitnesses, however, they attribute the poor quality of evidence to a lack of anthropological savvy on the part of reporters. Tuzin and Brown contend that before the publication of the 1892 edition of *Notes and Queries on Anthropology,* people did not know how to collect evidence for cannibalism. *Notes and Queries* contains a "well considered list of 19 questions to serve as a starting guide to the study of cannibalism" (Tuzin and Brown 1983). Unlike Sanday or Springer, Tuzin and Brown thus recognize that evidence for cannibalism is not to be obtained by simple observation, but is embedded in a context that may be inaccessible to an untrained observer.

Tuzin and Brown go on to say that by the time *Notes and Queries* appeared, "information was difficult to obtain" because "[u]nder conditions of empire, [cannibalism] was no sooner reported than it was forbidden, and the beliefs associated with it suppressed by influential colonial officials and missionaries" (1983, p. 2). One is awed by this power of British officials, especially considering the failure of the Soviets to obliterate religious customs during seventy years of Communist rule. Nevertheless, by the time that the 1951 edition of *Notes and Queries* was published, questions on cannibalism and human sacrifice were no longer included. Questions about headhunting, apparently either a more robust tradition or more easily tolerated by British officials and missionaries, survived in the later editions.

What Cannibalism Means versus Whether It Occurs

Symbolic anthropologists respond to Arens's complaint about lack of evidence for cannibalism by shifting attention from evidence for its occurrence to evidence for the meaning of claims about cannibalism in a society that attributes the practice to its forebears or its enemies. At least one anthropologist with experience in the New Guinea highlands says that he observed instances of cannibalism, but he attributes little significance to this (Poole 1983). Of his experiences Poole says, "All *observed* instances of Bimin-Kuskusmin anthropophagy . . . are *customary* aspects of mortuary rites" (p. 15, emphasis original). He describes the various mortuary customs and distinguishes occasions on which he sees merely the presentation of "morsels" of flesh from those on which he observes consumption. He reports that although the Bimin-Kuskusmin consider some of these rites "degrading or disgusting" (p. 16) or extraordinarily polluting (p. 17), they also believe that the rites are necessary to enhance fertility, strength, and the perpetuation of knowledge and power.

Poole maintains that he is not interested in the controversy provoked by Arens, because the *meaning* of cannibalism for the Bimin-Kuskusmin is far more important than whether it ever occurs. His essay accordingly discusses cannibalism's symbolic and mythical trappings, especially those that focus on gender identity and various notions about the human body. More than half his essay is taken up with the notorious Great Pandanus Rite, involving human sacrifice and exocannibalism. The rite is said to occur only about "once every generation" (p. 17) and may have last taken place during the late 1940s. No outsider has witnessed it (and lived to tell the tale). Poole's account is based entirely on interviews with ritual elders who discuss their memories of an event at least thirty years in the past.

Poole's co-participants in the symposium also focused on the symbolic aspects of cannibalism. MacCormack (1983) says that in the early days when Sierra Leone was a protectorate, groups frequently complained to the district commissioner that their rivals were cannibals, knowing that this accusation would be understood as marking the rivals as unfit to rule. Thus, MacCormack warns, "[i]t is a grave scholarly error for literal-minded historians to use [the Sierra Leone Government Archives] without any anthropological interpretation" (p. 53). While providing an interpretation for the Sierra Leone Protectorate context (cannibal = person unfit to rule), she offers no general guidelines for such interpretations. She obviously agrees, however, that accusations of cannibalism are politically motivated and are used to justify exclusion, oppression or denial of rights.

Sahlins (1983), who also took part in the symposium, so interweaves myth and history in his discussion of Fijian cannibalism that it seems beside the point to ask whether the Fijians—or any humans—ever actually engaged in cannibalistic acts. Sahlins wants to show that the "historical practice of cannibalism can alternately serve as the concrete referent of a mythical theory or its behavioral metaphor" (p. 91). He opens his essay with an observation by a nineteenth-century missionary to Fiji: "Cannibalism among this people is one of their institutions; it is interwoven in the elements of society; it forms one of their pursuits and is regarded by the mass as a refinement" (Williams and Calvert 1859). This quotation is reminiscent of similar remarks by Evans-Pritchard (1937) about Zande witchcraft, which carry the strong suggestion that the real object of anthropological study is not to determine whether cannibals or witches exist, but to understand the role that beliefs about them play in the lives of people.

In focusing on what cannibalism means to the cannibal rather than whether it occurs, symbolic anthropologists attempt to articulate the social and political relationships that are bound up with talk about cannibalism, and in this way they further the humanistic and scientific goal of understanding the nature of human social reality. Clearly, much can be learned about the culture by trying to understand its hopes and fears on their own terms. Yet by denying the importance of the factual question, anthropologists overlook a crucial consideration for grasping the social reality they seek to understand. Evans-Pritchard, for example, *knew* that witches, in the sense of persons with supernatural abilities to cause harm to others, did not exist. This knowledge significantly shaped his understanding of Zande accusations of witchcraft and their explanations involving the activity of witches. If anthropologists who study cannibalism talk do not know whether or not cannibalism occurs, their ability to understand the meaning of the talk is severely limited.

Material Evidence for Cannibalism

Inference to the Best Explanation. The symbolic approach would seem to hold little appeal for physical anthropologists, given their sources of information and their concern with material matters. Durham (1991), for example, holds that the function of cultural practices is to enhance reproductive fitness. Thus, he must provide explanations for any cultural practices that seem to undermine fitness. That is to say, he must either show that their failure to enhance fitness is only apparent, or give some other account of their persistence. Once convinced that cannibalism was practiced in New Guinea, Durham investigated the alleged link between mortuary cannibalism and the spread of kuru among the Fore people.

Durham claims that Arens made him skeptical about Fore cannibalism. He says that his doubts were overcome by lines of evidence that include a brief eyewitness report of Gajdusek, as well as convincingly detailed descriptions of several "endocannibalistic feasts" during the 1950s (Klitzman, Alpers, and Gajdusek 1984). These ethnographic lines of evidence, however, are not specific to physical anthropology. Durham appeals to the verisimilitude of detailed eyewitness accounts of "playlets" in which pre-1960s endocannibalistic feasts of the Fore's neighbors, the Gimi, are reenacted (Gillison 1983; Durham 1991, pp. 396–397). He also cites Meigs's (1984) account of Hua—another neighbor—beliefs about cannibalism. Whereas Poole claims to have *observed* Bimin-Kuskusmin rituals in which minute portions of the body were reluctantly consumed, Durham cites accounts that refer to *the women* going off to their houses to consume the entire body. They are said to call human flesh "good meat" and to have eaten quantities of it before the practice was outlawed.

The warlike Fore perceive deaths as threats to military strength, and counteract the threat by holding mortuary feasts for kin and allies. These feasts serve to strengthen social and military ties. Durham (1991) claims that the Fore were predisposed to gastronomic cannibalism by "older cultural values that advocated the use of body substances, such as hair and nail clippings for their nourishing and curative effects upon one's relatives" (p. 398, citing Meigs 1984, ch. 6). Yet Durham offers no evidence that medicinal cannibalism leads to gastronomic cannibalism in any other case.

To the contrary, one of the best-documented and long-lived practices of medicinal cannibalism did not lead to other forms. Gordon-Grube (1988) investigated the use of mummia or "mummy," consisting of prepared bits of mummified human body or blood, as a treatment for many types of ailments in post-Renaissance Europe. Paracelsus in the sixteenth century argued that in minute amounts—since only the essence was required—mummy was an excellent curative. He noted with regret the reluctance of physicians to prescribe it and of patients to take it (Gordon-Grube 1988). Mummy was listed as a remedy in the standard *London Pharmacopoeia* from 1618 to 1747. The same substance was even offered for sale in the 1908–1909 edition of the Merck manual under the title of "genuine Egyptian mummy—while supplies last" (Gordon-Grube 1988). The acceptance of mummia as a curative, however, did not predispose Europeans to further acts of cannibalism.

Another form of medicinal cannibalism in Europe is found in the widespread use of human blood to avert attacks by vampires. One recipe, reported by Barber (1988), involves dipping part of a shroud in the blood of a supposed vampire, leaching the blood out into brandy, and drinking the mixture. Barber remarks, "Whether or not vampires drank the blood of human beings, we have most persuasive evidence that human beings have drunk the blood of vampires" (p. 64). The reported reluctance with which Bimin-Kuskusmin consume small pieces of flesh at mortuary rites and their feelings of pollution as a result also argues against such forms of cannibalism predisposing a group to gastronomic cannibalism.

Aside from the ethnographic evidence for cannibalism, Durham (1991) believes that cannibalism *best explains* the pattern of the kuru epidemic. Here we are dealing with physical evidence, but it is problematic. The original agent of kuru is unknown, but apparently the disease first appeared after European contact. Those who believe that the Fore practiced cannibalism also believe that it was a relatively recent innovation, adopted within living memory of some older Fore. Despite the coincidental timing, the physical evidence that cannibalism *causes* kuru is not clear. Even Gajdusek and his coworkers agree that the clearest and perhaps only means of transmission of kuru is through direct invasion of the blood stream. They did not find any mechanism for transmission associated with consuming infected matter. (More recent work on the kuru-related "mad cow disease" has suggested such mechanisms, but the issue is far from settled.)

Arens argues against Durham's *inference to the best explanation* by showing that the pattern of kuru infection—which was far more prevalent among adult females than males—could be explained by adult females' role in mortuary practices. Men and women live separately. Sanitation is poor, hand washing rare. Women live with and care for children, prepare food, and take care of dead bodies. Small cuts and abrasions allow infection to enter their bodies. Durham, in opposition to Arens, while admitting that handling infected bodies is a possible mode of transmission, feels that "normal" or "traditional" mortuary practices during the precannibalistic period would not explain such contamination. Durham "reconstructs" the traditional practices, for which no direct evidence exists, on the basis of reported ethnographic data from neighboring tribes. Presumably, the bodies were exposed on platforms, left to decay, and—because of fear of pollution—handled as little as possible. Durham claims, though the evidence is hardly conclusive, that only the preparation of the bodies as food would have caused the widespread pattern of the disease.

Arens complains about internal inconsistencies in many accounts of cannibalism. Durham's account of how the Fore were predisposed to cannibalism while also fearing the pollution of dead bodies is vulnerable to this criticism. Durham tries to resolve the problem by accepting an implausible ethnographic account of the institution of cannibalism—the women decided it would be better to eat the entire bodies than to allow them to decay, he says, and the practice caught on very quickly. He then links this happenstance beginning to a functional account of a shortage of meat, and the perceived need to preserve the best meat (pork) for the male warriors in the tribe.

The dispute between Arens and Durham about how best to explain the kuru epidemic brings out another problem. With no clear criterion for what constitutes a best explanation, appeals to the standard of inference to the best explanation are defective. Moreover, we find that although Durham is a physical anthropologist whose expertise lies in handling material evidence, his physical evidence is not adequate to support the thesis of cannibalism. Durham depends heavily on ethnographic lines of evidence (alleged eyewitness accounts and detailed second-hand reports) for his "inference to the best explanation."

The Archaeological Record and Arguments from Analogy. In general, archaeologists have responded to Arens's critique by focusing on their own standards of physical evidence for cannibalism rather than by reexamining and reinterpreting ethnographic reports. In doing so, they admit that medicinal cannibalism and ritual consumption of small portions of human flesh will not be archaeologically recognizable. However, gastronomic cannibalism might be recognized archaeologically in the way in which the parts of bodies are treated and discarded. While no reliable ethnographic evidence for cannibalism exists to provide direct comparisons, analogies can be made with treatment of other large vertebrates eaten by humans. Rathje (1985), well known for his studies of what contemporary garbage tells us about human behavior, urges archaeologists to look more carefully at kitchen middens to try to find patterns of use of animal materials. Greater proportions of human bone in some sites, he says, might indicate cannibalism.

Despite examples like Choukoutien and Grotta Guattari, where evidence for cannibalism is so slight that it could be called whimsical, archaeologists generally have been more conservative than other anthropologists in their attributions of cannibalism. Archaeological findings of human bones in the American Southwest, for example, are simply referred to as "burials," no matter what their condition (White 1992). Recent reexamination of the physical evidence at Choukoutien (Zhoukoudian) and at Grotta Guattari strongly suggests that hyenas who used these caves as maternity dens were responsible for the conditions of the bones found there. Binford and Ho (1985) argue that the so-called unnatural breakage of bones at Zhoukoudian is the normal result of scavenger and other taphonomic processes, and that the items labeled "bone tools" are the result of rodent modification. Stiner (1991) questions whether the so-called ring of stones at Grotta Guattari was deliberately placed. She also notes that piercing the base of the skull—in the manner of the damaged cranium found at Grotta Guattari—is not an efficient way of extracting brain matter.

Motivated by the desire to recognize cannibalism in an archaeological context, White (1992) analyzed skeletal remains at Mancos, an Anasazi site in southwestern Colorado. The group burials there, unlike the single graves common to most of the Southwest, had age groups clustered in a suspicious way. White defined cannibalism broadly as "the conspecific consumption of human tissue" and credited the authors of the study of Neolithic cannibalism at Fontbrégoua (Villa et al. 1986, p. 431) with articulating four important lines of analogical evidence:

> 1. similar butchering techniques in human and animal remains: frequency, location, and type of verified cut marks and chop marks on human and animal bones must be similar, [with allowance for] anatomical differences between humans and animals;
> 2. similar patterns of long bone breakage that might facilitate marrow extraction;
> 3. identical patterns of postprocessing discard of human and animal remains;
> 4. evidence of cooking: if present, such evidence should indicate comparable treatment of human and animal remains. (White 1992, p. 9)

White subjected bones and their contexts to minute analysis to identify the bone parts and to determine their condition of preservation. Important clues are the color of bone (bones that have been defleshed before being exposed to weather are whiter), and such matters as whether fractures and other marks are the result of damage at the time of death or later, what sort of agent inflicted the marks, and whether the bone was subjected to burning or cooking. He used 98 different characteristics in the categories mentioned above, as well as a less labor-intensive, minimal "short list" of 24 characteristics relevant to cannibalism. One such characteristic is "pot polish." Abrasion of bones occurs as a result of tossing by wind or water, trampling, and being used as tools. White and his coworkers, experimenting with deer metapodials, found another source of polish: when flesh-covered bones are boiled in a ceramic pot of the type used at Mancos, the exposed ends of the bones are abraded through contact with the pot.

Neither pot polish nor any other feature in White's lists provides by itself the archaeological signature of cannibalism, but a number of features in conjunction can be persuasive. White says about the Mancos site:

> [T]he remains were deposited simultaneously. Most of the bones show little or no weathering, indicating that surface exposure time was minimal [this helps rule out animal damage to the bones]. The remains themselves bear substantial evidence of human-induced modification in the form of thermal trauma, cutmarks, chopmarks, pot polish, and fracture and crushing by hammerstone percussion. There are very few nonhuman bones associated with the human remains [perhaps an indication of starvation conditions]. There is no evidence that carnivores manipulated the bones. (p. 337)

While White feels that this evidence is very strong, he admits that it is not conclusive (p. 339). He says, however, that only finding traces of human bone in "demonstrably human coprolites" would be stronger. Moreover, he believes that such evidence might turn up in the dry caves of the Southwest. Even this latter criterion will not satisfy some critics of alleged evidence for cannibalism. Bahn (1993), for example, claims that the only conclusive evidence or "archaeological signature" would be human remains found in the human gut.

White's evidential standards for attributing cannibalism are, as he says, very high. Although his definition of cannibalism is the broadest possible, however, only gastronomic cannibalism will be detectable by his methods, and except for starvation conditions, such as those suspected at Mancos, even that may be elusive. Despite this, we may come to accept, on the basis of archaeological studies such as White's, a *negative* answer to a question raised by Sagan: Were our earliest ancestors all cannibals?

Present Status of the Cannibalism Controversy

In the aftermath of Arens's criticisms, ethnographic accounts and alleged physical evidence that human remains were "processed" as food have been closely reexamined. Some anthropologists discount the importance of the truth of cannibalism and the quality of evidence for it; others, including most archaeologists, have tried to reexamine physical evidence, or to seek new physical evidence, to see whether they can find an archaeological "signature" for cannibalism. (See Bower 1993; Turner 1983, 1999.) Durham depends heavily on ethnographic evidence but remains persuaded that no other hypothesis explains the observed physical evidence for the distribution of kuru. Alleged physical evidence for cannibalism from Zhoukoudian and some Neolithic sites, such as Grotto Guattari, has been reinterpreted as resulting from carnivorous scavengers, and these sites no longer support a cannibalistic hypothesis.

Cannibalism at other Neolithic sites, such as Fontebrégoua and Krapina, is not so easily dismissed. At these sites, as well as at Mancos, the human bones show the same marks as butchered animal bones. However, because some ethnographically recognized human burial practices involve defleshing bones, breaking them, exposing them, and treating them similarly to bones of animals that are eaten for food, evidence that bodies were "butchered" does not conclusively support cannibalism. White recognizes this, but draws on his extensive knowledge of burial practices in the Southwest to argue that the bones at Mancos are not just the result of an anomalous form of burial. He is persuaded by the physical evidence at Mancos that cannibalism is the hypothesis that best accounts for the condition of human bones and their manner of deposition.

As White and other archaeologists readily admit, even very strong archaeological evidence for cannibalism may give no hint as to why the cannibalism occurred. Was it a response to starvation, or some other form of cannibalism? White suspects survival cannibalism at Mancos because of the lack of other animal bones associated with the human remains. If White is correct, we can still ask whether survival cannibalism should be classified as psychotic behavior induced by severe stress or a culturally approved response to such stress. Thus, even the most convincing physical evidence may not be adequate to answer Arens's question of whether cannibalism has ever been an accepted cultural practice. Arens's continuing skepticism about the type and degree of evidence required to support a thesis of cannibalism infuriates many anthropologists and, according to Osborne (1997), helps drive physical anthropologists such as White from anthropology departments into biology departments.

Although European medicinal cannibalism has received less attention than other forms, the case for it seems very strong. However, Gordon-Grube's (1988) analysis of the use of mummia and the existence of catalogs offering it for sale over several centuries are tempered by the recognition that it was never a popular remedy. Much more work would be needed to discover how often mummia was used, in what quantities, and even whether its content was understood by those who prescribed and used it. Questions can be raised as well about the relevant differences between consuming "fresh" human flesh and minute quantities of the ancient mummified stuff that was once flesh.

Conclusion

On one level the controversy about cannibalism is a recalcitrant factual question about whether any culture has been cannibalistic. The controversy Arens generated has resulted in more meticulous definitions and classifications of the practice. While disagreement about the main question remains, anthropologists have agreed that the evidence does not support many formerly accepted attributions of cannibalism. Nevertheless, anthropologists disagree sharply about how much and what kind of evidence is required to say that cannibalism occurred. White, for example, has compared Arens's refusal to "admit all the implications of the hard scientific evidence" comparable to "flat-earthers denying the roundness of the earth" (quoted in Osborne 1997, p. 38). In other words, anthropologists, like other social scientists, find it easier to agree about what fails to constitute good evidence than to define their standards.

The disagreement between those for whom it matters whether cannibalism occurred and those who care only about what cannibalism means plays out the old dispute between scientific concern with facts and humanistic concern with meanings. The cannibalism controversy is at the heart of anthropology, which after all is concerned with who we are and how we came to be as we are. The controversy about cannibalism divides all subfields of the anthropological community. A resolution will require the best ethnographic and physical evidence and the most sensitive interpretation of that evidence. Examining the question of standards of evidence in this difficult context should further general understanding of the evidence in the social sciences.

Notes

I am grateful to Jeremy Sabloff, who made very helpful comments on several drafts of the manuscript. A version of this chapter was presented at the 1995 Spindel Conference on Explanation in the Human Sciences, organized by David K. Henderson and was published in *The Southern Journal of Philosophy* (vol. 34 suppl.), edited by D. K. Henderson. I am grateful to the journal for permission to use the material here. I also extend sincere thanks to Robert Feleppa, who commented on the paper at the Spindel Conference.

References

Arens, W. (1979), *The Man-Eating Myth: Anthropology and Anthropophagy.* New York: Oxford University Press.

Bahn, P. (1993), Letter. *Science News* 143(9):131.

Barber, P. (1988), *Vampires, Burial, and Death.* New Haven, Conn.: Yale University Press.

Binford, L., and C. K. Ho. (1985), Taphonomy at a distance: Zhoukoudian "the cave home of Beijing man"? *Current Anthropology* 26:413–442.

Bower, B. (1993), The cannibal's signature. *Science News* 143(1):12–14.

Clendinnen, I. (1991), *Aztecs.* Cambridge: Cambridge University Press.

Coon, C. (1963), *The Origin of Races.* New York: Knopf.

Durham, W. (1991), *Coevolution: Genes, Culture, and Human Diversity.* Stanford: Stanford University Press.

Evans-Pritchard, E. (1937), *Witchcraft, Oracles, and Magic among the Azande.* London: Oxford University Press.

Freeman, D. (1983), *Margaret Mead and Samoa.* Cambridge, Mass.: Harvard University Press.

Gillison, G. (1983), Cannibalism among women in the eastern highlands of Papua New Guinea, in P. Brown and D. Tuzin (eds.), *The Ethnography of Cannibalism.* Washington, D.C.: Society for Psychological Anthropology, pp. 33–50.

Gordon-Grube, K. (1988), Anthropophagy in post-Renaissance Europe: The tradition of medicinal cannibalism. *American Anthropologist* 90:405–409.

Harner, M. (1977), The ecological basis for Aztec sacrifice. *American Ethnologist* 4:117–135.

Harris, M. (1977), *Cannibals and Kings.* New York: Random House.

Henderson, D. K. (ed.). (1995), *Explanation in the Human Sciences. The Southern Journal of Philosophy,* vol. 34 (suppl.).

Klitzman, R. L., M. P. Alpers, and D. C. Gajdusek. (1984), The natural incubation period of kuru and the episodes of transmission in three clusters of patients. *Neuroepidemiology* 3:3–20.

MacCormack, C. (1983), Human leopards and crocodiles: Political meanings of categorical anomalies, in P. Brown and D. Tuzin (eds.), *The Ethnography of Cannibalism,* Washington, D.C.: Society for Psychological Anthropology, pp. 51–60.

Mead, M. (1928), *Coming of Age in Samoa*, New York: W. Morroe.
Meigs, A. S. (1984), *Food, Sex, and Pollution: A New Guinea Religion.* New Brunswick, N.J.: Rutgers University Press.
Montaigne, M. (1931), *Montaigne's Essays: John Florio's Translation,* ed. J. I. M. Stewart. London: Nonesuch Press.
Notes and Queries on Anthropology (2nd ed.), (1892), J. G. Garson and C. H. Read (eds.). London: The Anthropological Institute of Great Britain and Ireland.
——— (6th ed.), (1951), rev. and rewritten by a committee of the Royal Anthropological Institute of Great Britain and Ireland. London: Routledge and Paul.
Osborne, L. (1997), Does man eat man? *Lingua Franca,* April/May: 28–38.
Poole, F. (1983), Cannibals, tricksters, and witches, in P. Brown and D. Tuzin (eds.), *The Ethnography of Cannibalism.* Washington, D.C.: Society for Psychological Anthropology, pp. 6–32.
Rathje, W. (1985), The cannibal debate. *Wilson Quarterly,* Spring, 134–135.
Sagan, E. (1974), *Cannibalism: Human Aggression and Cultural Form.* New York: Harper and Row.
Sahagún, B. de. (1950–1982), *Florentine Codex.* Santa Fe: School of American Research; Salt Lake City: University of Utah, 13 vols.
Sahlins, M. (1983), Raw women, cooked men, and other "great things" of the Fiji Islands, in P. Brown and D. Tuzin (eds.), *The Ethnography of Cannibalism.* Washington, D.C.: Society for Psychological Anthropology, pp. 72–93.
Sanday, P. (1986), *Divine Hunger: Cannibalism as a Cultural System.* Cambridge: Cambridge University Press.
Springer, J. (1980), Review of Arens. *Anthropological Quarterly* 53:148–150.
Stiner, M. (1991), The cultural significance of Grotta Guattari reconsidered. 1. The faunal remains from Grotta Guattari: A taphonomical perspective. *Current Anthropology* 32:103–117.
Stocking, G. (April 1993), Lecture, "Reading the palimpsest of inquiry: Notes and queries and the history of British social anthropology," University of Pittsburgh.
Trinkaus, E. (1985), Cannibalism and burial at Krapina. *Journal of Human Evolution* 14: 203–216.
Turner, C. (1983), Taphonomic reconstructions of human violence and cannibalism based on mass burials in the American Southwest, in G. LeMoine and A. MacEachern (eds.), *A Question of Bone Technology.* Calgary: University of Calgary Archaeological Association, pp. 219–240.
——— (1999), *Man Corn: Cannibalism and Violence in the Prehistoric American Southwest.* Salt Lake City: University of Utah Press.
Tuzin, D., (1983), Cannibalism and Arapesh cosmology: A wartime incident with the Japanese, in P. Brown and D. Tuzin (eds.), *The Ethnography of Cannibalism.* Washington, D.C.: Society for Psychological Anthropology, pp. 61–71.
Tuzin, D., and P. Brown. (1965), Editors' preface, in P. Brown and D. Tuzin (eds.), *The Ethnography of Cannibalism.* Washington, D.C.: Society for Psychological Anthropology, pp. 1–5.
Vaillant, G. (1965), *Aztecs of Mexico.* Baltimore: Penguin Books.
Villa et al. (1986), Cannibalism in the Neolithic. *Science* 233:431–436.
Warner, M. (1994), *From the Beast to the Blonde: On Fairy Tales and Their Tellers.* London: Chatto and Windus.
——— (1997), Lecture, "Why do ogres eat babies?" University of Pittsburgh.
White, T. (1992), *Cannibalism.* Princeton, N.J.: Princeton University Press.
Williams, T., and J. Calvert. (1859), *Fiji and the Fijians.* New York: Appleton.

12

Multiple Personalities, Internal Controversies, and Invisible Marvels

IAN HACKING

Structures of Controversy

This chapter was first presented at the 1993 Conference on Scientific Controversies, and was revised a few days later to take into account what other participants had said. We all change our minds, or develop our ideas, so I cannot speak for what colleagues now believe, but what they said then seems to me important enough to recall it to mind. For my part, I used as an example of what I call internal scientific controversies—a label I explained—the then raging debates about multiple personality. I was at the time working on a book (Hacking 1995) that now documents the history I exploit below. But the specific use of that history as an example with which to think about controversy was dedicated to the conference and is not found in the book or elsewhere.

One valuable paper that helped organize the discussion at the conference was given by Aristides Baltas. He maintained that scientific controversies form a hierarchy of four tiers from the constitutive through the interpretive. My example, drawn from sensational and at the time popular psychology, is less easily structured, perhaps because things have even now not yet settled down. I can see only a nesting of levels. To begin abstractly, these are as follows:

X, controversies about global revisions of an entire taxonomic scheme;

Y, specific controversy about the existence of a basic class within a current scheme;

Z, controversy internal to those who insist on the reality of that class—disputes about how it is to be characterized (as my title indicates, my focus is at this level); and

O, silence: what is never discussed, an Althusserian *absence* (I shall argue that silences can be very important to the conduct of a controversy).

Philip Kitcher's paper argued that controversies have a beginning, a middle, and an end; their beginnings and middles are peculiar, but their final resolutions are similar in kind. I suspect my story is cyclic. Up until now, and for the foreseeable future, the history of my topic exhibits only spirals of knowledge. This is particularly evident with the regular recycling of global taxonomic schemes, my level X. We do not exactly repeat ourselves, but we do go round and round. I optimistically speak of spirals, with the implication that we are at least making a circular ascent, a progress.

Gideon Freudenthal'a paper proposed that controversies should be defined as disputes that nobody has any idea of how to settle. That is plainly true of the events at level Y that I shall describe—between clinicians in the multiple personality movement on the one hand, and the majority of psychiatrists on the other. But nothing in controversy is simple. I shall show that the internal argument at level Z is intended to do an end-run around the impasse at level Y. Thus, whereas Freudenthal had a sort of *aufhebung* in mind, an overcoming of the dispute by the production of deeper general ideas, in my example we may have the attempt to undermine the dispute by making the controversial category seem more innocuous (though if anything it is thereby made more pervasive, more dangerous).

Freudenthal urged, in addition, that controversies must have cognitive content. Mine do, but they differ in two ways from other controversies discussed at this meeting. First, they have immediate practical content: What, then, shall we do to help some very disturbed people? And, unlike the disputes that had been before our minds at the conference, mine are only secondarily about facts and theories. They are primarily about classification, about kinds of people, about the kinds of mental troubles that can afflict us. The taxonomies matter, because people are subjected to treatment according to the ways in which they are classified; moreover, the theories about what ails them depend largely on the organization of diagnostic categories. The theories do not so much explain the categories as legitimate them.

Most of my discussion here is at the lowest level on my scheme, Z. Controversy at level Y will be, if not familiar to you, at least readily intelligible: Does multiple personality exist? Or, better, is multiple personality disorder a viable diagnostic category for a large number of patients? However, the internal discussion on which I focus here, level Z, will not be familiar. In the course of the conference, William Wallace, an emeritus from Catholic University, remarked to me that he taught history of science for the University of Maryland because the department there was strong only in recent history, like that of quantum mechanics—some so new that it had hardly become history yet. Some of the history that I presented at the conference was so new that it had not even happened, at that time. Four years later, as I went through my paper again, I found that some of the events that I described were only just beginning to reveal themselves.

Psychiatric Patients

Before I fill in the blanks in my X, Y, and Z, I must make a general remark about psychiatry. It is not inevitable that doctors of medicine should be the segment of our society with authority and control over mentally disturbed people. Physicians and surgeons won that control. Their position was put in place during the Enlightenment and confirmed at the end of the eighteenth century. By now, the most widely read account of this gaining control is the early work of Michel Foucault (1965). There are many later versions of that story, such as Andrew Scull (1993). I want only to emphasize that the very concept of "psychiatry" stands for the organization and control of madness in Western culture. The central figure of Foucault's own account is Philippe Pinel. Like every great psychiatric revolutionary, he set forth a new nosology of madness. Yet although his classification was novel, and although it was replaced by many others—part of the recycling I mentioned above—there has at least since Pinel, and his contemporaries in other countries, been a fairly stable arrangement of Western insanity. I would not say that in front of an audience of mental health professionals, but for amateurs, I think we can stand far enough away from the topic to discern four chief configurations. This is not to say that symptoms, diseases, and injuries don't move from one to the other with each recycling, but the great prominences that we can make out are as follows:

A. Damage: head injury;
B. Defect or deficiency: among the derogatory labels are feeble-minded and idiot;
C. Reality problems: "psychoses" such as schizophrenia, paranoia, bipolar disorder (≈ manic-depressive ≈ *folie circulaire*), and catatonic states; and
D. Mental disorder, dysfunction: "neuroses" such as compulsions, obsessions, anorexia, alcoholism, neurasthenia, hysteria, conversion symptoms, depression, melancholia, hypochondria, dissociation, fugue, multiple personality, and chronic fatigue syndrome.

I deliberately mention categories from different diagnostic schemes, from different eras. I omit the ample borderline personality disorder, which occupies many beds in many wards and stands for the class of patients that doctors tend to hate the most. I also omit epilepsy, which for a long time was given the status of D, but would now be located in either A or B depending on symptoms and history—and yet in many cases, despite the power of modern neuroleptic drugs, epilepsy would require treatment in D, and may produce C-type symptoms, including epileptic automatism.

I do insist on the distinction between C and D. The unfashionable words "psychotic" and "neurotic" were by no means bad choices to mark a distinction. I am not claiming that schizophrenia is a good diagnostic category, but that there is a family of problems, often grouped under schizophrenia, which is far worse than mere dysfunction or disorder. It is characterized by a menu of hallucinations, delusions, terror, and catatonic states (before the advent of powerful psychotropic chemicals). Since I shall go on to speak about multiple personality disorder, I must avoid confusion with schizophrenia. Unfortunately the expression "split personality" rides, in popular discourse, on both backs. Eugen Bleuler (1908, 1911/1950) gave us the name "schizophrenia" as a replacement for Emil Kraepelin's *dementia praecox*. Kraepelin had emphasized the way in which the symptoms begin when the victim is age 17–24, hence his choice of name. Emergence in early adulthood remains a standard sign of schizophrenia. Bleuler did not think of a person splitting into different personalities; he chose the name because he believed that the core problem for his patients was that their will was split from their feeling.

Contrast this with multiple personality, which usually manifests itself later in life, 25–35, and is at present thought to be latent from the age of 5. Its epidemiology is almost the converse of schizophrenia, at least according to current wisdom. Schizophrenics are more commonly men than women, while 9 in 10 multiple personality patients are female. The main feature shared by the two, as diagnosed at present, is that both kinds of patients may report hearing voices. But the schizophrenic hallucinates voices from afar—God or a person at some distance behind him. The voices of other personalities are from within the head, or just outside the ear. Gossip among multiple personality patients begins with this advice: Never tell a doctor you hear voices, or he'll treat you like a schizophrenic and drug you out of your mind. Multiple personality does not respond specifically to chemicals. Schizophrenia does.

I have heard the author of the standard multiple personality textbook (Putnam 1989) say that patients with multiple personality aren't *mad!* And that goes across the board: very crudely, people in class C are crazy, and people in class D are troubled. There is at present also a profound practical difference between C and D. The behavior of psychotics responds (often for the better) quite specifically to individualized cocktails of psychotropic drugs. Hallucinations are terrifying, and you really can get rid of them by taking chemicals regularly. But there are no very specific medications for people in class D, aside from uppers and downers. Of course, you can influence people's behavior, diminish depression, make an alcoholic throw up after taking a drink, and on and on through several thousand types of tablets. Unfortunately, specific effective treatments are, despite the glossy advertisements in medical magazines, not in hand. Conversely, despite the legitimate agitation by those opposed to psychotropic drugs, they really do help psychotics get along with the everyday world and other people.

I began this section mentioning the medicalization of madness, and the formation of four distinct classes of mental disorder. That happened toward the end of the very period that was singled out in Peter Machamer's paper, the era when the individual comes to the fore, both philosophically and commercially. We may say that people in class C suffer from a problem with the epistemic ego, while those in class D suffer from a problem of the entrepreneurial ego. I allow myself a speculation, in line with Machamer's analysis: Why has the above arrangement (A–D) been relatively stable? There may seem to be no problem about A and B, but there is. Head injury may take up more of the space of C and especially D than has been acknowledged. But the interesting question is the distinction between C and D. Is that just an objective distinction about two biologically separate types of disorder? My remarks about chemicals show that the role of biology is real enough. Nevertheless, I would like to suggest a different type of analysis of the distinction between C and D—none other than the one at the heart of Peter Machamer's paper: between the *epistemically disabled* and the *entrepreneurially disabled.* That distinction makes sense chiefly in our type of culture—what anthropologist Mary Douglas (1992) calls the enterprise culture. I said the C/D distinction had long been stable in the West. I suggest that Douglas and Machamer may be combined to tell us why. Douglas tells us about the self in the enterprise culture; Machamer, about the self's imposed difficulties.

This section has been a gross oversimplification of the past and present of psychiatry. Forgive me. I use it only to separate class D, the disorders and dysfunctions. That is where my controversies are located, and where my X-Y-Z-O levels above are to be found. Thus, for example, I have nothing to say here about schizophrenia, although that is certainly a region for controversy, perhaps more familiar to some of you than the domain to which I now turn.

Three Levels of Controversy

I believe that my structure diagrammed as X-Y-Z-O is quite commonly exemplified in many scientific fields. One would not guess that from other contributions to this conference, but I shall not argue the point here. I am preoccupied by my own example, which is represented like this:

X, *global:* controversies about the "neuroses"—how are disorders of type D to be classified, treated, and explained?

Y, *specific:* within a current diagnostic scheme, is multiple personality disorder to have a significant place?

Z, *internal:* within the multiple personality movement—of clinicians and others who think the disorder is real, common, and significant for the theory and practice of psychiatry—what should be done about a crisis of confidence in the established etiology, diagnosis, and behavior of patients in therapy?

O, *silences:* about marvels ranging from trance and hypnosis to Satanic cults and abduction by aliens.

The Multiple Timeline

I would like to give a long-term sense of the controversy that I use as an example. There has been a raging epidemic of multiple personality disorder in North America, with a few bridgeheads in Europe, starting with Holland. A more complete history is examined in detail in Hacking (1995), so I omit scholarly citations here. A few salient dates suffice for the big picture:

1785. French commissions on mesmerism.

1791. First well-known detailed reports of double or split personalities more or less as such, although in retrospect one can find cases going back a long way in time (Germany and America).

1800–1875. Sporadic reports in medical literature of what was, in English, usually called double consciousness.

1853. Braid's scientific hypnotism, the successor idea to mesmerism, sharing many of its practices, but rejecting its language.

1870–1890. In France, hysteria is the dominant mental disorder of type D; reemergence of hypnotic treatment.

1875. Azam presents the first modern double personality, Félida X, who becomes the prototype for double consciousness.

1875–1895. French wave of multiples; Pierre Janet's assertion that every case of hysteria is at bottom a case of dissociation, of splitting of consciousness.

1885. The first truly multiple personality—that is, with more than two personalities; trauma, which used to mean a wound or physical lesion, comes also to denote psychological hurt.

1893. Freud's seduction hypothesis about the cause of hysteria, revised in 1897.

1895–1914. Dissipation of hysteria into a whole new arrangement of disorders—recycling at level X above; death blow to hysteria by Joseph Babinski in dealing with shell-shocked patients.

1900–1920. American multiples, a wave spreading out from Boston and Morton Prince, Connecticut.

1953. Eve (of the *The Three Faces of Eve*) diagnosed with multiple personality.

1961. Child abuse surfaces as a central American concern, in connection with battered baby syndrome.

1971–1976. Incest brought under the category of child abuse.

1974. The first thoroughly modern multiple: *Sybil,* described in a multobiography (novel published 1973; novel relaesed 1976), became the prototype for multiples of the 1970s and thereafter.

1980. American diagnostic manual (*DSM-III*) includes Multiple Personality Disorder as a legitimate diagnosis; MPD described as an epidemic two years later.

1984. Annual meetings of International Society for the Study of Multiple Personality and Dissociation established; journal *Dissociation* commenced four years later.

1985. Posttraumatic stress disorder takes off as a way of dealing with dysfunctional U.S. veterans of Vietnam War.

1986. Repeated childhood sexual abuse confirmed as the primary cause of multiple personality.

1987. MPD entrenched in next edition of American diagnostic manual (*DSM-III-R*); satanic ritual abuse surfaces as major ingredient in child abuse.

1989. First complete medical text book on MPD (Frank Putnam, *Diagnosis and Treatment of Multiple Personality Disorder*).

1990. Abduction by aliens increasingly reported by patients professing to be multiples and seeking therapy.

1992. November: theme for annual meeting of ISS MPD is how to deal with health insurance companies—MPD has joined the establishment, but storm clouds on the horizon. March: false Memory Syndrome (FMS) Foundation established, challenges therapeutically recalled memories of abuse, and takes off throughout the year.

1993. Final committee meetings for next edition of American diagnostic manual; first annual meeting of FMS Foundation.

1994. Publication of *DSM-IV. Multiple Personality Disorder renamed Dissociative Identity Disorder.*

1995. Backlash in full swing, accompanied by what is logically independent, causally irrelevant, but rhetorically valuable: Freud-bashing.

1996. Dissociative Identity Disorder hunkers down; PhD dissertations in sociology, anthropology, literature, and so on, are being written on the phenomenon of "repressed memory."

The Modern Multiple

I should very briefly present some stereotypical symptoms of multiple personality. They appear as a sort of template in the patient of a self-styled maverick psychoanalyst, Cornelia Wilbur. Wilbur could not get her case published in any of the leading journals, so she had a journalist write it up (Schreiber 1973). The author insisted that the patient, Sybil, had to be cured before the book would sell. A cure was achieved. The book sold well and became a movie. *Sybil* became a stereotype. Here are some of the features of multiple personalities from the 1970s to the present. I shall ignore the extent to which the symptomatology has evolved.

- Periods of lost time.
- Waking up in a strange place, not knowing how one got there.
- Strange credit card charges.
- Odd wardrobes, with clothes she would never wear in special parts of the wardrobe.
- "She"—90% of diagnosed multiples are female.
- Experiencing arguments in the head, or from just outside the head.
- Switching from one personality to an alter personality. If there are switches before therapy, or early in therapy, they commonly take place in a state of trance.
- Many alter personalities develop in therapy, usually between 16 and 25 personalities or person-fragments (some therapists get up to 100).
- Considerable mutual amnesia among alters, although in therapy they may become increasingly co-conscious.
- Child alters.
- Alters of different races, ages, sexes, sexual tastes.
- Cruel, vicious, persecutory alters, but also helpers.
- Experiences of antagonistic alters taking control in difficult situations, with a customer, with a boss, with the children; tricks of concealment.
- When a multiple is in the company of supportive people (perhaps other multiples in a self-help group), assumption of the personal pronoun "we"; talk of one personality withdrawing for a time "to another place."
- In memory therapy, recollections of abuse; different acts or types of abuse connected with different alters.
- *Etiology:* alters are dissociations from the personality, a response to early abuse; abuse causes "dissociation."
- *Treatment:* memory recall, often using hypnotism, even in some clinics sodium amytal; attempt to get warring alters to collaborate, form contracts, aim at co-consciousness; never try to "kill off" an alter.
- *Cure:* the usual model is co-consciousness and fusion of alters, together with an ability to become self-aware and put oneself together after reexperiencing past horrors; but not all multiples wish this—some would like to retain at least a few alters, as ways of giving expression to other aspects of themselves.

This prototype evolved during the period 1970–1990. In the early days, multiple personality was almost unknown and patients did not walk into clinics with symptoms except perhaps for confusion and periods of lost time. Many had already been diagnosed as having many other disorders and had been in the mental health system for an average of 8 years. There was much publicity in the popular media, tabloids, and

TV shows. After all this media action, disturbed people began walking in with quite florid symptoms. Yet even those who were already to some extent split developed a vast array of personalities or personality fragments only under therapy.

Marvels and Curiosities

After this brief immersion into the discourse of multiple personality, let us now stand back and examine what is *not* a part of this discourse. Students of scientific controversies seldom examine the absences, the silences, the topics that are excluded from science. What is missing from my story?

First, I remark on curiosities and marvels. Scientific curiosities are topics that are acknowledged by scientists but about which they can do nothing. Brownian motion was a curiosity for a century, and the photoelectric effect was a curiosity for 80 years. They were scientific because only scientists with a certain amount of instrumentation could observe them, and they were curiosities because they were isolated phenomena that fit with no vision of the world. There is a continuum from curiosities to marvels. In the early pages of the *Philosophical Transactions of the Royal Society of London* (1664–1800), there are all those marvels that Peter Machamer mentioned at the conference, three-headed sheep and the like. They became curiosities only because they were written up for fellows of a select society. In more recent times marvels lived in circuses and freak shows; now they inhabit tabloids and afternoon TV. The greater the human interest, the more likely it is that a curiosity, first reported by a scientist, will become a marvel.

One way to silence, to expel, a topic of research is to turn it into a marvel. One range of human behavior that fluctuates between marvel and curiosity is loosely grouped under trance: hypnotism, demonic possession, types of what were once called hysteria and are now called dissociation. Now you may begin to suspect why I put hypnotism and its ilk into the multiple timeline above. These phenomena of industrialized Western cultures are paralleled in most societies. Our anthropologists speak of trance states, which include, for example, Shamanism. I am not saying that these practices in other societies are all "the same," only that our experts lump them as the same. Many peoples have cultural roles for these types of behavior. We have none. They are important only to fringe parts of our anticulture, meditation or New Age movements, for example.

Science abhors a marvel not because marvels are vacuous, empty of meaning, but because they are too full of meaning, of hints, of suggestions, of feeling. Marvels are meanings out of control. You can expel a topic from science by making it a marvel. Conversely, if you are forced to look a marvel in the face, the thing to do is to bring it into the laboratory. There it will languish and die until the laboratory itself is cast out of science. Then it will become a marvel again, but it has been somehow rendered less potent for having declined a laboratory niche. Take, for example, the way in which psychical research was made a laboratory science in 1882, sponsored by some of the best scientific minds of London and Cambridge. It has tried to locate at the greatest laboratory centers of the world, Stanford, and recently Princeton. In each place it has become not marginalized but marvelized, cast out, not without being first stripped of its cash assets. I'm not talking metaphorically, although to talk literally and truthfully would arouse the interest of libel lawyers.

Trance phenomena have shared much the same fate. That is remarkable, for there are many people of a scientific bent who would be willing to state as a matter of fact that there are no parapsychological phenomena except the phenomenon of parapsychology itself—while asserting that trance phenomena are real enough. But trance is constantly being cast into the shadowy realm of marvel.

Philosophers like to talk about "the aims of science." If ever there was a time that science acted with concerted aim, it was in the two commissions that worked in 1785 to determine the validity of animal magnetism. One commission was established by the Academy of Medicine in Paris, while the other was in effect a royal commission over which Lavoisier presided, and which numbered Benjamin Franklin among the five commissioners. Mesmer had claimed the rank of science for his practice and had proposed a new theoretical entity, the magnetic fluid; he had laboratory practice, he had cures. But it was determined that there was no substance to his claims. Mesmerism was consigned to the level of popular marvel, where it played a significant role in underground anti-establishment movements leading up to 1789.

In the 1850s James Braid tried to restore animal magnetism to science. He abandoned all talk of the fluid and renamed the practice neurhypnology or scientific hypnotism. But scientific it never became. It did briefly flourish in France at the time of Jean-Martin Charcot and *la grande hystérie,* around 1885. By 1892 Pierre Janet was propounding a general therapeutics of hypnotism for restoring past memories and then resolving them. That had indeed already been done with Josef Breuer's patient Anna O. of 1882, the woman whose hysteria led Freud to the assertion that hysterics suffer from reminiscences. That was the beginning of what Breuer called the "talking cure," which later mutated into psychoanalysis. Freud first followed in Charcot's footsteps, but then renounced hypnotism and developed other techniques for getting in touch with memories (Léon Chertok and Isabelle Stengers 1992). Psychoanalysis has in this respect remained true to Freud, particularly in France during the dominance of Jacques Lacan, where hypnotism was the greatest taboo of all.

America, always more attuned to popular movements and ill-disposed to authority, has been much more eclectic about hypnotism. Yet almost no funds from the overall budgets for research psychology are dedicated to trance. I don't deny grounds for skepticism. Undoubtedly some behaviors that we loosely lump under the heading of trance states are readily induced and involve *very* peculiar happenings. In science, the suffix "-ian" is very often the sign of a marginal cult or sect, and so it is in America, where a major school of hypnotic research is called Eriksonian. You learn Eriksonian hypnotic induction in much the same way you learn psychoanalysis and become a Freudian. You have to become a novice and learn from an approved school. Scientific sects are often marked by the way in which initiation proceeds following the rubrics of established Christendom—new initiates are confirmed by laying on of hand in the ceremony of confirmation conducted by a bishop. If you are not confirmed by the episcopacy, then you are a heretic. And so it is with so much hypnotism. In practice, however, there are many therapists who use approaches tantamount to hypnotism. In my own jurisdiction (the Canadian province of Ontario) no one may practice hypnotism except a doctor of medicine or dental surgery—hypnotism is allowed for anesthesia. The therapists evade the law by using what they call guided imagery, which has the effect of producing a trance state in which the guide (the hypnotist) has control.

So hypnotism is alive and well, at least in North America? No, it is either a marvel, contained in a sect, or traveling incognito. Every branch of science nowadays cries for lack of funding, but there are still enormous treasure chests for psychological research financed by the taxpayers of numerous nations. Yet try to get funding for research on hypnotism! Yes, there is some, as long as you can dress it up in statistics. Or if you can hook up some wires to the head of a person in hypnosis and analyze the resulting squiggles, you'll be given some quite costly hardware. But serious thinking and research on hypnotism? That has happened chiefly in private foundations, whose patrons are generous but often classed as daft. When public funding has been lavishly made available in retrospect, we tend to think that the government was daft. For example, at the time of the Korean war, it was believed that American prisoners of war were being "programmed" by their captors. The film *The Manchurian Candidate* (1962), based on this premise, was a great success. It told of how the evil Asian reds had programmed an American to disrupt the presidential election.

There was a great flurry of disreputable investment in America on mind-bending drugs. These have recently become scandalous, as it turns out that unknowing patients in mental hospitals were used as guinea pigs. A companion body of research was dedicated to suggestion and hypnosis. That is the only time during the twentieth century that there has been ample funding for the investigation of hypnosis. We know that trances are easily induced. Memories, behavior, and character are all deeply affected. But what is a trance? By and large, the question is excluded as a marvel.

Hypnotism, Trance, and Multiplicity

What has all this to do with multiple personality? The disorder has always gone hand in hand with hypnotism—I wrote the timeline above to illustrate this. Mesmer had emphasized the crisis, the frenzy, into which magnetized patients were thrown. After 1785 the crisis was dropped by the magnetizers, who now emphasized the trance state, which they called artificial somnambulism, *somnambulisme provoquée.* To us, somnambulism means sleepwalking, but it once meant a sleeplike state in which a person had most or perhaps all of the abilities she had when awake. But somnambules, artificial or spontaneous, could not remember their trances when awake, although when once again in trance they could remember both the waking and previous trance states. There were two states of consciousness, and this furnished the original model for double personality, called double consciousness when the switching from one state to another was spontaneous. There was a steady dribble of double consciousnesses until 1875, usually but by no means exclusively studied by magnetizers or hypnotists. Whenever a theory of the phenomenon was produced—for example, in 1844 when the dual structure of the brain was mooted as an explanation of double consciousness (the two halves were not communicating)—there would be a cascade of cases.

Multiple personality emerged in full in 1875. The first case was produced by E. E. Azam. It was he who had tried to introduce scientific hypnotism to France in 1858. He returned to one of his two demonstration cases in 1875. His Félida became the prototype for turn-of-the-century multiplicity, just as Cornelia Wilbur's Sybil was the late twentieth-century prototype. Félida had two states, in one of which she was amnesic for the second state. Her switches were spontaneous and interrupted by a period of deep sleep. She followed the double consciousness tradition and was more lively, sexual, in the second state. She conceived in the second state and denied her pregnancy in her first state. According to the diagnostic canons of the day, she suffered from hysteria and much else. She was highly hypnotizable. After her there was a cascade of cases, first in France and then in New England; all were hysterics, and all were highly hypnotizable.

Multiple personality has been a bit like a parasite that requires a host, a cultural milieu in which to thrive. Like most of the mental disorders that are transient and have their ups and downs, it needs to find an ecological niche. (See Hacking 1998 for a theory of transient disorders and ecological niches.) In 1875 the host was part positivism and part hysteria. The positivism may seem surprising—on the contrary, multiples were taken to refute the claim that there is a transcendental, metaphysical, religious, or Kantian self. It is often said that hysteria disappeared. Not exactly. The syndrome dissipated into a diverse collection of symptoms and diagnoses. When both hysteria and positivism went their way into oblivion in France, multiple personality had no place to go. In New England a slightly different host survived for a while, namely, spiritism and mediums, for what better niche is there for an alter personality than a ghost?

With the passing of these hosts, multiple personality faded away and became something between a scientific curiosity and a marvel. As curiosity, cases went on being reported in the medical literature (far more frequently than is commonly made out in the multiple literature today). But they reverted to the status of curiosities, like Brownian motion. And they lived on as marvels; my favorite quotation is a newspaper headline of 1926: "That Scarlet Demon of the Modern Mind" (Hacking 1991). And always multiple personality has had an assistant, namely, hypnotism. I do not mean that every patient has been subject to hypnotism or like procedures. But multiple symptoms have flourished only when there was a host, and only when hypnotism had some sort of scientific credibility, at least in some scientific subculture.

Despite the lack of funding for pure research on hypnotism today, the practice is back among therapists who work with dissociation. It is a good assistant to multiplicity. I do not say every clinician uses hypnosis or some variant. I do believe, however, that using trance states of the patient is essential to almost every practitioner. But this is barely mentioned and seldom discussed. Since trance is a curiosity or a marvel, a respectable practice must keep it behind closed doors.

Hypnotism has always been associated with multiple personality. But the disorder flourishes only when there is a host. For the past two decades multiple personality has thrived in an emotional and political climate that (rightly) emphasizes the prevalence of child abuse. This has furnished an etiology. A child was severely sexually abused; it survived this trauma by dissociating, by developing a fantasy personality that experienced or reacted to this trauma, of which the main personality thereby had no knowledge. Dissociation was a coping mechanism. For a long time the focus on child abuse was completely safe. To expose child abuse was to be on the side of the angels, even to be a ministering angel. Then child abuse began to generate marvels: satanic ritual abuse, alien abduction, and the like. Science, as I have said, abhors a marvel.

Child abuse activism was sorely threatened, which probably harmed a lot of children in need of protection. But also the lawsuits began against the therapists who had encouraged a multiplicity of alters, and memories of satanic abuse.

The Multiple Personality Controversy

Multiple personality has always been controversial, and I had thought to use it as an example for this conference. But in itself it provides little novelty except that it is such an extreme example of a scientific controversy, decked out in bizarre stories. In brief, as my timeline indicates, multiple personality came in on the coattails of child abuse, established itself as an official diagnosis in 1980, and confirmed its professional niche in 1984. Whereas the first meeting of the professional society in 1984 was beleaguered, trying to establish that it had a disorder to discuss, in 1992 it was teaching its members how to make claims for health insurance—not a welcome type of claim, as therapy might last up to 6 years. But the mighty insurance companies took the issue seriously, sending vice presidents to treat with the society about what claims would be allowed. At the meeting dozens of private clinics distributed glossy brochures advertising their costly treatments of multiple personality and dissociation. By this time it seemed that at least the entire middle class of North American society knew all about multiple personality, its causes, and the often bizarre claims that were being reported about the practices of some families.

Nevertheless, multiple personality was embattled. The majority of psychiatrists rejected it as a diagnosis. The authors of *The Three Faces of Eve* (1957), a sort of predecessor to *Sybil* that was also made into a movie, were appalled in 1984 when they saw the sudden epidemic of MPD. They wrote that there were probably no more than a dozen "real" cases then current in the United States (Thigpen and Cleckley 1984).

There is a ten-story psychiatric institute half a mile from my home. Never, on its whole ten stories, is anyone ever diagnosed with MPD. When a patient walks in and claims to have multiple personalities, the doctor proceeds as follows. "Show me your OHIP card," a card possessed by everyone in Ontario bearing only the words Health-Santé, a person's name, and two sets of numerals. It guarantees full free medical treatment by a practitioner chosen by the patient. The patient produces the piece of plastic. "Whose name is on the card?" The patient reads the name, say, "Ian Hacking." "Well then, that is the person whom I am treating. Tell me about yourself." I should say that another ten-story building, equidistant from my office but in the opposite direction, was the multiple personality capital of Ontario. There used to be weekend workshops of 50 clinicians whose client list consisted mostly of dissociative patients, many of whom have florid multiple personalities. I was a paid-up member of a number of those workshops, and learned of many more marvels than are discussed in departments of philosophy.

That's the practical side. On the more theoretical side, one psychiatrist, Harold Merskey (1992), as it happens in a city neighboring mine, published diagnoses for each well-known historical case of multiple personality in the literature, starting in 1791, diagnosing them as having something other than multiple personality. He published in a leading psychiatric journal, although not without controversy—the leading American journal sent the paper for peer review to members of the multiple movement, and it was rejected. So it was published in Britain, an island relatively immune to the disorder, but the editor had to publish two dozen angry letters maintaining the reality of multiple personality. At a less recondite level, in December 1992 the man who wrote the textbook, Frank Putnam, wrote a semipopular essay for *The Sciences,* published by the New York Academy of Science. He used Breuer and Freud's Anna O as an exemplar of multiple personality; in the next issue Merskey vehemently replied. Controversy, anyone?

Believe the Alters?

In the 1960s claims about the prevalence of physical child abuse and neglect were widely disseminated. At first they were greeted with horror, then by outrage, but as the extent of the claims grew, a certain skepticism set in. In the 1970s child abuse passed from physical abuse to sexual abuse, and the same cycle was repeated, but the issue became politicized. A large number of activists, counting many feminists in their ranks, proclaimed that child abuse was a product of patriarchy and that its cardinal vice was incest. The established authority figures, judges and doctors, doubted that incest was widespread. "Believe the children" became the motto. Because I shall shortly mention some remarkable events that have occurred in the multiple movement, I should make one point clear. In my opinion the child abuse movement was one of the most valuable consciousness-raising agents of the late twentieth century. The activists were right to insist on the prevalence, and the authority figures were wrong to deny it.

The multiple movement used the child abuse agitation as a host. In therapy among movement members, patients recalled garish sexual events of childhood, events that they themselves had repressed, and which, in many cases, few outsiders could credit. The multiple movement followed the child abuse movement: "Believe the alters." If an alter describes abuse that occurred thirty years ago, from a hitherto trusted parent, believe the alter.

During the 1980s, allegations of child abuse increasingly involved bizarre abuse of children at the hands of satanic cults. The events described involved torture, cannibalism, ritual sacrifices, and much more, including events that had to be fantasies, since they were contrary to all physical law, in effect, people flying around. In the most famous cases, whole schoolyards were dug up to find the skeletal traces of horrendous rites that must have left bones, but none were found. The multiple movement, only a few years behind, was confronted by a problem. Alters confessed to horrendous tales of ritual abuse by cults. Alters remembered having been used as serial baby breeders for infants whose blood they had to drink during the sacrifice. And much, much worse. But confirmation was impossible. Police forces could not establish anything. Believe the alters?

Worse, at the end of the 1980s, satanic ritual abuse gave way to abduction by aliens. Alters remembered these things, too. Since the multiple movement had operated on the "believe the alters" principle, there was a crisis. And on an external front, accused families began to organize in 1992, as the False Memory Syndrome Foundation. At the 1992 multiple personality congress, the speakers were saying this organization was funded by a rich unknown abuser who had to be exposed. Meanwhile in April 1993 (at Valley Forge!) the FMS Foundation had its first annual congress.

The Internal Controversy

There is a schism within the multiple movement. The movement is itself a pyramid, with psychiatrists at the top. Somewhat lower down are the "Ph.D. psychologists," and then a spreading out into Masters of Social Work and various types of clinicians with qualifications of several sorts. It is a grassroots movement, rather unlike standard psychiatry, and the roots have a far stronger voice than do ordinary psychiatrists. But the top layer of the movement was running scared. At the 1992 congress, the theme, as I have said, was how to deal with insurance companies, with plenary sessions scheduled for a cavernous amphitheater. The sessions on ritual abuse were scheduled for quite small rooms, which were overflowing. "Talk about denial!" people said, meaning denial by the bigwigs, the organizers, that there was a great danger abroad—ritual abuse. Here is controversy at a completely different level, between the people with power, who want to keep aligned with mainstream psychiatry, and grassroots, who couldn't care less about the medical mainstream.

Numerous conciliatory moves were made. One of importance was a change in name. Satanic ritual abuse is SRA, in code. Let's change this to Sadistic ritual abuse—ritual cruelty to children. That takes away part of the air of horror. Yet the SRA wing had evolved a whole new technology, patterned on the "programming" that came to the fore in the 1960s, when people talked about the sect of Reverend Moon programming young people. There were secret sects who programmed children to adopt alter personalities later, even to the extent that they were programmed to spy on their therapists and report back to home base, the cult. Problem: law enforcement authorities could seldom locate cults. In fact, if one listened to reports, the only identifiable members of cults were members of the victim's families. Oddly enough, the historically minded observer noticed that in many ways there was a turn away from the sexual fixation of the movement, to plain old-fashioned cruelty to children, perpetrated by parents and cloaked in fantasy by the victims—but also probably cloaked in fantasy by the perpetrators themselves, thereby making the victims' memories more veridical and less fantastic.

But things were moving too fast. Not just cruelty, not just Satan, but alien abduction. The FMS Foundation could have a field day! What were the doctors, the leaders to do? There is a good deal of rhetoric among sociologists of science about the "rhetoric of science." I doubt that those sociologists have seen serious rhetoric of the sort that is inflicted on multiple personality discussions. There's curious aggrandizement at work. Thus, in 1991 those who denied the reality of cult abuse were compared to the "good Germans" who lived beside the gas chambers and denied their existence. In a subsequent issue of the movement newsletter, its president, in an article about the FMS Foundation, says we must learn from the end of the Soviet empire that the curtailment of criticism is not the way to succeed in keeping the high ground. It was widely said by newspaper pundits that the work of Loftus and Ketcham (1994) and of Ofshe and Waters (1994) brought science to the rescue. Really? Does science need, as in each of these two cases, a second author who is a journalist?

You see that the response to crisis was radical. I interviewed one important figure in the midst of these goings on. Two years before, he had published a supportive essay on satanic rituals. He had recanted and wished his article could be suppressed. When a prospective patient talked to him of abduction by aliens, he got rid of her quite easily; anyone abducted for torture by aliens who said they would whisk her away next time she went shopping, terrified of being picked up any time anywhere, would certainly be fearful, ergo paranoia, and the patient could be referred to a hospital where she would have her delusions removed by chemicals. In short, don't believe the alters after all.

The same psychiatrist told me that when he takes on a patient for memory therapy today, he insists that it will be horribly painful. At the end, if the patient bears up, she will have memories all right. But she must understand right now, and throughout treatment, that no matter how real those memories seem to her at the end, she will have no guarantee that what she remembers really happened. In short, this is one doctor who has completed Freud's journey 1893–1897. The memories of seductions may be fantasies. But in our televisual world, far more grotesque fantasies may surface than in troubled fin-de-siècle Vienna. There is a certain irony in this, because Freud has been much unloved by feminists and the child abuse movement. Hence, he was ignored by the multiple movement, even as his doctrine of infantile sexual trauma was accepted. But now Freud, or rather the simplest, vulgar, pre-1900, pre-Oedipus complex Freud is back.

Beleaguered by Marvels

Multiple personality managed to contain the marvels that presented themselves by finding hospitality within child abuse. All the excluded marvels of hypnotism and trance states could be put on one side before the overpowering diagnosis of dissociation caused by child abuse. If the therapist ventured into the marvellous, it was all in aid of good cause. The marvels could be silenced because they were drowned out by the cries of the anguished children, the alters bearing witness to crimes of long ago.

But then the cries became too marvellous, and there arose internal dissension. Damage control was attempted. But still the implicit motto, "believe the alters" had all its power. Ever since 1875 alters had different personalities—not just different consciousnesses, different memories, but different lives. These were cultivated after 1970. In therapy one was told never to eliminate an alter, which would be akin to murder, killing a person. Persecuting alters were said to have minds filled with lust for murder; they would force the host personality to commit suicide, thereby achieving their goals. In fact, there was a deliberate personification of alters. A first precaution, in treating a multiple, was to establish contracts among alters, to ensure that a persecutor never arranged a suicide/murder. A useful move was to elicit an all-knowing and benign alter, or else what was called an internal self-helper. These entities were persons whom the therapist should trust, believe, and use to negotiate with evil alters.

How could the top of the movement, the leading psychiatrists, control their flock of patients, of alters, but above all their flock of movement activists, clinicians? They do have some control, for they determine the knowledge, the official diagnoses. They write (after intense negotiation) the clinical description to appear in the official diagnostic manual. As soon as the revised third edition of the *Diagnostic and Statistical Manual* of the American Psychiatric Association appeared in 1987, experts began negotiating for the fourth edition, which came out in 1994. The experts consulted everyone, but in the end a small committee wrote the entries, the diagnoses, and their criteria. One way to resolve the problem of the strange memories of alters might be this: *the alters are not, after all, persons or personalities at all!* They are at most personality fragments, incomplete dissociated memories and dispositions. If so, one does not have the same moral relationship to them as one has to persons. They don't have to be believed.

But how can one say the alters aren't persons, if they are manifestations of an individual who suffers from multiple personality disorder? What's in that name? Well, personality, for one thing. Solution: change the name of the disorder. This is extraordinary. The society has fought tooth and nail for the legitimacy of multiple personality disorder. The one thing to insist on was this: multiple personality disorder is *real!* Yet the decision was made: change the name. Multiple personality disorder leaves. Enter a new name, covering much the same phenomena. Enter a new diagnosis: *Dissociative Identity Disorder.*

The patients are dissociated, they split into fragments, they are unsure of their identity, they assume different identities. The causes of their confusion remain the same. They were hurt in childhood. That produced the dissociation and the confusion. Dissociation remains a coping mechanism. The whole structure of the disorder remains in place, but false memories have been institutionally vanquished. As psychiatrists we are always supportive; we tell the patient that we are always in sympathy, always listening. But we don't have to believe what we sympathize with. That was just a dissociated personality fragment speaking. Perhaps we can even work at a meta-level with the patient to make her realize that although her memories will be intense, they may just be "memories."

Many philosophers of science in recent years have discussed what they call the problem of closure, that is, the question of how a scientific controversy gets resolved. How are disagreeing positions brought into harmony, or losers excluded? An open question becomes closed—hence, philosophers have asked, what makes closure possible? I have not, myself, used "closure" as a technical term, or even mentioned it in print until the present paragraph. But now perhaps I can use it in a nontechnical way. If it works, the multiple movement may have achieved closure—it may have come full circle. It has come full circle to Freud's position of the late 1890s, preceding the arrival of psychoanalysis. No longer always believe the patient's memories.

Yet has the movement not thereby lost everything, lost multiple personality? No, for the future remains promising, though the center turns to dissociation. In the United States, as some plan of universal health care is worked out, mental illness will divide in two. It will be the old division, my C and D above, the psychoses and the neuroses. But the key division splits off the drug-specific maladies, which are relatively cheap and which are in the interests of the drug companies. Then there is the residue, the neuroses, as people used to say. These will be the disease of the middle classes who will pay, as best they can. Dissociation takes up the space that hysteria and neurasthenia occupied a century ago. It will fill the clinics. Multiple personality, once a faithful ally, can be jettisoned. All the cases will be covered by dissociative identity disorder, part of a larger framework of classification and treatment. Thus I foresee a rewriting of the taxonomy at level X with which I began. For example, dissociation reaches out for anorexia (one personality fragment binges, another one hates that one, etc.). Here we have controversy indeed, but the stakes are patent. Who controls that space of mental maladies that are not helped by specific chemical medication? One thing is certain: you will not control this valuable high ground if you allow marvels. So some must be excluded: scotch the worst memories. Others are allowed as long as they attract no notice. Hypnosis, by many names, becomes a private consensual practice between clinician and client, behind closed doors. When the marvels are thus controlled, the controversy can then proceed in a more dignified way. To discuss my levels of controversy X-Y-Z, while ignoring the absences, level O, is to forget that if level O were not silenced, then the stately fabric of a controversy would be torn to shreds.

That is a suggestion worth pondering. Without silence, without the unspoken, without a hidden mass of marvels about which the respectable will not speak, there can be no scientific controversy at all.

References

Bleuler, Eugen (1908). Die Prognose des Dementia Praecox: Schizophreniengruppe. *Allgemeine Zeitschrift für Psychiatrie* 65: 436–464.

——— (1911/1950). *Dementia Praecox, Or, The Group of Schizophrenias,* trans. Joseph Zinkin. New York: International University Press.

Chertok, Léon, and Isabelle Stengers (1992). *A Critique of Psychoanalytic Reason: Hypnosis as a Scientific Problem from Lavoisier to Lacan.* Stanford: Stanford University Press.

Douglas, Mary (1992). The Person in an Enterprise Culture. In *Understanding the Enterprise Culture: Themes in the Work of Mary Douglas,* ed. S. H. Heap and A. Ross, pp. 41–62. Edinburgh: Edinburgh University Press.

Foucault, Michel, (1965). *Madness and Civilization.* New York: Pantheon.

Hacking, Ian (1991). Two Souls in One Body. *Critical Inquiry* 17: 838–867.

——— (1995). *Rewriting the Soul: Multiple Personality and the Sciences of Memory.* Princeton, N.J.: Princeton University Press.

——— (1998) Mad Travelers: Reflections on the Reality of Transient Mental Illnesses, Charlottesville: University Press of Virginia.

Loftus, Elizabeth, and K. Ketcham (1994). *The Myth of Repressed Memories: False Memories and Allegations of Sexual Abuse.* New York: St. Martin's Press.

Merskey, Harold (1992). The manufacture of Personalities. The Production of Multiple Personality Disorder. *The British Journal of Psychiatry* 160: 327–340.

Ofshe, Richard, and E. Waters (1994). *Making Monsters: False Memories, Psychotherapy and Sexual Hysteria.* New York: Scribners'.

Putnam, Frank (1989). *Diagnosis and Treatment of Multiple Personality Disorder.* New York: Guilford Press.

Schreiber, F. R. (1973). *Sybil.* Chicago: Regnery.

Scull, Andrew (1993). *The Most Solitary of Afflictions: Madness and Society in Britain 1700–1900.* New Haven, Conn.: Yale University Press.

Thigpen, C. H., and H. Cleckley (1984). On the Incidence of Multiple Personality Disorder. *International Journal of Clinical and Experimental Hyphosis* 32: 63–66.

13

The Theory of Punctuated Equilibria

Taking Apart a Scientific Controversy

MICHAEL RUSE

In this chapter, I look at a controversy that has raged in recent evolutionary biology: that which arose in paleontology about the so-called theory of "punctuated equilibria." My story begins in the 1960s with two young would-be paleontologists, Niles Eldredge and Stephen Jay Gould, in graduate school at Columbia University in New York City. Students of evolution's history will know that these were the full summer days of neo-Darwinism, the "synthetic theory" blending natural selection and Mendelian (already becoming molecular) genetics, supposedly explaining phenomena right across the biological spectrum. The key to understanding was "adaptation," seeing organic features molded by selection to serve life's needs. Teleology (although most preferred not to use that term) therefore reigned supreme, and the questions asked and answers given were consequently phrased in terms of "function" or "purpose."

It is important to stress that both Eldredge and Gould fell into this adaptationist picture. This can be seen fully in a review of the literature on the topic written by Gould—a simply remarkably detailed and mature piece of work for all that it was a review—on problems of relative growth (allometry; Gould 1966). Encouraged by the father figure of neo-Darwinism, Sir Julian Huxley, who had himself mapped out the conceptual space of an evolutionary approach to growth (Huxley 1932), Gould stressed again and again in the review that one must see the living world in terms of adaptive function, and that this is the key to the problems of biology that one such as he would consider: "As a paleontologist, I acknowledge a nearly complete bias for seeking causes framed in terms of adaptation" (p. 588).

Punctuated Equilibria: Claims and Counterclaims

Move next to the early 1970s. Eldredge was on staff at the American Museum of Natural History, already a trilobite expert. Gould was at Harvard, attached to the Museum of Comparative Zoology. A student of fossil snails, he was starting to show broader literary interests, as well as a fondness for the history of his subject. In 1971, Eldredge published a paper in the evolutionists' leading journal, *Evolution,* on one of the most puzzling questions facing the paleontologist: Why is it, if evolution is true, that we rarely if ever see in the fossil record the evidence of a smooth transition from one form (species, genera, and above) to another? Why do we rather see one fixed and defined form ("stasis") and then as we move through the strata an abrupt change, a jump, to another equally defined form? Denying the usual face-saving move, supposing that one has a gradual process of "phyletic" change, and that the gaps are due to an incomplete record, Eldredge argued rather that the record as we see it is precisely what one would expect if the synthetic theory were true! Eldredge drew attention to what students of living organisms (especially Ernst Mayr 1963) believe to be the major form of speciation, so-called "allopatric speciation," where new species are the result of the isolation of small subpopulations from the main group (with an inevitable atypical genetic constitution and consequent rapid evolution). Eldredge claimed that this would create exactly the stepwise record found.

I should stress that, as becomes a junior scientist publishing in a prestigious journal, Eldredge's tone was modest and respectful. A normal scientist doing normal science, to use Kuhnian language. The same tone of deference cannot be found in the next publication, co-authored by Eldredge and Gould (1972). Appearing in a volume intended to push a more biologically informed approach to paleontology, the authors threw down the gauntlet. Distinguishing between "phyletic gradualism," where fossil lineages change gradually and smoothly, and "punctuated equilibria," where change comes in spurts, they announced boldly:

> The history of life is more adequately represented by a picture of "punctuated equilibria" than by the notion of phyletic gradualism. The history of evolution is not one of stately unfolding, but a story of homeostatic equilibria, disturbed only "rarely" (i.e., rather often in the fullness of time) by rapid and episodic events of speciation. (Eldredge and Gould 1972, p. 84)

There is small wonder that, having commissioned the piece, the editor of the volume had to be pressured to accept it. He did, and the rest is history.

What I want to emphasize is the extent to which the authors were now actively in the business of moving out from the adaptive constraints of strict neo-Darwinism. Although Eldredge and Gould did not want to argue that new species appear entirely without regard to adaptive needs, they did argue that what is immediately adaptive is not necessarily long-term adaptive. Species might appear according to the needs of the moment (set within the context of the randomness imposed by allopatric speciation), but the overall pattern (trend) may well be pointing in other directions. At least as significant is the treatment of stasis. Why do we not get ongoing morphological change in evolving lines of organisms? Why is there stability—equilibrium—between spurts of rapid change? Here there was a turning to notions of "homeostasis," where this is to be understood as meaning that there are certain in-born constraints buffering against the external world and its immediate effects.

The all-importance of selection as a day-to-day mechanism is diminished. Selection may have been important in the past, but once its work is done, stability becomes, in its own right, "an inherent property of both individual development and the genetic structure of populations." Moreover, "its power is immeasurably enhanced, for the basic property of homeostatic systems, or steady states, is that they resist change by self-regulation" (Eldredge and Gould 1972, p. 114).

Gauging the Controversy

So much for the initial expression of the theory. It was not long before the Eldredge/Gould papers started to attract attention, first among paleontologists and then more widely. Some liked their ideas; others did not. I will turn in a moment to reactions and counterresponses. First, however, I want to start by trying to measure the effects of the punctuated equilibria hypothesis. How much attention, favorable and unfavorable, did it attract? How big a controversy was it, or was it going to be? Was it really significant, or was it all a storm in a teacup?

When faced with a problem like this, there is one tool of inquiry that springs to mind, namely, the *Science Citation Index.* I have surveyed the *Index* for a quarter century (in fact, 26 years) from 1965 to 1990 inclusive, the findings of which I present in the Appendix to this chapter. My aim has been to compare Eldredge and Gould against their peers and others, judging their influence. Beginning at the beginning, let us see what we can learn about the two initial periods 1965–1969 and 1970–1974, that is, the half decade before the theory was presented and the half decade when it was first presented.

First, a couple of benchmarks. By the 1960s, molecular biology was the really hot area in the life sciences, and there was nothing more important than the Jacob-Monod operon theory of the gene. This is reflected in the fact that, in the period 1965–1969, Francois Jacob got almost 3,500 references, of which about a thousand were to his classic paper (Jacob and Monod 1961) announcing his find. In a somewhat less prestigious science, although one with obvious connections with our inquiry, geology, there had been a move forward that everyone did recognize as revolutionary (Ruse 1981): the theory of continental drift and its mechanism of plate tectonics. One of the major figures here, Fred Vine, garnered (overall) about an order of magnitude fewer references than did Jacob, and this held true also of *his* key paper (Vine and Matthews 1963).

Turning now to evolutionary biology, the dominant figure—especially in America—was Theodosius Dobzhansky. Overall, he got about half the references as did Jacob, and like Vine the references to his key work (*Genetics and the Origin of Species* [1937]) was about an order of magnitude less than for the molecular biologist (although one should note that the last edition of this book had appeared way back in 1951). Of younger evolutionists, Richard Lewontin was surely among—if not the best and brightest of—the new crop. Overall he was about level with Vine, although his key papers, applying molecular techniques to traditional problems (using so-called "gel electrophoretic" techniques: Lewontin and Hubby 1966; also Hubby and Lewontin 1966), got fewer references than did Vine's key paper. (It did appear somewhat later, only at the beginning of our period, so there would have been a lag here)

If, with an eye to our discussion, we ask about people just starting to make a mark in the evolutionary field, one thinks first of George Williams, author of the stimulating critique *Adaptation and Natural Selection,* published in 1966. He was very small beer, with only about twenty-five references to his book. The same is true also of England and of its really creative thinkers, especially of William Hamilton, who had just published (what was essentially) his thesis, including the idea of kin selection—a breakthrough many would consider the most significant since the work of Ronald Fisher and Sewall Wright, if not back to Darwin (Hamilton 1964a,b). And completing our background survey, mention must be made of Edward O. Wilson, especially on the ground that he too was to be embroiled in controversy, one which certainly entangled Gould. In our period, he got about the same number of references as Lewontin.

Against this background, we find that in 1965–1969, Eldredge got two references (one by Gould!) and Gould got about forty references (about half those of Hamilton and Williams), including fourteen to his *Biological Reviews* paper (Gould 1966, his most cited publication). Although these are modest figures, I take it that they are absolutely no surprise whatsoever. After all, we have merely two young men, just out of graduate school.

Move straight on to the five-year period 1970–1974. Jacob's work is still very important, but no longer quite so innovative in such a fast-moving field as molecular biology. Vine's work is fully recognized, but still one has the feeling that geology is not a science in the same league as that of Jacob. Dobzhansky holds solid (in 1970, at the beginning of this period, he published his *Genetics of the Evolutionary Process,* essentially the fourth edition of his great book). Lewontin's significance is being appreciated (although his major book, *The Genetic Basis of Evolutionary Change,* did not appear until 1974), and the same is true of Wilson (whose 1971 *Insect Societies* got immediate attention, and whose 1967 *Theory of Island Biogeography,* jointly authored with R. W. MacArthur, is getting solid attention). Note, however, that both of these men are still more in the geology than the molecular biology category, and the same is true of Williams and Hamilton, although they too are starting to get attention.

What of our two paleontologists? It certainly seems fair to say that Eldredge and Gould have careers that are solidly on track. Gould in particular is attracting attention, in no small part because of his 1966 *Reviews* article, although not exclusively because of it. Already their punctuated equilibria papers are being noted, and interestingly the joint paper (Eldredge and Gould 1972) is more of a hit than the strictly scientific publication of Eldredge (1971)—although the figures are small and may not be very significant. What surely is significant is that the smallness of the figures points to the fact that punctuated equilibria hardly arrived to a major crashing of symbols. Compare, for instance, the far greater effect in the earlier half-decade of the Lewontin-Hubby study. Punctuated equilibria is under way, but this is not yet really the stuff of controversy.

The Controversy Builds: 1975–1979

Let me again pick up the story for the next five-year period. It was this time that most would remember as immediately preceding the high point of the punctuated equilibria debate. There were some, either in series or in parallel, who were arguing for a picture of evolution very much in line with that of our paleontologists. One was Steven Stanley, a young paleontologist at Johns Hopkins University, who had been the junior author of the then basic text in paleontology (Raup and Stanley 1971). It was he who coined the term "species selection" for the supposed process whereby the overall pattern of change (between species) might display epiphenomena quite independent of the immediate adaptive needs of individual species, either their origins or their survivals (Stanley 1975, 1979). Thus, individually, the adaptive pressure might be toward increased size, but overall the trend might show reduced size. This could happen, for instance, if there were an appropriate differential rate of extinction.

There were also identifiable and articulate critics. One of these was a young vertebrate paleontologist, Yale educated and at that time teaching at the University of Michigan, Philip Gingerich (1976, 1977). Relying on incredibly detailed studies, Gingerich argued that now we do have a record of microevolution in action. And the message is gradualism. Punctuated equilibria is just plain false.

Expectedly, Eldredge and Gould responded to these criticisms, but what made punctuated equilibria *controversial,* as opposed to simply a disagreement between professionals, was the fact that the major response, appearing in 1977 and written now mainly by Gould, was larded with provocative musings (Gould and Eldredge 1977). What really proved to be gasoline over flickering flames was the suggestion that those who accept traditional Darwinian gradualism are still stuck with nineteenth-century laissez-faire liberalism. Perhaps, they suggested, there is an alternative, better philosophy:

> Hegel's dialectical laws, translated into a materialist context, have become the official "state philosophy" of many socialist nations. These laws of change are explicitly punctuational, as befits a revolutionary transformation in human society. One law, particularly emphasized by Engels, holds that a new quality emerges in a leap as the slow accumulation of quantitative changes, long resisted by a stable system, finally forces it rapidly from one state to another (law of the transformation of quantity into quality). Heat water slowly and it eventually transforms to steam; oppress the proletariat more and more, and guarantee the revolution. . . .
>
> In the light of this official philosophy, it is not at all surprising that a punctuational view of speciation, much like our own, but devoid (so far as we can tell) of reference to synthetic evolutionary theory and the allopatric model, has long been favored by many Russian paleontologists. . . . It may also not be irrelevant to our personal preferences that one of us learned his Marxism, literally at his daddy's knee. (Gould and Eldredge 1977, pp. 145–146)

It is important to point out that this claim was made at the height of another controversy in evolution, that over the supposed biological basis of human thought and behavior, epitomized by opposition to Wilson's *Sociobiology: The New Synthesis* (1975). Gould was one of the co-signatories to a notorious letter to the *New York Review of Books* (Allen et al. 1975) faulting Wilson as a genetic determinist and racist/sexist/

capitalist. This really was a nasty dispute, especially given that Gould (and fellow signatory Lewontin) were in the same department at Harvard as Wilson. There were really raw emotions at stake here far transcending biology. Apart from anything else, many of Wilson's opponents, as Jews, felt very threatened by what they saw as a return to Germanic thought of the 1930s. As can be imagined, any other controversy in which Gould was embroiled was going to be examined with great care, and his sociobiological opponents (especially Wilson) were not about to miss an opportunity to attack Gould on grounds of his (supposed) Marxism. (See also Wilson 1978.)

Note that, beneath the inflammatory remarks of the Gould-Eldredge response, one can see clear evidence that the screw is being turned a little more tightly on adaptationism. More explicit was a belief that organisms have basic blueprints, *Baupläne,* and that these could constrain and cause stasis. Significant change can occur only as one switches from one *Baupläne* to another, and this could require a relaxation of selection. Close functional tracking was being questioned:

> At the higher level of evolutionary transition between basic morphological designs, gradualism has always been in trouble, though it remains the "official" position of most Western evolutionists. Smooth intermediates between *Baupläne* are almost impossible to construct, even in thought experiments; there is certainly no evidence for them in the fossil record (curious mosaics like *Archaeopteryx* do not count). (Gould and Eldredge 1977, p. 147)[1]

Without at all belittling the initial contributions of Eldredge, it was becoming plain for all to see that it was Gould who had taken over punctuated equilibria as *his* theory. If more were needed to reinforce the point that it was Gould leading the flight from adaptationism, there was the fact that in this same year he published both a very popular work on evolution, *Ever Since Darwin* (Gould 1977a), and a major scholarly book, *Ontogeny and Phylogeny* (Gould 1977b). This latter was a combination history and conceptual analysis of problems of relative growth as expressed through time, making explicit the ways in which constraints on and effects of growth might be expected to create problems for the strict Darwinian, that is, for one who sees all organic features all of the time as direct functions of immediate adaptive needs.

However, 1979 was the year in which Gould really made overt his hostility to panselectionism. Together with Lewontin, Gould contributed to a symposium under the auspices of England's Royal Society, where he seized the opportunity to take the fight against adaptationism right into the home of the belief (Gould and Lewontin 1979). With brilliant use of example, metaphor, and simile, the two American evolutionists argued that much in the organic world has but an indirect connection to adaptive necessity. *Baupläne,* constraints on growth, incidental effects, and more are major shapers of organic form. Natural selection is all very well, but it has a limited and hobbled effect on the processes and products of life: "[C]onstraints restrict possible paths and modes of change so strongly that the constraints themselves become much the most interesting aspect of evolution" (p. 594)

The distaste for pure Darwinism was now explicit, and the way was prepared to push the punctuated equilibria controversy to its highest pitch. But first, let us take people's temperatures through the *Citation Index.*

Gauging the Controversy, 1975–1979

In this time period, sociobiology was the hot issue, and this fact seems certainly to be recorded in the index. Overall, Wilson explodes into major-figure status, and his 1975 *Sociobiology* is a significant factor. Moreover, whether as cause or effect, related work is getting more attention, that of Hamilton, for instance. Yet, before turning to punctuated equilibria, perhaps the sociobiology controversy does help make one important point. We must draw a careful distinction between a *scientific* controversy and a controversy involving *scientists.* To be perfectly candid, the sociobiology controversy was in major part the latter, with the battle being fought in the media, and with the full and happy participation of people who were not professional evolutionists. These included philosophers like myself, for instance! (See Caplan 1978; Ruse 1979; Kitcher 1985 was much admired by those who do not like human sociobiology, but it did not really appear until after the controversy had died down somewhat.)

I do not make this point in order to downgrade the significance of the controversy; indeed, the debate was about as significant an issue as you could have: the status of humankind. But it was not strictly one within science—and though the *Index* shows that there was a scientific controversy, it reflects the extrascientific effects also. There may have been somewhat more discussion and use of Wilson's *Sociobiology* than of Lewontin's 1974 *Genetic Basis,* and even more general discussion if you include the work to Hamilton, Dawkins, and others. But overall, Wilson really scored no more than Lewontin, or Dobzhansky for that matter.

Against this background, what can we say about punctuated equilibria? Gould as a scientist is certainly gaining respect and appreciation in the scientific world, although note that he is certainly not up with Lewontin or Wilson, or Williams and Hamilton, for that matter. The other participants—Eldredge, Stanley, Gingerich—are establishing solid careers, although they in turn lag behind Gould. If we think about punctuated equilibria in particular, there seem to be two main conclusions to be drawn. First, without Gould there would not have been much of a controversy. The solo-authored Eldredge article, for instance, got very little attention. If one looks ahead to the next decade (that is, the period 1980–1990), this is a conclusion that is confirmed strongly. I am not saying that this makes Gould a better scientist or anything; rather, I am saying that it was he who drew attention to punctuated equilibria. Second, within the scientific community, it is still not that big a controversy. It is nowhere as controversial as Wilson's *Sociobiology.* If we compare, say, the original Eldredge-Gould paper with Hamilton's seminal papers, we see that the scientific (for which I would read "evolutionary") community does not seem to rate the paleontologists' ideas as of comparable importance with kin selection.

What one does start to see, however, specifically in the case of Gould, is that people in the scientific community are showing an interest in his views challenging strict adaptationism—which include, but also extend beyond, his punctuated equilibria writings. The *Reviews* article gets lots of attention, and note that *Ontogeny and Phylogeny* is also getting noticed (see appendix to this chapter). Highly pertinent here are the breakdown figures from *Paleobiology* and *Evolution.* They suggest strongly that punctuated equilibria as such has made few inroads to the evolutionary community taken as a whole. Even the few references tend to be neutral. However, more general work by Gould does tend to get some (more) attention. This all fits with the commonsense observation that scientists are likely to be much more attracted to something they can use, rather than to more general ideas. One can use (check, test, experiment with, etc.) kin selection, non-adaptation through growth, and so forth. One cannot, if one is a regular evolutionist, use hierarchies and macrotheories and the like.

The Controversy Boils Over, 1980–1914

The beginning of the new decade, 1980, saw the peak of the punctuated equilibria controversy. It really became a matter of interest in the scientific community at large and with the general public. This was thanks particularly to a symposium on the topic at the Field Museum in Chicago, and a provocative report in *Science* by the journalist Roger Lewin (1980), telling us that the old way of doing evolution is past. Evolution by fits and starts is the new orthodoxy—perhaps we do indeed have a new "paradigm," in the Kuhnian sense. Gould himself, playing his role, published in 1980 his most extreme discussion of punctuated equilibria, declaring that neo-Darwinism is "effectively dead"—that none of its major tenets remain standing. Moreover, he began flirting with the idea that perhaps species change can occur in one or two generations. He began championing the reputation of the geneticist Richard B. Goldschmidt (1940), who was a saltationist and an arch-opponent of the synthetic theorists.[2]

Yet, even by microevolutionary scales the moment of extremism was short-lived. The geneticists were now truly stirred to action, and the criticism started to flood in. It was one thing for a paleontologist to presume in his own field. It was another if he presumed in neontological studies. In response, Gould quickly moved back into line (Gould 1982). We learn that he had never meant to be read as a radical. He was much more interested in "expanding" conventional Darwinian evolutionary theory than in refuting it. At the microlevel, it may well be that natural selection rules are OK. It is at the macrolevel—where the paleontologist is monarch—that we must come to appreciate the force of punctuated equilibria. Evolutionary thinking must therefore be hierarchical, with different ideas respectively appropriate for different levels:

> Each level generates variation among its individuals: evolution occurs at each level by a sorting out among individuals, with differential success of some and their progeny. The hierarchical theory would therefore represent a kind of "higher Darwinism" with the substance of a claim for reduction to organisms lost, but the domain of the abstract "selectionist" style of argument extended. (Gould 1982, p. 386)

The implication is that it was not so much that Gould was unsound as it was that those who would not go with him were unduly conservative, blinkered by the constraints of their own narrow discipline.

So much for what Gould was doing and thinking. Always more interested than Gould in classification for its own sake, Eldredge was (like most others at the American Museum of Natural History) an enthusiast for the school of cladism, trying to do taxonomy on strict phylogenetic lines (Eldredge 1972). This interest had extended into work on biogeography, a pursuit that led to fruitful collaboration with others of a similar bent (e.g., Eldredge and Cracraft 1980). And, for all that the two men have always remained loyal to each other, one can now detect some divergence of interests and emphasis between Eldredge and Gould. Most particularly, with Eldredge one does not sense the deep ambivalence about adaptation found in Gould. At heart, Eldredge really likes Darwinism, and he has truly internalized natural selection.

Probably by the 1980s most people still thought (as I am sure they still do) of punctuated equilibria as the Eldredge-Gould theory, but the debate had spread out somewhat. Stanley had now published a book (Stanley 1979). Also important was the fact that the attempt to bring empirical evidence to bear on the issue was now in full bloom. Most noteworthy was the work of a young invertebrate paleontologist, Peter Williamson (1981, 1985), who apparently had found solid evidence of punctuated evolution among mollusks of East African lakes. This work was widely hailed by True Believers, and the fact that Williamson was a member of Gould's department at Harvard meant that there was full opportunity to spread the word to the world at large.

Gauging the Controversy, 1980–1984

The *Index* in this time period strongly confirms the extent to which Gould was rising to be a public man, although generally there seems to be a heightened interest/activity in matters evolutionary. It is not true yet that, within the scientific community, Gould is *the* major evolutionist. Wilson, if anybody, has that honor, but Gould alone among the paleontologists has achieved major status. What is becoming increasingly evident is the extent to which people are referring selectively to Gould's work. (His 1977 *Ever Since Darwin* has been referenced but once in *Evolution.*) It was Gould's 1980 paper in *Paleobiology,* "Is a New and General Theory of Evolution Emerging?" that really got shocked gasps, and yet its scientific effect was hardly overwhelming. The direction in which people still preferred to turn was toward his earlier, more moderate (in scientific claim) punctuated equilibria work. What was quickly judged inadequate did not receive continued major attention.

However, if we look at the work of Gould that did get attention, an interesting effect starts to emerge. People are interested in punctuated equilibria, but they are as interested—*if not more!*—in Gould's other work, where there is a more general attack on ubiquitous adaptationism. Even back to the 1966 *Biological Reviews* paper, and then up through *Ontogeny and Phylogeny* (1977), and on to the jointly authored "Spandrels" paper (Gould and Lewontin 1979), there are solid sets of references. For now, I simply note this fact, although there are some obvious explanatory hypotheses to which I turn shortly.

Looking briefly at the others involved in the punctuated equilibria controversy, the *Index* does support the claim that, in general, nonpaleontological critics got in and made their points and then went on to other things. There is solid interest in the work of paleontologists involved in the punctuated equilibria controversy, with individual items that correspond to Gould's work receiving about the same reduced proportion of citations as the people themselves compared overall to Gould. But the outstanding fact to emerge is just how much slighter the punctuated equilibria controversy remains compared to sociobiology. Nothing in the punctuated equilibria controversy seems to have had the impact, one way or the other, of Wilson's 1975 *Sociobiology: The New Synthesis.*

The Concluding Years, 1985–1990

Of course, these years were not really concluding. The story goes on. But from our viewpoint, speaking especially now of *controversy,* this completes our tale. Although both Eldredge and Gould have remained prolific authors, frenetic even, conceptually (and empirically) I see no significant innovations. And I think this is reflected in the fact that no one now seems inclined to argue in a heated fashion for or against punctuated equilibria. Basically, commitments have been made, and people think about other things.

There is certainly no triumph, say, as happened with plate tectonics, and no disaster, as with cold fusion. Looking at the information yielded by the *Index,* the professional Gould seems to have caught up with the public Gould, and he is now the most frequently cited evolutionist. But despite this, and despite the solid careers of the other punctuationists, the conclusions derived earlier still stand. Punctuated equilibria does not seem to have had the same impact as sociobiology. Wilson's *Sociobiology* is still more cited than punctuated equilibria in our survey.

Likewise, if we look at the actual pieces published on and around punctuated equilibria and compare them with Gould's more general material, the same pattern as before prevails. Nobody could say, for instance, that Eldredge or Williamson dominates the field. And Gould himself has more effect away from his theory, considered directly. Twenty years after it was published, his *Reviews* paper still gets more citations than any single piece he wrote on punctuated equilibria, including the original article co-authored with Eldredge. More generally, it is his book *Ontogeny and Phylogeny* (1977) and his "Spandrels" paper co-authored with Lewontin that get the fullest attention. The major interest in Gould's work seems to be in his general attack on panselectionism. The basic punctuated equilibria papers just hold steady or decline, and the same seems true of more recent general statements.

Analysis: Mere Words?

What can we now say about the punctuated equilibria dispute? There are three lines I wish to pursue, beginning first with the fact that many critics insist that the punctuated equilibria dispute was no real dispute at all. It is dismissed as a "wrinkle" on evolutionary thought (Dawkins 1986), as "mere hand waving by the paleontologists," as it was characterized by one eminent evolutionist when I asked for his opinion.

We know that, at one level, this is simply not true—if, that is, the criticism is meant to imply that there was no dispute at all, or even that there was no *scientific* dispute at all. I accept that much of the attention came from outside the professional scientific community, and that within the community it certainly does not seem to have been that big a matter. But judging from the interest paid, both to the key pieces of advocacy and to the major critiques, there was a dispute within the evolutionary community. This cannot be denied.

At another level, pointing to the subjective and personal dimension in some respect, there is surely truth in that claim. If anything is certain, it is that the punctuated equilibria controversy would never have existed without Gould's brilliant rhetorical skills. His use of apt example and provocative metaphor; his folksy casual style, which adds rather than detracts from the very serious concern about science and its morality; his lawyerlike nose for the weakness in the other's argument—all these are quite without equal. It is no wonder that he is a favorite with literary theorists (Lynne and Howe 1986; Selzer 1993).

Let me try to dig a little more deeply here. I am not denying that the "facts" at some level were important, for and against punctuated equilibria: the work of Gingerich and of Williamson spring to mind, as does the initial work of Eldredge on trilobites. The point is that the punctuated equilibria debate was more than just a question of the facts, for apart from anything else, everybody accepted that some facts swung one way and some facts swung the other way. And the initial Eldredge paper backs this belief. One's sense is that, had Gould not been around, it would have been considered a pretty minor piece of work. It has simply not been in the same league as, say, Hamilton's papers, even if one ignores the fact that it probably benefited from the halo effect of what was to come.

This conclusion is just what the content analysis supports. Eldredge's paper was good, but it was just an application of Mayr to the fossil record. Without more being involved, there was just no reason why punctuated equilibria should take off as an idea in its own right, let alone be particularly controversial. That something more was the Gouldian component. And though (with respect) Gould has done empirical work (on Bermuda snails) and is happy to refer both to his own work and to that of others, his contribution is not basically from the rock face; our conclusion that the rhetoric played a major role is confirmed (For early empirical work, see Gould 1969; Eldredge 1972).

Standing High for Paleontology

I am starting to make Gould sound like a grotesque egotist. We have a rather minor conceptual point, with some limited empirical evidence, which may or may not have come from Gould himself, which was then blown up through rhetoric into a supposed major theory and a controversy. Yet, although there is truth in this, if we were to leave matters at this point, we would be committing a sin of omission. For Gould, it has always been very clear that there is much more at stake than personal glory. Again and again he stresses that what really concerns him is the hope of finding a significant place for paleontology within the evolutionary family. For too long the field of evolution has been dominated by the geneticists and experimentalists. Despite the general public identification of evolution with the fossil record, among professional evolutionists paleontologists have been despised, told when and where to jump. Now, however, with punctuated equilibria these hewers of wood and drawers of water (breakers of rock?!) can play their part in the full development—the articulation and the justification—of evolutionary thought. I draw your attention to the title of a presentation when Gould was on display before his fellow evolutionists (Gould 1983): "Irrelevance, submission, partnership: the changing role of paleontology in Darwin's three centennials and a modest proposal for macroevolution." The desire to be treated like a grownup is really rather touching.

In fact, as the dispute developed, Gould got bolder in his claims. At first, the feeling was joy that one no longer has to twist and turn—invoking an incomplete fossil record—to fit paleontology into causal thinking about evolutionism. It can indeed support it. Then confidence grew to the point where it was thought that paleontology can contribute in places where genetics and the rest must fall silent. It is true that there was a nasty reminder of limitations, when Gould was rapped sharply on the scientific knuckles over his perceived saltationism. But this was more a setback than a defeat. The claim for essential status was revised in such a way that there is no danger of overlapping and contesting the beliefs of already established areas of evolutionary biology, except where they themselves are clearly out on a limb. Thus the strong emphasis in recent years on the hierarchical nature of evolutionary thought, and on the significance of paleontology for understanding macroevolution (Gould 1982, 1989, 1990; Eldredge 1985; Eldredge and Salthe 1984; and many, many more).

As we now know, Gould and his fellow paleontologists were successful, but only partially so, in what they set out to do. They certainly got attention within the scientific community. But the fact is that they simply did not get the attention that other areas get, sociobiology in particular. Even if we ignore the fact that, when there was a clash with genetics, it was the paleontologists who had to step smartly back in line, the overall recognition gained by punctuated equilibria was respectable rather than stunning (as measured by references). Eldredge and Gould (1972), for instance, got less than half the recognition that Hamilton (1964a,b) got for kin selection. The simple fact is that we are still left with the feeling that we are dealing with a science of the second order (See the data on Gould references in *Evolution* in the appendix to this chapter).

Rival Visions?

Once again, I seem to be drawing a rather negative conclusion. First I suggested that everything was rhetoric and Gould's personal pitch for glory. Now I am suggesting that no one very much succeeded anyway. But I want to claim more than this. I see punctuated equilibria theory as an idea through which Gould significantly has been stimulated to think about life's processes, and in particular the true causes of evolutionary change. In particular, it has been a stimulus to move away from a strict Darwinian perspective, a stimulus that has had an effect on the community at large. This has not been entirely a one-way causal process, with paleontology the dog that wagged the tail of everything else. But the general critique on adaptationism owes much to the experiences and theorizing in paleontology. And if this is so, then punctuated equilibria has had a wider influence on evolutionary thought.

This is perhaps the way that one might have expected things to be. As noted, the reason why Hamilton's work has been so significant is that he gave people models that they could use—a bright graduate student could take kin selection into the field and see if it applies to some interesting species of organism. In the other corner, however, hierarchies and jumps are all very well, but after you have finished talking about them at a conference, what can you do with them?[3] Yet, this impotence does not hold true of a general critique of adaptationism, particularly when it is linked to ontogeny (embryology, genetics including the molecular variety, and so forth). Such a critique offers something that impinges on the work of everyone, even offering the chance of doing something experimentally or in the field, hence the increased interest, reflected in the frequency of the references. (Consider, by way of confirmation of this point,

John Endler's [1986] overview of natural selection and its operation in nature. Six of Gould's papers are referenced, and the emphasis is on those that stress nonadaptive features of the organic world.)

We are led straight to another point of interest. Whatever the major direction of the causal links, it is clear that at some level the punctuated equilibria enthusiasts, Gould particularly, managed to sound a responsive chord in the biological community. Apparently, the supposed ubiquitous adaptationism of the synthetic theory was sufficiently unstable that, within twenty years of its triumph (the centenary of the *Origin of Species* in 1959), people were listening with attention to those prepared to argue that neo-Darwinism is an impoverished view of the evolutionary process. My findings do suggest that because most evolutionists are not paleontologists, they preferred to work from nonpaleontological writings. But this in no way detracts from the fact that, as punctuated equilibria theory has evolved, it has become the epitome (certainly in Gould's version) of the move away from adaptationism—constraints, randomness, *Baupläne,* and so forth.

What does this all mean? You might claim that people have simply been convinced by the beauty and power of the arguments put forward by Gould and company. Once, evolutionists were all happy adaptationists; now, thanks to disinterested science, they see the error of their ways. But I would add another element. Punctuated equilibria has had the success that it has had because it is in tune with a deep strain of already existing (although perhaps for a time submerged) nonadaptationism in evolutionary thought, or rather, in evolutionists' thinking. The adaptationism of the synthetic theory was a veneer over rather different sympathies—at least, it was a veneer for some evolutionists, although for others (especially English evolutionists) adaptationism was solid oak all of the way down.

In particular, I am inclined to argue that what we have are two different "paradigms" or conceptual frameworks. We have two different visions of the evolutionary process and product (Russell 1916). Although the one may have been submerged for a while, the paleontologists have been at the forefront of reviving it. To talk in terms of *metaphor,* one set of evolutionists (the Darwinians) regard the organic world in terms of adaptation, which is to say as though it is *functioning* or *designed.* The other set of evolutionists, of which Gould is a prime representative, do not. They think of the organic world in terms of form, which means that there are certain basic *structures* or *blueprints* according to which organisms are *constructed.*

Historically, there has always been this dichotomy in the history of evolutionism, with those such as Darwin pushing function and those such as the German *Naturphilosophen* arguing rather that transcendental laws of form are what govern the processes of transformation. (Gould himself, especially in his *Ontogeny and Phylogeny,* has long stressed this point.) Punctuated equilibria theory stands in the second, transcendentalist tradition, and this (at least in part) accounts for its success: it is simply doing what many evolutionists have always been doing. American evolutionism has always deeply internalized nonadaptationism in a way that British evolutionism never has (Richards 1987). *The* crucial figure dominating American evolutionism in the last century was never Charles Darwin. Beyond doubt, it was Herbert Spencer (Russett 1976; Pittenger 1993). And the crucial point to note here is that, unlike Darwin, Spencer was ever a man with a casual attitude toward adaptationism—an attitude he passed on to others. It was not that Spencer was against adaptation. It was rather that, for him, it was not the overriding mark of organic nature. Far more significant was the upward rise of organic life, the move from homogeneity to heterogeneity, as organisms increasingly complexify, down through the course of time.

Spencer's ongoing influence can be traced through his enthusiast L. J. Henderson to *his* student, none other than the eminent evolutionist Sewall Wright and so on down the line to the present (Ruse 1996). Simply, then, punctuated equilibria stands in this American tradition. It succeeds as nonadaptationist because this has always been the way of American evolution. It is as American as applepie and (Gould will appreciate this) baseball.

Appendix of Empirical Data

I present in this appendix my raw data from the *Citation Index* together with other pieces of information used in this chapter. I should say that, having come to the end of this, my most empirical paper, I have considerably more respect for scientists, including social scientists, than when I went in.

Citations during the Period 1965–1969.

Figures are given in parentheses after each reference.

Dobzhansky, T. (1951) *Genetics and the Origin of Species* (New York: Columbia University Press). (111)

Eldredge, N. (1968) Convergence between two Pennsylvanian gastropod species: a multivariate mathematical approach. *Journal of Paleontology,* 42: 186–196. (2)

Gould, S.J. (1966) Allometry and size in ontogeny and phylogeny. *Biological Reviews of the Cambridge Philosophical Society,* 41: 587–640. (14)

Hamilton, W.D. (1964a) The genetical evolution of social behaviour I. *Journal of Theoretical Biology,* 7: 1–17. (19)

——— (1964b) The genetical evolution of social behaviour II. *Journal of Theoretical Biology,* 7: 18–36. (12)

Hubby, J.L., and R.C. Lewontin (1966) A molecular approach to the study of genic heterozygosity in natural populations I. The number of alleles at different loci in *Drosophila pseudoobscura. Genetics,* 54: 577–594. (53)

Jacob, F., and J. Monod (1961a) Genetic regulatory mechanisms in the synthesis of proteins. *Journal of Molecular Biology,* 3: 318–356. (955)

——— (1961b) On the regulation of gene activity. *Cold Spring Harbor Symposia on Quantitative Biology,* 26: 193–211. (263)

Lewontin, R.C., and J.L. Hubby (1966) A molecular approach to the study of genic heterozygosity in natural populations II. Amount of variation and degree of heterozygosity in natural populations of *Drosophila pseudoobscura. Genetics,* 54: 595–609. (87)

MacArthur, R., and E.O. Wilson (1967) *The Theory of Island Biogeography* (Princeton, N.J.: Princeton University Press). (29)

Vine, F.J. (1966) Spreading of the ocean floor: new evidence. *Science,* 154: 1405–1415. (151)

Vine, F.J., and D.H. Matthews (1963) Magnetic anomalies over oceanic ridges. *Nature,* 199: 947–949. (139)

White, J.F., and S.J. Gould (1965) The interpretation of the coefficient in the allometric equation. *American Naturalist,* 99: 5–18. (8)

Williams, G.C. (1966) *Adaptation and Natural Selection* (Princeton, N.J.: Princeton University Press). (23)

Citations during the period 1970–1974

Dobzhansky, T. (1970) *Genetics of the Evolutionary Process* (New York: Columbia University Press). (127)
Eldredge, N. (1971) The allopatric model and phylogeny in paleozoic invertebrates. *Evolution,* 25: 156–167. (9)
Eldredge, N., and S.J. Gould (1972) Punctuated equilibria: an alternative to phyletic gradualism. In *Models in Paleobiology,* edited by T.J.M. Schopf (San Francisco: Freeman Cooper). (6)
Wilson, E.O. (1971) *The Insect Societies* (Cambridge, Mass.: Belknap Press). (54)

Dobzhansky (1951) (79)
Gould (1966) (51)
Hamilton (1964a) (42)
Hamilton (1964b) (22)
Hubby and Lewontin (1966) (107)
Jacob and Monod (1961a) (543)
Jacob and Monod (1961b) (95)
Lewontin and Hubby (1966) (185)
MacArthur and Wilson (1967) (177)
Vine (1966) (205)
Vine and Matthews (1963) (147)
White and Gould (1965) (8)
Williams (1966) (76)

Citations during the period 1975–1979

Dawkins, R. (1976) *The Selfish Gene* (New York: Oxford University Press). (100)
Gingerich, P.D. (1976) Paleontology and phylogeny: patterns of evolution at the species level in early tertiary mammals. *American Journal of Science,* 276: 1–28. (16)
Gould, S.J. (1977a) *Ever Since Darwin* (New York: Norton). (3)
——— (1977b) *Ontogeny and Phylogeny* (Cambridge, Mass.: Belknap Press). (40)
Gould, S.J., and N. Eldredge (1977) Punctuated equilibria: the tempo and mode of evolution reconsidered. *Paleobiology,* 3: 115–151. (27)
Lewontin, R.C. (1974) *The Genetic Basis of Evolutionary Change* (New York: Columbia University Press). (401)
Oster, G., and E.O. Wilson (1977) *Caste and Ecology in the Social Insects* (Princeton, N.J.: Princeton University Press). (5)
Stanley, S.M. (1975) A theory of evolution above the species level. *Proceedings of the National Academy of Sciences,* 72: 646–650. (53)
Williams, G.C. (1975) *Sex and Evolution* (Princeton, N.J.: Princeton University Press). (147)
Wilson, E.O. (1975) *Sociobiology: The New Synthesis* (Cambridge, Mass.: Belknap Press). (501)
——— (1978) *On Human Nature* (Cambridge, Mass.: Harvard University Press). (6)

Dobzhansky (1951) (88)
Dobzhansky (1970) (205)
Eldredge (1971) (24)
Eldredge and Gould (1972) (110)
Gould (1966) (80)
Hamilton (1964a) (245)
Hamilton (1964b) (85)
Hubby and Lewontin (1966) (89)
Jacob and Monod (1961a) (381)
Jacob and Monod (1961b) (64)
Lewontin and Hubby (1966) (142)

MacArthur and Wilson (1967) (519)
Vine (1966) (50)
Vine and Matthews (1963) (84)
White and Gould (1965) (11)
Williams (1966) (262)
Wilson (1971) (333)

Citations during the period 1980–1984

Charlesworth, B. (1982) A NeoDarwinian commentary on macroevolution. *Evolution,* 36: 474–498. (31)

Eldredge, N. (1980) *Phylogenetic Patterns and the Evolutionary Process: Method and Theory in Comparative Biology* (New York: Columbia University Press). (96)

Gould, S.J. (1980a) The promise of paleobiology as a nomothetic, evolutionary discipline. *Paleobiology,* 6: 96–118. (24)

——— (1980b) Is a new and general theory of evolution emerging? *Paleobiology,* 6: 119–130. (81)

——— (1982) Darwinism and the expansion of evolutionary theory. *Science,* 216: 380–387. (47)

Gould, S.J., and R.C. Lewontin (1979) The spandrels of San Marco and the Panglossian paradigm: a critique of the adaptationist program. *Proceedings of the Royal Society of London, Series B: Biological Sciences,* 205: 581–598. (173)

Gould, S.J., and E. Vrba (1982) Exaptation—a missing term in the science of form. *Paleobiology,* 8: 4–15. (31)

Lumsden, C., and E.O. Wilson (1981) *Genes, Mind, and Culture: The Coevolutionary Process* (Cambridge, Mass.: Harvard University Press). (49)

——— (1983) *Promethean Fire: Reflections on the Origin of Mind* (Cambridge, Mass.: Harvard University Press). (4)

Stanley, S.M. (1979) *Macroevolution: Pattern and Process* (San Francisco: Freeman). (169)

Stebbins, G.L., and F.J. Ayala (1981) Is a new evolutionary synthesis necessary? *Science,* 213: 967–971. (36)

Williamson, P.G. (1981) Paleontological documentation of speciation in cenozoic molluscs from Turkanan Basin. *Nature,* 293: 437–443. (70)

Dawkins (1976) (261)
Dobzhansky (1951) (89)
Dobzhansky (1970) (230)
Eldredge (1971) (20)
Eldredge and Gould (1972) (181)
Gingerich (1976) (38)
Gould (1966) (125)
Gould (1977a) (10)
Gould (1977b) (253)
Gould and Eldredge (1977) (151)
Hamilton (1964a) (376)
Hamilton (1964b) (105)
Hubby and Lewontin (1966) (60)
Jacob and Monod (1961a) (197)
Jacob and Monod (1961b) (32)
Lewontin (1974) (491)
Lewontin and Hubby (1966) (60)
MacArthur and Wilson (1967) (738)
Oster and Wilson (1978) (130)
Stanley (1975) (62)
Vine (1966) (50)
Vine and Matthews (1963) (51)

White and Gould (1965) (24)
Williams (1966) (351)
Williams (1975) (291)
Wilson (1971) (389)
Wilson (1975) (682)
Wilson (1978) (44)

Citations during the period 1985–1990

Dawkins, R. (1982) *The Extended Phenotype: The Gene as the Unit of Selection* (Oxford: Freeman). (110)

Dawkins, R. (1986) *The Blind Watchmaker* (New York: Norton). (47)

Eldredge, N. (1986) *Unfinished Synthesis: Biological Hierarchies and Modern Evolutionary Thought* (New York: Oxford University Press). (50)

Gould, S.J. (1985) The paradox of the first tier: an agenda for paleobiology. *Paleobiology,* 11: 2–12. (29)

Gould, S.J. (1989) *Wonderful Life* (New York: Norton). (5)

Holldobler, B., and E.O. Wilson (1990) *The Ants* (Cambridge, Mass.: Harvard University Press). (10)

Vrba, E.S., and S.J. Gould (1986) The hierarchical expansion of sorting and selection: sorting and selection cannot be equated. *Paleobiology,* 12: 217–228. (19)

Charlesworth (1982) (90)
Dawkins (1976) (210)
Eldredge (1971) (10)
Eldredge (1980) (161)
Eldredge and Gould (1972) (161)
Gingerich (1976) (25)
Gould (1966) (215)
Gould (1977a) (20)
Gould (1977b) (377)
Gould (1980a) (18)
Gould (1980b) (60)
Gould (1982) (53)
Gould and Eldredge (1977) (169)
Gould and Lewontin (1979) (350)
Gould and Vrba (1982) (32)
Hamilton (1964a) (470)
Hamilton (1964b) (119)
Hubby and Lewontin (1966) (41)
Lewontin (1974) (380)
Lewontin and Hubby (1966) (67)
Lumsden and Wilson (1981) (44)*
Lumsden and Wilson (1983) (18)
MacArthur and Wilson (1967) (695)
Oster and Wilson (1978) (197)
Stanley (1975) (43)
Stanley (1979) (151)
Stebbins and Ayala (1981) (26)
White and Gould (1965) (21)

* There were many more references to this book in the *Social Sciences Citation Index.* (A similar pattern was found for Gould's *Mismeasure of Man,* which between its appearance in 1981 and 1990 recieved 74 references in the *Science Index* and 196 references in the *Social Sciences Index.*)

Williams (1966) (371)
Williams (1975) (295)
Williamson (1981) (55)
Wilson (1971) (500)
Wilson (1975) (443)
Wilson (1978) (35)

Total Overall Citations by Publication

Reference	Citation Period	Total Number of Citations
Charlesworth (1982)	1982–1990	121
Dawkins (1976)	1976–1990	471
Dawkins (1982)	1982–1990	110
Dawkins (1986)	1986–1990	47
Dobzhansky (1951)	1965–1984	367
Dobzhansky (1970)	1970–1984	562
Eldredge (1971)	1971–1990	63
Eldredge (1980)	1980–1990	257
Eldredge and Gould (1972)	1972–1990	458
Gingerich (1976)	1976–1990	79
Gould (1966)	1966–1990	485
Gould (1977a)	1977–1990	33
Gould (1977b)	1977–1990	670
Gould (1980a)	1980–1990	42
Gould (1980b)	1980–1990	141
Gould (1982)	1982–1990	100
Gould (1985)	1985–1990	29
Gould and Eldredge (1977)	1977–1990	347
Gould and Lewontin (1979)	1979–1990	523
Gould and Vrba (1982)	1982–1990	63
Hamilton (1964a)	1965–1990	1152
Hamilton (1964b)	1965–1990	343
Hubby and Lewontin (1966)	1966–1990	350
Holldobler and Wilson (1990)	1990	10
Jacob and Monod (1961a)	1965–1984	2076
Jacob and Monod (1961b)	1965–1984	454
Lewontin and Hubby (1966)	1966–1990	541
Lewontin (1974)	1975–1990	1272
Lumsden and Wilson (1981)	1981–1990	93
Lumsden and Wilson (1983)	1983–1990	22
MacArthur and Wilson (1967)	1967–1990	2158
Oster and Wilson (1978)	1978–1990	332
Stanley (1975)	1975–1990	158
Stanley (1979)	1980–1990	320
Stebbins and Ayala (1981)	1981–1990	62
Vine (1966)	1966–1984	456
Vine and Matthews (1963)	1965–1984	421
Vrba and Gould (1986)	1986–1990	19
White and Gould (1965)	1965–1990	72
Williams (1966)	1966–1990	1083
Williams (1975)	1975–1990	733
Williamson (1981)	1981–1990	125
Wilson (1971)	1971–1990	1276
Wilson (1975)	1975–1990	1626
Wilson (1978)	1978–1990	85

Total Overall Citations, by Author and Time Period

	1965–1969	1970–1774	1975–1979	1980–1984	1985–1990
Dawkins			254	734	874
Dobzhansky	1219	1027	1448	1560	
Eldredge	2	44	231	500	616
Gingerich		26	260	501	688
Gould[a]	33	145	610	1876	3330
	(41)	(159)	(731)	(2081)	(3531)
Hamilton[b]	74	250	870	1560	1839
	(62)	(228)	(785)	(1455)	(1720)
Jacob	3060	2040	1472	1088	
Lewontin	470	830	1350	1820	1593
	(523)	(937)	(1439)	(2053)	(1984)
Stanley[c]		93	410	728	836
Vine	420	605	278	155	
Vrba				136	354
Williams	85	384	750	1260	1284
Williamson				105	104
Wilson[a]	390	650	1550	1880	2102
	(419)	(827)	(2074)	(2801)	(3066)

[a]The figures in parentheses represent the cumulative totals if one includes all of the mentioned publications on which they appeared as second authors. I appreciate that there are other such publications that are unrecorded here, but I doubt that their omission makes much difference at the level of accuracy that I require.

[b]The figures in parentheses represent the totals if one regards (as I am inclined to think one should) Hamilton's seminal kin selection papers as but one continuous argument.

[c]I suspect that Stanley's figures would have been much inflated if I had included the textbook that he co-authored with Raup; but given that others had also published textbooks, I decided to not to include any such figures.

Citation Patterns for Hubby and Lewontin (1966) and Lewontin and Hubby (1966)

	Hubby and Lewontin (1966)		Lewontin and Hubby (1966)	
	Genetic	Evolutionary	Genetic	Evolutionary
1965–1969	27	0	39	3
1970–1974	34	7	70	12
1975–1979	30	12	48	12
1980–1984	18	4	39	8
1985–1990	9	2	24	5

"Genetic" includes *Genetics, Canadian Journal of Genetics, Biochemical Genetics, Genetika, Heredity, Journal of Heredity, Genetical Research, Annual Review of Genetics, Japanese Journal of Genetics,* and *Journal of Medical Genetics.* "Evolutionary" includes *Evolution, Journal of Molecular Evolution, Evolutionary Biology, Journal of Human Evolution,* and *Genetics, Selection, and Evolution.*

Total Citations Made to Gould in *Paleobiology*

	Total Citations[a]	Total Articles Citing[b]	Total Self-Citations
1975	6	3	—
1976	6	3	—
1977	6	6	11
1978	37	9	—

1979	27	5	—
1980	24	8	—
1981	49	14	13
1982	41	8	—
1983	12	7	2
1984	25	4	—
1985	33	4	11
1986	32	7	10
1987	13	3	5
1988	20	2	—
1989	18	—	—
1990	12	2	—
Total	361	85	52

Total number of articles in *Paleobiology* in the years 1975–1990: 723

Total number of articles making reference to punctuated equilibrium: 85[c]

References favoring punctuated equilibrium: 31[c]

References disfavoring punctuated equilibrium: 20[c]

References having a neutral stance: 34

Circulation of *Paleobiology* as of 1993: 2200

[a]Total citations to Gould, excluding self-citations.

[b]Total number of articles (by all authors) citing at least one punctuated equilibria article (defined as Eldredge and Gould 1972; Gould and Eldredge 1977; Gould 1980b). Eldredge 1971 was never cited alone, without one of these three articles.

[c]These figures were determined by a content analysis. About 12% of articles refer to punctuated equilibria; about 4% favor it.

Total citations made to Gould in "Evolution"

	Total Citations[a]	Total Articles Citing[b]	Total Self-Citations
1970	—	—	—
1971	3	—	—
1972	2	—	—
1973	8	1	—
1974	7	4	—
1975	2	1	7
1976	3	1	—
1977	3	—	—
1978	4	2	—
1979	11	2	3
1980	8	1	1
1981	13	1	—
1982	37	14	—
1983	15	3	—
1984	28	3	—
1985	27	1	9
1986	34	2	—
1987	32	1	6
1988	23	1	—
1989	24	2	12
1990	28	1	—
Total	309	41	38

Total number of articles in *Evolution* in the years 1975–1990: 2016

Articles referring to punctuated equilibria: 1970–1990, 41; 1975–1990, 36

Articles favorable to punctuated equilibria (including one by Stanley): 3[c]

Articles unfavorable to punctuated equilibria: 7[c]

Articles neutral to punctuated equilibria: 31[c]

Total articles referring to Gould: 1970–1990, 282; 1975–1990, 261

Articles referring to other Gould (non-punctuated equilibria) writings: 1970–1990, 241, 1975–1990, 225

Circulation of *Evolution* 1992–1993: 4500

[a]Total citations to Gould.

[b]Articles (by all authors) referring to punctuated equilibria articles (defined as Eldredge and Gould 1972; Gould and Eldredge 1977; Gould 1980b).

[c]These figures were determined by a content analysis. Roughly 15% of articles referring to Gould refer to his punctuated equilibria papers (1970–1990). Absolutely, roughly 2% of articles refer to his punctuated equilibria papers, and the number accepting the theory is insignificant (1975–1990).

Most Common Non-Punctuated Equilibrium Citations to Gould in *Evolution*

Gould (1966) (35)

S. J. Gould (1971) "Geometric similarity in allometric growth: a contribution to the problem of scaling in the evolution of size" (10)

S. J. Gould and R. F. Johnson (1972) "Geographic variation" (7)

S. J. Gould (1975) "Allometry in primates, with emphasis on scaling and the evolution of the brain" (10)

Gould (1977b) (47)

P. Albrech, S.J. Gould, and D. Wake (1979) "Size and shape in ontogeny and phylogeny" (19)

Gould and Lewontin (1979) (28)

Gould (1980b) (13)

Gould (1982a) (7)

All other citations appear less often (usually much less often). *Ever Since Darwin* (1977) and *The Mismeasure of Man* (1981) have but one citation each.

Citations Made to Eldredge

	Cites to Eldredge and Gould	Self-Citation	Other Citations
Paleobiology			
1975–1979	19	3	27
1980–1914	33	0	30
1985–1990	15	0	17
Evolution			
1970–1974	1	5	3
1975–1979	6	0	2
1980–1914	15	0	6
1985–1990	8	0	10

These figures are for Eldredge as first author.

Citations Made to Stanley

	Self-Citations	Other Citations
Paleobiology		
1975–1979	7	59
1980–1984	23	93
1985–1990	27	134
Evolution		
1970–1974	3	1
1975–1979	0	3
1980–1984	5	27
1985–1990	0	31

These figures are for Stanley as first author.

Notes

1. I cannot overemphasize how significant gradualism is for Darwinians. Selection can only work if variation is slight, otherwise one gets out of adaptive focus. Darwin, who at one early point contemplated "hopeful monsters," quickly stepped back in line (Ruse 1979b). Fisher's *The Genetical Theory of Natural Selection* (1930) makes the definitive Darwinian case for gradualism.

2. Gould has always stressed (to me) that he was never a saltationist. But, as with the Marxism, he did trail his coat somewhat. The notorious "reconsidering" paper (Gould and Eldredge 1977) was prefaced by two quotations by that arch-saltationist Thomas Henry Huxley: "You have loaded yourself with an unnecessary difficulty in adopting *Natura non facit saltum* so unreservedly" (to Darwin) and "I see you are inclined to advocate the possibility of considerable 'saltus' on the part of Dame Nature in her variations. I always took the same view, much to Mr. Darwin's disgust" (to William Bateson, the saltationist's saltationist). See also Dennett (1995).

3. Jean Gayon makes the strong claim that thinking or not thinking in terms of hierarchies actually makes no difference whatsoever to one's scientific theorizing. "I strongly recommend that any naturalist who would believe that being 'for' or 'against' hierarchy is of primordial importance read Kant's Second Antimony of Pure Reason and its "solution" as given in *The Critique of Pure Reason*" (Gayon 1990, p. 35).

References

Allen, E., et al. (1975), Letter to the editor. *New York Review of Books,* 22(18), 43–44.
Caplan, A. (ed.). (1978), *The Sociobiology Debate.* New York: Harper and Row.
Dawkins, R. (1986), *The Blind Watchmaker.* New York: Norton.
Dennett, D. C. (1995), *Darwin's Dangerous Idea.* New York: Simon and Schuster.
Dobzhansky, T. (1937), *Genetics and the Origin of Species.* New York: Columbia University Press.
——— (1970), *Genetics of the Evolutionary Process.* New York: Columbia University Press.
Eldredge, N. (1971), The allopatric model and phylogeny in paleozoic invertebrates. *Evolution* 25: 156–167.
——— (1972), Systematics and evolution of "Phacops rana" (Green, 1832) and "Phacops iownesis" Delo, 1935 (Trilobita) in the Middle Devonian of North America. *Bulletin of the American Museum of Natural History* 47: 45–114.
——— (1985), *Unfinished Synthesis: Biological Hierarchies and Modern Evolutionary Thought.* New York: Oxford University Press.

Eldredge, N., and J. Cracraft. (1980), *Phylogenetic Patterns and the Evolutionary Process.* New York: Columbia University Press.

Eldredge, N., and S. J. Gould. (1972), Punctuated equilibria: an alternative to phyletic gradualism. *Models in Paleobiology,* ed. T. J. M. Schopf, pp. 82–115. San Francisco: Freeman, Cooper.

Eldredge, N., and S. N. Salthe. (1984), Hierarchy and evolution. *Oxford Surveys in Evolutionary Biology* 1: 182–206.

Endler, J. (1986), *Natural Selection in the Wild.* Princeton, N.J.: Princeton University Press.

Fisher, R. A. (1930), *The Genetical Theory of Natural Selection.* Oxford: Oxford University Press.

Gayon, J. (1990), Critics and criticisms of the modern synthesis: the viewpoint of a philosopher. *Evolutionary Biology* 24: 1–49.

Gingerich, P. D. (1976), Paleontology and phylogeny: patterns of evolution at the species level in early tertiary mammals. *American Journal of Science* 276: 1–28.

——— (1977), Patterns of evolution in the mammalian fossil record. *Patterns of Evolution, as Illustrated by the Fossil Record,* ed. A. Hallam, pp. 469–500. Amsterdam: Elsevier.

Goldschmidt, R. (1940), *The Material Basis of Evolution.* New Haven, Conn.: Yale University Press.

Gould, S. J. (1966), Allometry and size in ontogeny and phylogeny. *Biological Reviews of the Cambridge Philosophical Society* 41: 587–640.

——— (1969), An evolutionary microcosm: Pleistocene and Recent history of the land snail "P. (Poecilozonites)" in Bermuda. *Bulletin of the Museum of Comparative Zoology* 138: 407–532.

——— (1977a), *Ever Since Darwin.* New York: Norton.

——— (1977b), *Ontogeny and Phylogeny.* Cambridge, Mass.: Belknap Press.

——— (1980), Is a new and general theory of evolution emerging? *Paleobiology* 6: 119–130.

——— (1982), Darwinism and the expansion of evolutionary theory. *Science* 216: 380–387.

——— (1983), Irrelevance, submission, partnership: the changing role of paleontology in Darwin's three centennials and a modest proposal for macroevolution. *Evolution from Molecules to Men,* ed. D. S. Bendall, pp. 347–366. Cambridge: Cambridge University Press.

——— (1989), *Wonderful Life.* New York: Norton.

——— (1990), Speciation and sorting as the source of evolutionary trends, or "Things are seldom what they seem." *Evolutionary Trends,* ed. K. J. McNarmara, pp. 3–27. London: Belhaven.

Gould, S. J., and N. Eldredge. (1977), Punctuated equilibria: the tempo and mode of evolution reconsidered. *Paleobiology* 3: 115–151.

Gould, S. J., and R. C. Lewontin. (1979), The spandrels of San Marco and the Panglossian paradigm: a critique of the adaptationist program. *Proceedings of the Royal Society of London, Series B: Biological Sciences* 205: 581–598.

Hamilton, W. D. (1964a), The genetical evolution of social behaviour I. *Journal of Theoretical Biology* 7: 1–16.

——— (1964b), The genetical evolution of social behaviour II. *Journal of Theoretical Biology* 7: 17–32.

Hubby, J. L., and R. C. Lewontin. (1966), A molecular approach to the study of genic heterozygosity in natural populations I. The number of alleles at different loci in *Drosophila pseudoobscura. Genetics* 54: 577–594.

Huxley, J. S. (1932), *Problems of Relative Growth.* London: Methuen.

Jacob, F., and J. Monod. (1961a), Genetic regulatory mechanisms in the synthesis of proteins. *Journal of Molecular Biology* 3: 318–356.

Kitcher, P. (1985), *Vaulting Ambition.* Cambridge, Mass.: MIT Press.

Lewin, R. (1980), Evolutionary theory under fire. *Science* 210: 883–887.

Lewontin, R. C. (1974), *The Genetic Basis of Evolutionary Change.* New York: Columbia University Press.

Lewontin, R. C., and J. L. Hubby. 1966. A molecular approach to the study of genic heterozygosity in natural populations. II. Amount of variation and degree of heterozygosity in natural populations of "Drosophila pseudoobscura." *Genetics* 54: 595–609.

Lynne, J., and H. F. Howe. (1986), "Punctuated equilibria": rhetorical dynamics of a scientific controversy. *Quarterly Journal of Speech* 72: 132–147.

MacArthur, R. W., and E. O. Wilson. (1967), *The Theory of Island Biogeography.* Princeton, N.J.: Princeton University Press.

Mayr, E. (1963), *Animal Species and Evolution.* Cambridge, Mass.: Harvard University Press.

Pittenger, M. (1993), *American Socialists and Evolutionary Thought, 1870–1920.* Madison: University of Wisconsin Press.

Raup, D., and S. M. Stanley. (1971), *Principles of Paleontology.* San Francisco: Freeman.

Richards, R. J. (1987), *Darwin and the Emergence of Evolutionary Theories of Mind and Behaviour.* Chicago: University of Chicago Press.

Ruse, M. (1979a), *Sociobiology: Sense or Nonsense?* Dordrecht: Reidel.

——— (1979b), *The Darwinian Revolution: Science Red in Tooth and Claw.* Chicago: University of Chicago Press.

——— (1981), What kind of revolution occurred in geology? *PSA 1978,* ed. P. Asquith and I. Hacking, vol. 2, pp. 240–273. East Lansing, Mich.: Philosophy of Science Association.

——— (1996), *Monad to Man: The Concept of Progress in Evolutionary Biology.* Cambridge, Mass.: Harvard University Press.

Russell, E. S. (1916), *Form and Function: A Contribution to the History of Animal Morphology.* London: John Murray.

Russett, C. E. (1976), *Darwin in America: The Intellectual Response. 1865–1912.* San Francisco: Freeman.

Selzer, J. (1993), *Understanding Scientific Prose.* Madison: University of Wisconsin Press.

Stanley, S. M. (1975), A theory of evolution above the species level. *Proceedings of the National Academy of Sciences of the USA* 72: 646–650.

——— (1979), *Macroevolution, Pattern and Process.* San Francisco: Freeman.

Vine, F. J., and D. H. Matthews. (1963), Magnetic anomalies over oceanic ridges. *Nature* 199: 947–949.

Williams, G. C. (1966), *Adaptation and Natural Selection.* Princeton, N. J.: Princeton University Press.

Williamson, P. G. (1981), Paleontological documentation of speciation in Cenozoic molluscs from Turkana basin. *Nature* 293: 437–443.

——— (1985), Punctuated equilibrium, morphological stasis and the paleontological documentation of speciation. *Biological Journal of the Linnaean Society of London* 26: 307–324.

Wilson, E. O. (1971), *The Insect Societies.* Cambridge, Mass.: Harvard University Press.

——— (1975), *Sociobiology: The New Synthesis.* Cambridge, Mass.: Harvard University Press.

——— (1978), *On Human Nature.* Cambridge, Mass.: Harvard University Press.

14

Quasars, Causality, and Geometry

A Controversy that Should Have Happened but Didn't

WESLEY C. SALMON

Quasars, originally called quasi-stellar radio sources, were discovered thirty years ago; the discovery was announced to the world at large in the December 1963 issue of *Scientific American* (Greenstein, 1963). The discovery occurred at a time when radio astronomy was beginning to achieve fairly high resolution, so radio sources could in some cases be identified with visible objects. In fact, the first identification of a quasi-stellar radio source was made in 1960, but at the time it was taken to be a visible star that had the rather unusual property of emitting radio waves.

The Discovery

The key to the 1963 discovery was the careful spectral analysis of three sources, in which several known spectral lines could be identified if one assumed very large redshifts, suggesting that these sources are not stars in our galaxy, but rather are extragalactic objects. The chief observational data available at the time were the following:

1. In photographs they look like faint stars; they emit in both the optical and the radio regions of the spectrum.
2. Their spectra show very large redshifts.
3. Their brightness varies rapidly, for example, by as much as 30 percent in a year.

These "observations" were, of course, made with the aid of the most sophisticated instruments of observation available at the time.

The first problem concerns the redshifts. If they are cosmological—that is, results of the overall expansion of the universe—their sources must be very far away, for example, five billion light years. There are, of course, other types of redshifts, and for a time some astronomers denied that these were cosmological, but that denial seems by now to have been pretty generally abandoned. This was not a major scientific controversy; it was a relatively short-lived disagreement about the interpretation of the data. It follows, then, that these sources must be extremely bright—perhaps 100 times as bright as our galaxy. Otherwise, given their enormous distances, we would not be able to see them.

Item 3, the variability of the sources, has been crucial in the minds of many astrophysicists, who have used a causal argument in an attempt to show that the relatively rapid variability in brightness implies that the sources are extremely compact. This conclusion was drawn in 1963, and it has been frequently repeated ever since, right up to the present. Moreover, since 1963, many other quasi-stellar radio sources—now usually called quasars or QSOs—have been discovered with *much* greater redshifts and *much* more rapid variation (on the order of days). According to the standard line of reasoning, they must be *much* brighter and *much* more compact. It is this causal argument on which I wish to focus attention.

The Argument

The argument in question is based on what might be called the "$c\Delta t$ size criterion," where c is the speed of light and Δt is the time in which the variation occurs. It goes as follows:

> An overall change in brightness can be achieved only by propagating signals throughout the region of variation.
>
> No signal can travel faster than light.
>
> ∴ The region of variation cannot be larger than distance light travels in its time of variation.

It should be added that the variation need not be periodic, and that it may be either an increase or a decrease in brightness.

I shall show that this argument is fallacious—indeed, egregiously fallacious. The question is why it has hung on for thirty years without noticeable dissent (the only exception I know of is Dewdney, 1979), during which time it has been applied to a wide variety of other fluctuating objects, for example, BL Lacertae Objects (BL Lacs), pulsars, X-ray bursters, Seyfert galaxies, and active galactic nuclei (AGN), including that of our very own Milky Way. A number of examples are given in the appendix of this article. As a matter of fact, the fallacious character of the $c\Delta t$ size criterion argument was pointed out the following year by Banesh Hoffmann (1964) in a brief article in *Science*. Like any other fallacious argument, this one can, of course, be made valid by furnishing additional premises, but if this strategy is to be adopted we deserve at least a hint of what these other premises might be. A different response to Hoffmann (Terrell, 1964), offered in the same year and in the same journal, advanced a geometrical argument to which I shall return below.

My interest in these issues was first aroused by an article on BL Lacs in which the following claim was made:

> A successful model must account for the operation of what may be the most powerful engine in the universe, and it must fit that engine into a compartment of trifling size. A *crucial test of all models* is the fastest variation in luminosity that can be accommodated, since that period corresponds to the time required for a signal to traverse the emitting region. *Some models have had to be discarded already* because they do not allow rapid enough variations. (Disney and Véron 1977, p. 39, emphasis added)

A similar claim occurs in the recent *Astronomy and Astrophysics Encyclopedia* (Maran, 1992): "At present observations only give upper limits on the sizes of these objects [the central engines]. . . . Some AGN [active galactic nuclei] are strongly variable; in these, *causality limits the size* to the distance light can travel in a characteristic variability time" (p. 8, emphasis added). I do not know whether this encyclopedia qualifies as a serious technical reference work or merely a coffee-table display piece. Be that as it may, a similar argument is offered in the massive treatise *Gravitation* (Misner et al., 1973, p. 634), which is *very* far removed from the category of coffee-table fluff.

It should be noted that in the quotation from Disney and Véron the $c\Delta t$ size criterion is applied to the "emitting region" of the object in question, not to its overall size. The same point arises in the encyclopedia quotation. This statement explicitly applies the $c\Delta t$ size criterion to the nuclei of galaxies; it is not used to ascertain the overall sizes of the galaxies themselves.

As things have turned out, astrophysicists now have a rather generally accepted model of quasars—matter falling into a black hole from an accretion disk—and it does satisfy the $c\Delta t$ size criterion. Although there are still technical problems to be solved, I am *not* rejecting the model; the object of my criticism is the *argument.* It might be said that the argument is irrelevant; the aim of astrophysists is to construct a satisfactory model rather than to support a statement—a theory or hypothesis—as the conclusion of an argument. But even if model building is the aim, one can reasonably ask what constraints should be placed on the model. Surely no model that involved violation of the law of conservation of angular momentum could be accepted; the same would be true, I should think, of a model that violated special or general relativity or the laws of optics. But even though many authors seem to claim the same status for the $c\Delta t$ size criterion, suggesting that it is a consequence of special relativity, it cannot be put into the same category. It is *neither* a basic principle of *nor* does it follow from special relativity. So whether we are dealing with theories and their supporting arguments or with models and their constraints, the same fundamental issues concerning the $c\Delta t$ size criterion remain.

The Fallacy

To see the invalidity of the argument based on the $c\Delta t$ size criterion, it is essential to understand the distinction between genuine causal processes and pseudo processes. Consider a simple example (see figure 14.1). Suppose a large circular building, such as the Astrodome, is fitted out at its center with a rotating beacon that sends out a beam of white light. When the light is on, and the interior is otherwise dark, the beacon casts a white spot on the wall. As the beacon rotates the spot of light moves around the wall. A pulse of light traveling from the beacon to the wall is clearly a causal process, and

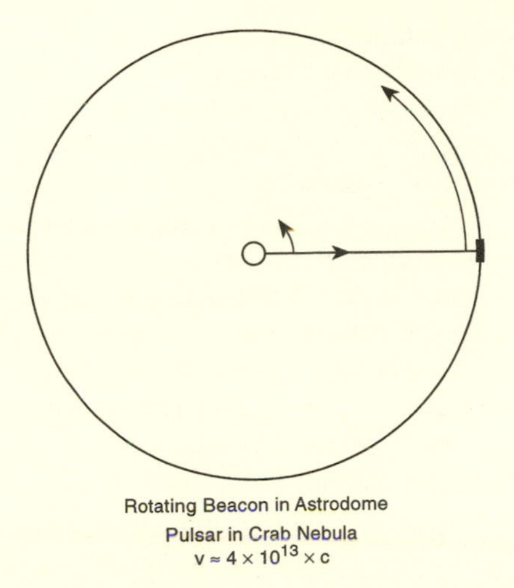

Figure 14.1 Causal vs. Pseudo-Processes

it transpires at the speed of light. No causal process can travel faster than light (in vacuo). Its causal character is revealed by the fact that it can transmit a mark; for example, if a red filter is placed in its path anywhere between the beacon and the wall, the white light changes to red *and it remains red* from that point on until it reaches the wall.

The spot of light that moves around the wall is a pseudo process. It is possible to mark the spot at any given place in its path—for example, by placing a piece of red cellophane at the wall where the spot passes—but when it travels past that point it will revert to its white color. It will not continue to be red as a result of a single local intervention in its path. Pseudo processes can be marked, but they do not transmit marks.

Suppose that our Astrodome has translucent walls, and that it is being observed at night by a distant observer. It will appear to get brighter and dimmer as the spot of light sweeps around the walls. Moreover, there is no finite limit on the speed at which the spot can travel. Imagine that as the beacon continues to rotate at the same rate the size of the building increases. The time required for the spot to traverse the entire circumference will remain constant, but the distance traveled in that time will increase as the circumference does.

Let us consider, instead of the foregoing fictitious example, a real physical system. There is a well-known pulsar in the Crab Nebula that is believed to be a rotating neutron star that beams radiation toward us much as the fictitious beacon in the Astrodome beams radiation to the walls of the building. The pulsar rotates 30 times per second and is located 6,500 light years from us. Look at figure 14.1 again, but this time suppose that the beacon is the pulsar and that Earth is located at some point on the circumference of the circle. A light pulse would require 13,000 years to cross the diameter of the circle, but the spot of radiation requires one-thirtieth of a second to sweep around the circumference. As this spot passes us it is traveling at about $4 \times 10^{13} \times c$. Faster and more distant pulsars are known, but if 4×10^{13} is not a big enough factor to be convincing, I doubt that a few more orders of magnitude would do the trick.

One possible objection to the example of the Astrodome with translucent walls is that, while the *emitting region* has the dimensions of the whole building, the *source of emitted energy* is much more compact, namely, the beacon at the center. This source would satisfy the $c\Delta t$ size criterion. Applying the criterion to the energy source instead of the emitting region makes sense because, from the beginning, a major problem about quasars has been to explain how such prodigious quantities of energy could be radiated by highly compact sources. But even though it requires us to furnish a slightly more complicated example, this shift does not save the criterion. Imagine, instead of a beacon in a building with translucent walls, a celestial object surrounded by a cloud of atoms or molecules in a metastable excited state. Suppose that this central object emits a quick burst of radiation that propagates isotropically toward the surrounding cloud, and that this light causes a burst of radiation by stimulated emission. The central source emits a relatively small quantity of radiant energy; the major part of the energy radiated by the cloud resides in the cloud; the central light is only a trigger. Although it could not be seriously entertained as a model of a quasar, because its spectrum would be totally unsuitable, this physically possible example shows that the $c\Delta t$ size criterion does not necessarily apply even to the size of the energy source.

The foregoing examples are mine, but in an article on Quasar 3C 273, one of the three discovered in 1963, Courvoisier and Robson (1991) appeal explicitly to the the $c\Delta t$ size criterion, and they offer the following analogy to explain its application:

> As a simple example, consider a line of 10 light bulbs. If one wishes to decrease the total luminosity significantly, a large number of the bulbs must be switched off, say, at least six. To do this, one must send a signal that instructs the bulbs to turn off. . . . The dimming process will therefore take at least the time light needs to cross the distance from the center of the line of bulbs to the most distant bulb to be turned off. (p. 54)

In a letter (20 June 1991) to the editor of *Scientific American,* in which the article appeared, I wrote:

> Far from supporting their contention, this example clearly shows that the size of the array cannot be inferred from the time required for dimming. Suppose that the bulbs are arranged along an arc of a very large circle. At the center of this circle place a powerful laser on a pivot. Aim it at the line of bulbs, turn it on, and rotate it rapidly. This device can zap the entire group of 10 light bulbs in an arbitrarily small time interval. No part of the laser needs to travel at a speed greater than that of light. To decrease the amount of time required for any given speed of rotation of the laser, simply move the laser farther away. Better yet, set up ten lasers, one aimed at each bulb, and turn them on simultaneously, wiping out the whole array of bulbs instantaneously.

On 5 July 1991 I sent a postscript to the foregoing letter in which I added, "I thought you might be interested to note that the very argument I criticize occurs in [the July] issue as well. . . . In this case it is applied to the Great Annihilator, but it is precisely the same argument." My letter was sent to one of the authors (Courvoisier); the entire text of his reply follows:

> The flight time arguments brought by Professor Salmon are correct. They are indeed used in the description of light echo phenomena (for example, in the context of SN 1987A). The relevance of these arguments in the context of quasars and AGN is, however, not convincing. The following points can be made:

> 1. The arguments we used can be applied to the distance between the laser or whatever controls the apparatus and the bulbs.
> 2. One can imagine many kinds of particular geometry with alignments along the line of sight in which the light travel time arguments can be defeated. They all suffer from being peculiar and contrived. Consider, for example, one single line of bulbs aligned with the line of sight and have the switch signal start at the farthest bulb. Since the switch signal travels at about the speed of light along the wire, we will have the impression that the process of intensity decrease (or increase) takes less time than the length of the array divided by *c*.
>
> The time of flight argument is not watertight, and we know that very well; it is nonetheless a very reasonable estimate of sizes that does not presuppose specific geometries. (Quoted by the kind permission of Professor Courvoisier)

Courvoisier's first point—that the $c\Delta t$ size criterion can be applied to the operation of the laser I proposed—is true but irrelevant. In the case of the quasars we observe the fluctuation on the surface; we do not observe the mechanism that produces it. We must keep clearly in mind the fact that the entity emitting the radiation is a three-dimensional object, whereas all that we can observe is part of its two-dimensional surface. The problem is to infer the size of the object from its observed period of variation. Because we cannot observe the internal mechanisms, the $c\Delta t$ size criterion does not solve that problem.

In his second point Courvoisier complains that examples like mine are "peculiar and contrived." Regarding this criticism I have two responses. First, the example cited is theirs, not mine. Second, if someone produced an intricate and complex device that turned out to be a genuine perpetual motion machine of the first kind—one that could actually do work without any input of energy—I doubt that anyone would complain that it was contrived. It would indeed be a contrivance, but one that would be extraordinarily interesting and useful.

Courvoisier concludes by remarking that their argument is not "watertight," and they are perfectly aware of that fact. This constitutes an explicit recognition that the $c\Delta t$ size criterion does not have the status of a law of nature or a consequence thereof. The editors of *Scientific American* informed me that my letter and the coauthor's reply did not merit publication.

An Actual Counterexample

Enough of these fictitious setups. In 1986 I sent the following technical report to *Science:*

> Charles V. Shank's article, "Investigation of Ultrafast Phenomena in the Femtosecond Domain" (*Science,* 19 September 1986), contains a fascinating discussion of the generation and uses of extremely brief pulses of light. I was, however, astonished at what seems a glaring omission—that is, any reference to the minute size of the apparatus that produces these pulses. Indeed, the article contains no hint of the miracle of miniaturization that has apparently been achieved.
>
> My knowledge of this feature of Shank's work is not derived from direct acquaintance; it comes from an application of a principle of astrophysics. In discussions of such

fluctuating sources as quasars, BL Lacs, X-ray bursters, and pulsars, appeal is often made to what might be called "the $c\Delta t$ size criterion." According to this criterion, an upper limit on the size of a source that fluctuates over an interval Δt is given by the product of Δt and the speed of light (3×10^{10} cm/sec). It is often presented as a rigorous consequence of special relativity, and hence as an inviolable law of nature.

For example, the very next issue of *Science* (26 September 1986) contains the article by K. Y. Lo, "The Galactic Center: Is It a Massive Black Hole?" (pp. 1393–1403). Writing about radiation from active galactic nuclei in general, he says, "Such radiation is sometimes found to vary on time scales as short as days, implying that the source is $<10^{17}$ cm in extent" (p. 1395). Although Lo does not explicitly invoke the $c\Delta t$ size criterion—probably because it is too well known to require mention—it would appear to be the basis of his calculation. An upper limit on the size of a source that fluctuates in one day is about 2.6×10^{15} cm, and one that fluctuates in 100 days has the approximate limit given by Lo. In a 1982 report in *Science* on the same topic, M. Waldrop, explicitly invoked the $c\Delta t$ size criterion, taking it as an unexceptionable law of nature (Waldrop, 1982). V. Trimble and L. Woltjer, in the recent survey article "Quasars at 25" (*Science,* 10 October 1986, pp. 155–161) [they took 1960 as the date of birth], also seem to make repeated appeals to this size criterion (pp. 155–157).

Shank reports that optical pulses with durations less than 8 femtoseconds (1 fsec = 10^{-15} sec) have been produced. Applying the $c\Delta t$ size criterion to the sources of such ultrabrief pulses, we find that the upper limit on their size is 2.4×10^{-6} m. [This is roughly the length of a human chromosome; such an apparatus would fit comfortably within a human cell.]

In all seriousness, I profoundly doubt that Shank's apparatus (including such equipment as tunable dye lasers) has dimensions of a couple of microns. So how are we to reconcile laser theory and astrophysics? The answer was given by Banesh Hoffmann in the pages of this journal more than 20 years ago. The $c\Delta t$ size criterion is not a law of nature; it is not a consequence of special relativity. What is its status? At best it is a heuristic device that has been used to assess the plausibility of physical theories pertaining to quasars, black holes, and other celestial objects.

This technical report was not published; it was returned to me without comment. The editors of *Science* apparently had no sense of humor. Also, I think, they had no sense of history. They seemed to see no problem in an Aristotelian brand of physics that has one set of laws for terrestrial phenomena and an entirely different set for celestial phenomena.

The Geometric Argument

Hoffmann (1964) pointed out that the surface of a sphere could brighten instantaneously as a result of a causal process that propagated uniformly from the center, reaching all parts of the surface simultaneously. A concrete illustration of this idea was published in the *New York Times* (Browne, 1993). It involves a new theory regarding the nature of supernova explosions. I am neither endorsing nor rejecting this theory; it is simply an example of Hoffmann's basic point. Standard supernova theory holds that when a star has used up almost all of its supply of hydrogen, a series of nuclear transmutations occurs creating iron and lighter elements. This is followed by a violent implosion in which heavier elements are created. According to the new

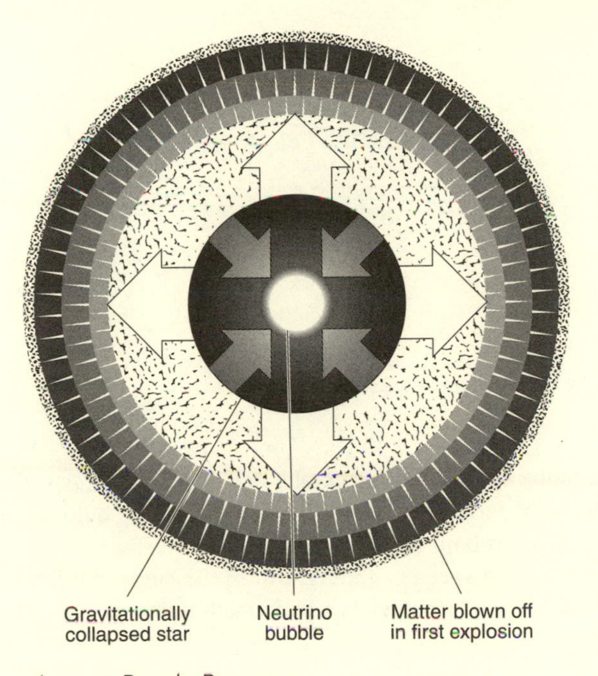

Figure 14.2 Causal versus Pseudo-Processes

theory, "[A] bubble of neutrinos forms in the implosion, lasts for 10 seconds, then ripples outward. As it reaches the surface a flash of light bursts out. Billions of miles out into space, shock waves from the star collide with gas ejected years before, generating radio signals that accompany supernova explosions" (see figure 14.2).

Given the obvious physical possibility of this sort of phenomenon, Terrell (1964) asks what a distant observer will see if it occurs. His answer is that the brightening will appear to occur, not instantaneously, but over a span of time. In a nice rhetorical flourish he appeals to relativity theory as the basis for asserting that the speed of light is finite (a fact we have known since Rømer's observations of the eclipsing of the moons of Jupiter in the seventeenth century). If we are observing a large spherical object the light from the parts nearest to the observer will arrive earlier than light from the periphery because that light has farther to go to reach us (see figure 14.3). Moreover, the difference in distance is roughly equal to the radius of the sphere, so the result is similar to the conclusion drawn from the $c\Delta t$ size criterion. I have noticed this argument in the semi-popular literature on quasars only once, in Paolo Maffei's *Monsters in the Sky* (1980, p. 263).

Notice the relationship between the $c\Delta t$ argument and the geometrical argument. According to the former, actual instantaneous brightening is physically impossible. According to the latter, even if actual instantaneous brightening occurs, it will appear to be noninstantaneous. In fact, given certain particular geometrical configurations, these two considerations can cancel one another out. The point (which I had not noticed before) was illustrated in Courvoisier's reply to my letter. Recall the case of a

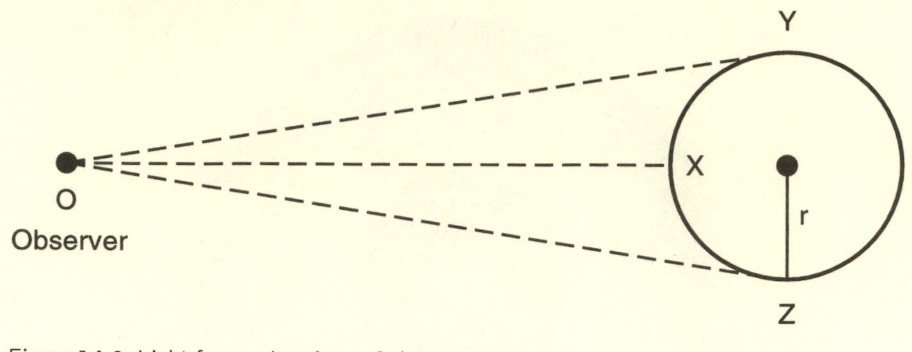

Figure 14.3 Light from a Luminous Sphere

line of 10 light bulbs arranged almost along our line of sight. A switch is flipped at the most distant bulb, and a signal travels along the line of bulbs at about the speed of light, turning on each bulb as it reaches it. Since the signal controlling the lights travels at about the same speed as the light from the bulbs, the light from all of the bulbs will reach us almost simultaneously, no matter how far apart the bulbs happen to be.

The striking feature of Terrell's geometrical argument is its dependence on the approximately spherical shape of the emitting object. Suppose, instead, that the emitting object is a flat disk oriented perpendicularly to our line of sight (see figure 14.4). Let X be the center of the disk, Y a point on its edge, and O the position of the observer. The Pythagorean theorem along with some trivial high school algebra shows that the absolute difference between the length of OX and that of OY approaches zero as OX increases in length and XY remains fixed. Indeed, if we were looking at a ring with the same orientation instead of a disk, all of the light would take the same amount of time to reach us, no matter how large OX might be.

The question we must ask, therefore, concerns the shapes of objects that we find in the sky. Our own galaxy is a spiral; the ratio of its thickness to its diameter is approximately equal to that ratio in an ordinary phonograph record (see figure 14.5). Of course,

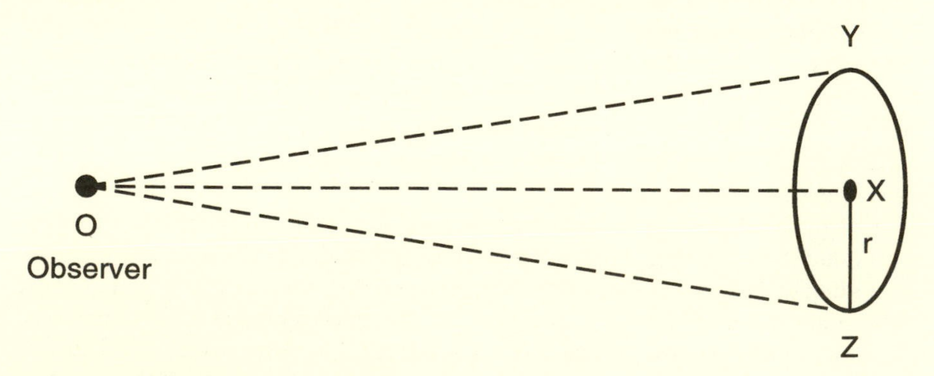

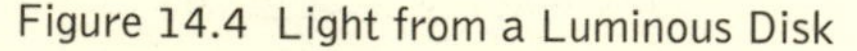
Figure 14.4 Light from a Luminous Disk

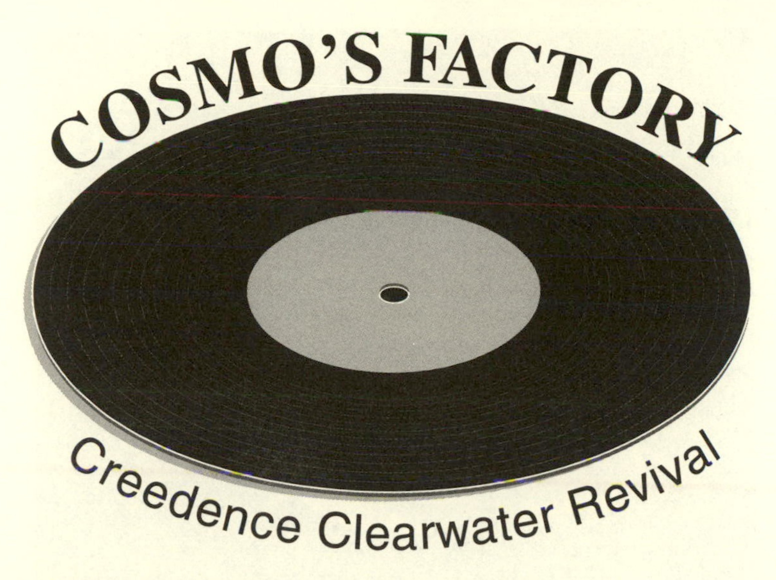

Figure 14.5 The Shape of the Milky Way

the edges are more ragged and there is a bulge at the center, but the approximation is pretty good. In order to display this shape I looked through our home collection of old LP records and serendipitously came upon "Cosmo's Factory" by Creedence Clearwater Revival, a happy discovery given that we are interested in various types of engines in the cosmos, a clearing of the waters muddied by invalid arguments, and in a revival of credence in theories or models of such engines. Spiral galaxies are numerous in the universe, and there is no reason to think ours is particularly special. It is, of course, the homeland of humans, but this fact seems to me to have little cosmic importance.

Consider another familiar celestial object, the planet Saturn (see figure 14.6). Although the planet itself is roughly spherical, the entire object, including the rings, is not. Now imagine that the planet disappears, leaving only the rings visible to us. Imagine that Saturn began as a much more massive object than it is, and that it had undergone gravitational collapse, becoming a black hole. The result would be quite similar to the above-mentioned model astrophysicists currently favor for quasars—a black hole surrounded by an accretion disk, with matter from the disk falling into the black hole. Since a black hole is in principle invisible, what we see is a ring radiating prodigious quantities of energy. Of course, these rings are not all oriented in the same way with respect to our line of sight, and this is an important point. But not all quasars have the same rate of observed fluctuation, so the argument does not depend on any general assumption about the orientations of quasars.

Some Methodological Points

When quasars were first discovered, and for a long time thereafter, they presented extremely puzzling phenomena. No reasonably satisfactory model was available. Un-

Figure 14.6 Saturn

der such circumstances the most fruitful avenue for astrophysicists would be to conceive and develop as many models as possible, hoping to find one that at least comes close to fitting the facts. It would be counterproductive to impose artificial and unnecessary constraints, such as the $c\Delta t$ size criterion, and to rule out immediately any proposed model that would violate it—as reported by Disney and Véron in the quotation given above. Perhaps what actually happened was that a proposal was made—for example, that quasars are large dense collections of stars (see Misner et al., 1973, p. 634)—that could be rejected on other grounds. In this case, for instance, there seemed to be no mechanism by which the brightening and dimming of members of the group could be orchestrated to produce the overall fluctuation that was observed. That is the real objection. It had nothing to do with the fact that the model violated the $c\Delta t$ size criterion.

As I noted above, a fluctuation violating the $c\Delta t$ size criterion could be produced by a huge shell of matter in an excited state induced to radiate by stimulated emission from a central source. However, this kind of entity would not yield a spectrum anything like those of observed quasars. Again, the $c\Delta t$ size criterion has nothing to do with the inadequacy of the model.

It may be that various models can be constructed that violate the $c\Delta t$ size criterion, and that every one of them can be rejected on completely different grounds. If that were to happen it would be perfectly reasonable to treat the $c\Delta t$ size criterion as a plausibility claim. An astrophysicist might say, in effect, "I don't believe you can construct a satisfactory model that violates this criterion, but if you have any such model let's take a look at it to see whether it's viable on other grounds. I seriously doubt that it will survive careful scrutiny." If this is what is involved, it seems to me, astrophysicists should say so explicitly, rather than invoking a criterion as a consequence of a law of nature when it is no such thing. And if the $c\Delta t$ size criterion is adopted as a plausibility principle, we may reasonably ask on what basis its plausibility rests. I have not found any answer to this question in the literature.

Perhaps the vast majority of authors feel queasy about appealing to plausibility considerations, thinking that such appeals fall short of scientific objectivity and rigor. Anyone who adopts a Bayesian approach to scientific confirmation can point directly to the fact that prior probabilities occur in Bayes's theorem; plausibility arguments can be identified with assessments of prior probabilities. According to the Bayesian approach, plausibility considerations not only are admissible but also are indispensable to scientific confirmation. Bayesianism has the important virtue of calling explicit attention to plausibility considerations and to the grounds for their evaluations. When the *Astronomy and Astrophysics Encyclopedia,* cited above, says that "causality limits the size," this sounds like an inviolable principle. It would have been far better to say that models that violate the $c\Delta t$ size criterion do not stand much of a chance of meeting the other requirements that models of quasars must satisfy, and why. Before astrophysicists had found fairly satisfactory models, the motto should have been, "Let a thousand flowers bloom, and let us look at them all."

The Rhetoric of the Argument

My attention was drawn to the $c\Delta t$ size criterion by articles in such journals as *Science* and *Scientific American.* I have subsequently found similar arguments in a number of other journals, such as *Nature, American Scientist,* and *Physics Today.* These publications share an important characteristic, namely, that they are widely read by scientists and other scientifically literate people who want to find out what goes on in scientific areas outside of their own specialities. *Nature* and *Science* are two of the most prestigious scientific journals in the world, but they are not confined to any single narrow speciality. *Science* and *American Scientist* are organs of scientific societies that purposely lack narrow disciplinary boundaries. But these are not publications like *Time* and *Newsweek* that are addressed to the general public. My complaint, then, concerns an apparent failure of scientists to level with their fellow scientists in other areas of research. Thoughtful astrophysicists apparently realize that the $c\Delta t$ size criterion has to be used with care, and only if certain conditions are fulfilled. Unfortunately, it is difficult for "outsiders" who have a genuine interest in the subject to discover any explicit formulation of the conditions that need to be taken into account.

In *The Remarkable Birth of the Planet Earth* (1972), Henry M. Morris, the leading proponent of "creation science" in the United States, says, "Cosmogony seems to be a sort of game that astronomers play, a tongue-in-cheek charade in which only the initiates know the rules and the spectators stand in awe" (p. 57). The basis on which

Morris makes this statement is, of course, entirely different from my concerns about the $c\Delta t$ size criterion and the geometrical argument. Nevertheless, it seems to me that bona fide scientists should do their best not to give potential ammunition to influential practitioners of pseudo science.

Recent Developments

In 1993 I learned of two mechanisms that had recently been postulated as causes of fluctuations in apparent brightness of celestial objects. The first of these is gravitational lensing. It has long been realized that a gravitational lens could intensify the radiation we receive from a distant source, but it has recently been suggested that brief pulses of intensified radiation from stars in one of the Magellanic Clouds—a nearby galaxy—were due to gravitational lensing by dark bodies in the halo of the Milky Way. The pulses were brief because the dark bodies passed relatively quickly between us and the more distant stars. (See Alcock, et al., 1993 and Augbourg, et al., 1993 for technical reports.) Since as much as 90 percent of the matter in the universe may be dark, it seems to me that we know very little about the frequency and circumstances of fluctuations caused by this sort of gravitational lensing.

The second mechanism involves relativistic jets emitted by quasars. Conservation of angular momentum suggests that the trajectories of these jets will be spirals rather than straight lines. If the axis of the spiral makes quite an acute angle with our line of sight to the quasar, it follows from basic geometrical considerations that the main body of the jet will sometimes be traveling more directly toward us and sometimes more directly away from us. As a result, we will observe a brightening and dimming associated with the jet. I do not know whether this sort of fluctuation has any bearing on the size of the source of the jet. The technical report on this proposal is Schramm et al. (1993).

Conclusion

A major theme of this chapter, as suggested by its subtitle, is why no genuine scientific controversy emerged over almost the past 40 years regarding the sizes of variable celestial objects. I do not have an answer to this question. Even though some thoughtful astrophysicists are aware that the $c\Delta t$ size criterion has limited applicability, I fear that in general the connection between the time taken for variation to appear and the size of the object may become a dogma—one whose basis for the rest of us remains obscure.

Appendix

The following are a few more examples of use of the $c\Delta t$ size criterion from the literature. This sample is given not only for further documentation but also to illustrate the variety of contexts in which the criterion is applied.

Morrison, P. (1973), "Resolving the Mystery of the Quasars?" *Physics Today 26* (Mar.), 25:

> Many quasars vary their *optical* intensity on a time scale which is characteristically a tenth of a year. . . . It follows from this variability that quasars must be very compact. If they weren't compact, they couldn't vary in a tenth of a year. No object can double in brightness in a time much smaller than the light-transit time across an appreciable part of the object. So we conclude that one-tenth of a light year is a characteristic maximum dimension for the optical heart of a bright variable quasar.

GBL (1974), "Evidence Accumulates for a Black Hole in Cygnus X-1," *Physics Today 27* (Feb.), 17, 19:

> Within the past year many observers have become convinced that Cygnus X-1 contains a black hole. The most recent evidence, reported at the December [1973] meeting of the American Astronomical Society in Tucson, is from an X-ray detector aboard a rocket; a group at the Goddard Space Flight Center reported seeing millisecond variations in intensity, suggesting a compact object. . . . The most recent evidence that Cygnus X-1 contains a black hole . . . is that its X-ray output is flickering with variations as short as a millisecond, a behavior characteristic of a very small object.

Schaefer, B. E. (1985), "Gamma-Ray Bursters," *Scientific American 252* (Feb.), 55:

> [T]heoretical reasons show it is plausible that a gamma-ray burster might contain a neutron star; certain observational facts make it probable that it does. One such fact is the very short time within which bursts change their intensity. Some bursts have been as short as .01 second, whereas a burst that occurred on March 5, 1979, rose in intensity in .0002 second. Since a source cannot significantly change brightness in a time shorter than the time it takes light to travel across the source region, the size of the March 5 burster must be smaller than .0002 light-second, or about 40 miles. There are few astronomical objects that meet the size limitations or have enough available energy to power a burst. A neutron star satisfies both of these requirements.

Hutchings, J. B. (1985), "Observational Evidence for Black Holes," *American Scientist 73* (Jan.–Feb.), 52:

> For the 22 years since their discovery, quasars have occupied the attention, time, and resources of many of the world's astronomers. We are now essentially certain that they are the most luminous single objects in the universe and also very small—often significantly changing their vast output of energy within days, and in some instances within minutes. (This limits the size of the radiating region to the distance that light can travel in that time.)

Miller, H. R. et al. (1989), "Detection of Microvariability for BL Lacertae Objects," *Nature 337,* 627:

> Large-amplitude, rapid optical variability is a well-known identifying characteristic for BL Lacertae objects ("blazars"). Although large-amplitude variations on timescales ranging from days to decades have been well documented, considerable controversy surrounds the nature of microvariability, that is, optical variations on timescales significantly shorter than a day. Here we report observations of BL Lacertae in which rapid changes were detected in the total optical flux from this object. These variations occurred on timescales as short as 1.5 hours. Although their structure is complex, the minimum

timescale for the variations may be used to place constraints on the size of the emitting region.

"Galactic Center Antimatter Factory Found at Last?" (1990) *Sky & Telescope 79,* 363:

From the time-scale of these variations and the finite speed of light, researchers argue that the radiation arises in a source less than 1 light-year across.

Remillard, R. A. et al. (1991), "A Rapid Energetic X-ray Flare in the Quasar PKSO558–504," *Nature 350,* 591:

The flaring timescale (Δt) provides, from the causality argument, an upper limit for the size of the emitting region $R < c\Delta t$.

Lin, Y. C., et al. (1992), "Detection of High-Energy Gamma-Ray Emission from the BL Lacertae Object Markarian 421 by the EGRET Telescope on the *Compton Observatory," Astrophysical Journal 401,* L61:

Mrk 421 exhibits significant time variability in many wave-length bands, particularly in X-rays [references] and in optical wavelengths [references]. Most recently, Fink et al. [reference] observed a 20% change of the soft X-ray flux (0.1—2.4 keV) in 2 hr. The rapid variability reported in these references strongly suggests that Mrk 421 contains a compact object.

Dermer, C. D., and R. Schlickeiser. (1992), "Quasars, Blazars, and Gamma Rays," *Science 257,* 1645:

We now know that there can be rapid variability in the gamma-ray emission of 3C279, which seems to require an emission site less than about a light-week away from the central black hole for 3C279. . . .

Bignami, G. F. (1992), "Gamma-ray Power from 3C279," *Nature 355,* 299:

Photon-photon absorption limits the amount of energy that can escape from a bright source if the density of photons at different energies is high enough. Theorists have already had to cope with this limit in explaining the γ-rays from the weaker 3C273 source. The difficulty all depends on the size of the source region, which can be inferred from the timescale for the variability (a source can change only with a maximum rate determined by the light transit time over its dimensions).

Baring, M. G. (1992), "Ignition of X-ray Flares," *Nature 360,* 109:

The X-ray emission from active galactic nuclei is thought to emanate from volumes barely greater than the Solar System (10^{10} km across) around a supermassive black hole at the hub of each galaxy. Such compact sources are inevitably involved if the rapid variations in intensity are to be explained.

Powell, C. S. (1993), "Inconstant Cosmos," *Scientific American 268* (May), 111:

"When you look at the sky at high energies, it's an amazingly inconstant place," reflects Neil Gehrels, the project scientist for GRO [Compton Gamma Ray Observatory]. On time scales ranging from weeks to thousandths of a second, objects brighten and dim, flicker and oscillate. Such rapid changes imply that the sources of the radiation are minuscule on a cosmic scale (otherwise it would take far to long for a physical change to affect a large part of the emitting region). Yet those same objects are emitting tremendous quantities of energetic radiation.

Kniffen, Donald A. (1993), "The Gamma-Ray Universe," *American Scientist, 81* (July–Aug.), 342–349:

> Because gamma-ray bursts fluctuate over very brief periods of time (less than one-1,000th of a second in some instances), the region emitting some of the gamma rays must be quite small (less than 100 kilometers in diameter). (p. 344)
>
> Some active galactic nuclei release energy in all parts of the spectrum, from radio waves to gamma rays. They are the brightest objects in the universe. . . . Remarkably, some of these objects appear to be releasing most of their energy at gamma-ray wavelengths. One of these, a quasar identified as 3C 279, lies about 6 billion light-years away and may release as much as 10 million times more gamma rays than our own galaxy. Curiously, four months after 3C 279 was discovered it ceased flaring almost entirely. Such tremendous variation in output appears to be a common feature of these objects. In some cases they vary their output in less than a day, suggesting that the region of emission is relatively small (less than a light-day across). (p. 346)

Notes

I express my sincere thanks to the following individuals for helpful suggestions and discussions: Frederic Chaffee, T. Courvoisier, Donald Kniffen, James Small, and Raymond Weymann. Each of them would, I believe, have serious objections to the present chapter; their generosity does not entail agreement with my main theses.

References

Alcock, C., et al., 1993. "Possible Gravitational Microlensing of a Star in the Large Magellanic Cloud," *Nature* 365, pp. 621–623.

Aubourg, E., et al., 1993. "Evidence for Gravitational Microlensing by Dark Objects in the Galactic Halo," *Nature* 365, pp. 623–625.

Browne, M. W. (1993), "Strange, Violent Physics Born in the Death of Stars," *New York Times,* 27 April, B5.

Courvoisier, T., and E. I. Robson. (1991), "The Quasar 3C 273," *Scientific American 264* (no. 6): 50–57.

Dewdney, A. (1979), "A Size Limit for Uniformly Pulsating Sources of Electromagnetic Radiation," *Astrophysical Letters 20:* 49–52.

Disney, M., and P. Véron. (1977), "BL Lacertae Objects," *Scientific American 237* (no. 2): 32–39.

Greenstein, J. L. (1963), "Quasi-Stellar Radio Sources," *Scientific American 207* (no. 6): 54–62.

Hoffmann, B. (1964), "Fluctuating Brightness of Quasi-Stellar Radio Sources," *Science 144:* 319.

Lo, K. Y. (1986), "The Galactic Center: Is It a Massive Black Hole?" *Science 233:* 1393–1403.

Maffei, P. (1980), *Monsters in the Sky.* Cambridge, Mass.: MIT Press.

Maran, S. P., ed. (1992), *Astronomy and Astrophysics Encyclopedia.* New York: Van Nostrand.

Misner, C. W, K. S. Thorne, and J. A. Wheeler. (1973), *Gravitation.* San Francisco: Freeman.

Morris, H. (1972), *The Remarkable Birth of the Planet Earth.* San Diego: Creation-Life Publishers.

Schramm, K. J. et al., 1993. "Recent Activity in the Optical and Radiofrequency Lightcurves of Blazar 3C 345: Indications for a 'Lighthouse Effect' Due to Jet Rotation," *Astronomy and Astrophysics* 278, pp. 291–405.

Shank, C. V. (1986), "Investigation of Ultrafast Phenomena in the Femtosecond Domain," *Science 233:* 1276–1280.
Terrell, J. (1964), "Quasi-Stellar Diameters and Intensity Fluctuations," *Science 145:* 919–919.
Trimble, V., and L. Woltjer. (1986), "Quasars at 25," *Science 234:* 155–161.
Waldrop, M. (1982), "A Hole in the Milky Way," *Science 216:* 838–839.

Index